T0281201

DISINFECTION BY-PRODUCTS in WATER TREATMENT

The Chemistry of
Their Formation and Control

The Editors

Roger A. Minear, Director of the Institute for Environmental Studies and Professor of Civil Engineering at the University of Illinois, received his B.S. (1964) in Chemistry, M.S.E. (1965) in Sanitary Engineering, and Ph.D. (1971) in Civil Engineering (specializing in environmental chemistry) from the University of Washington. He served as an Instructor at Oregon State University, 1966–1967, an Assistant Professor at Illinois Institute of Technology, 1970–1973, joined the University of Tennessee, Knoxville, as an Associate Professor in 1973, became Professor in 1977, and was named the first Armour T. Granger Professor in 1983.

Dr. Minear holds memberships in the American Chemical Society, the American Society of Civil Engineers, the American Society of Limnology and Oceanography, the American Association for the Advancement of Science, the American Water Works Association, the Association of Environmental Engineering Professors, the International Association on Water Quality, the Society of Environmental Toxicology and Chemistry, the Water Environment Federation, and Sigma Xi.

Dr. Minear's major areas of scientific interest involve the nature, origin, transport, and transformation of organic and inorganic compounds in natural and wastewaters; chemistry of aqueous solutions and chemical processes of water and wastewater treatment; and trace and environmental analysis. He has generated more than 110 professional paper presentations, 42 invited seminar presentations, 105 published articles, chapters, reports, and 35 graduate theses in the Environmental Science and Engineering field.

DISINFECTION BY-PRODUCTS in WATER TREATMENT

The Chemistry of Their Formation and Control

Edited by

Roger A. Minear and Gary L. Amy

CRC Press
Taylor & Francis Group
Boca Raton London New York

CRC Press is an imprint of the
Taylor & Francis Group, an **informa** business

CRC Press, Inc.
Taylor & Francis Group
6000 Broken Sound Parkway NW, Suite 300
Boca Raton, FL 33487-2742

©1996 by Taylor & Francis Group, LLC
CRC Press is an imprint of Taylor & Francis Group, an Informa business

First issued in paperback 2019

No claim to original U.S. Government works

ISBN-13: 978-0-367-44872-1 (pbk)
ISBN-13: 978-1-56670-136-5 (hbk)

Visit the Taylor & Francis Web site at
http://www.taylorandfrancis.com

and the CRC Press Web site at
http://www.crcpress.com

Library of Congress Cataloging-in-Publication Data

Disinfection by-products in water treatment : the chemistry of their formation and control /
    edited by Roger A. Minear, Gary L. Amy.
          p. cm.
    Includes bibliographical references and index.
    ISBN 1-56670-136-8 (alk. paper)
    1. Water--Purification--Disinfection--By-products.   I. Minear, R.
A.   II. Amy, Gary L.
TD459.D582   1995
628.1'662--dc20

                                                                      95-35535
                                                                         CIP

Library of Congress Card Number 95-35535

Gary L. Amy received his Ph.D. in Civil/ Environmental Engineering from the University of California at Berkeley in 1978. From 1978 through 1990, he was a faculty member in the Environmental Engineering Program within the Department of Civil Engineering at the University of Arizona. He is presently a Professor of Environmental Engineering within the Department of Civil, Environmental, and Architectural Engineering at the University of Colorado-Boulder. At Arizona and now at Colorado, he has been involved in research pertaining to aquatic chemistry, water treatment, and hazardous waste  treatment, and in teaching graduate and undergraduate courses in Environmental Engineering. His aquatic-chemistry research has largely been in the area of subsurface contaminant transport, with a focus on humic-facilitated transport of trace metals and polynuclear aromatic hydrocarbons (PAHs) in groundwater. His water-treatment research has focused on (i) characterization of natural organic matter (NOM) and implications thereof for treatment process selection, (ii) removal of NOM by ozone oxidation, chemical coagulation, activated carbon adsorption, and membranes processes, and (iii) evaluation of disinfection by-product (DBP) formation and control. He has directed the research of more than 30 graduate students. He has been the recipient of research grants from NSF, USEPA, USGS, USDOE, as well as various water and wastewater utilities. Water utility support has included the Metropolitan Water District of Southern California (MWD) and the Orange County Water District (OCWD). He is an active member of the American Water Works Association, the American Chemical Society, and the Association of Environmental Engineering Professors. Dr. Amy has served as an Environmental Consultant to various engineering firms and water utilities on various water quality issues. He is the author of 80 technical articles/papers.

Contributors

Robert C. Andrews
Department of Civil Engineering
University of Toronto
Toronto, Ontario, Canada

Susan A. Andrews
Department of Civil Engineering
University of Waterloo
Waterloo, Ontario, Canada

Lina S. Chin
Department of Civil Engineering
University of Illinois at
      Urbana-Champaign
Urbana, Illinois

R. Chinn
Water Quality Division
Metropolitan Water District of
      Southern California
La Verne, California

Z.K. Chowdhury
Malcolm Pirnie, Inc.
Phoenix, Arizona

M. Robin Collins
Department of Civil Engineering
University of New Hampshire
Durham, New Hampshire

William J. Cooper
Drinking Water Research Center
Florida International University
Miami, Florida

J. De Laat
Laboratoire Chimie Eau et Nuisances
Poitiers, France

Alicia C. Diehl
University of Texas
Austin, Texas

M. Doré
Laboratoire Chimie Eau et Nuisances
Poitiers, France

M.J. Ferguson
Water Branch
City of Edmonton
Edmonton, Alberta, Canada

William H. Glaze
Department of Environmental
      Science and Engineering
University of North Carolina
Chapel Hill, North Carolina

Charles N. Haas
Environmental Studies Institute
Department of Civil and
      Architectural Engineering
Drexel University
Philadelphia, Pennsylvania

Shannon Harrell
Department of Civil Engineering
University of Illinois at
      Urbana-Champaign
Urbana, Illinois

Jürg Hoigné
Department of Chemistry
Swiss Federal Institute for
    Environmental Science
    and Technology
EAWAG
Dubendorf, Switzerland

P.M. Huck
Department of Civil Engineering
University of Waterloo
Waterloo, Ontario, Canada

James N. Jensen
Department of Civil Engineering
State University of New York
    at Buffalo
Buffalo, New York

Stuart W. Krasner
Water Quality Division
Metropolitan Water District of
    Southern California
La Verne, California

Charles N. Kurucz
University of Miami
Coral Gables, Florida

Richard A. Larson
Institute for Environmental Studies
University of Illinois at
    Urbana-Champign
Urbana, Illinois

Solomon W. Leung
College of Engineering
Idaho State University
Pocatello, Idaho

Fei T. Mak
Drinking Water Research Center
Florida Internaional University
Miami, Florida

Larry McCollum
Contra Costa Water District
Concord, California

Jennifer Miller
Department of Civil Engineering
University of Illinois at
    Urbana-Champaign
Urbana, Illinois

Michael G. Nickelsen
High Voltage Environmental
    Applications, Inc.
Miami, Florida

D.M. Owen
Malcolm Pirnie, Inc.
Carlsbad, California

Jeffrey L. Oxenford
AWWA Research Foundation
Denver, Colorado

David A. Reckhow
Department of Civil and
    Environmental Engineering
University of Massachusetts
Amherst, Massachusetts

Khatereh L. Sawal
Drinking Water Research Center
Florida International University
Miami, Florida

Michael J. Sclimenti
Metropolitan Water District of
    Southern California
La Verne, California

Hiba Murad Shukairy
Drinking Water Research Division
U.S. Environmental Protection
    Agency
Cincinnati, Ohio

Mohamed S. Siddiqui
Department of Civil Engineering
University of Colorado at Denver
Denver, Colorado

Louis A. Simms
University of Houston
Houston, Texas

RoseAnn Slifker
Drinking Water Research Center
Florida International University
Miami, Florida

Vernon L. Snoeyink
Department of Civil Engineering
University of Illinois at
        Urbana-Champaign
Urbana, Illinois

Rengao Song
Institute for Environmental Studies
University of Illinois at
        Urbana-Champaign
Urbana, Illinois

Harvey Wayne Sorensen, Jr.
University of Houston
Houston, Texas

Gerald E. Speitel, Jr.
University of Texas
Austin, Texas

R. Scott Summers
Civil and Environmental
        Engineering Department
University of Cincinnati
Cincinnati, Ohio

H. Suty
ELF-ATOCHEM
Levallois Perret, France

James M. Symons
Department of Civil and
        Environmental Engineering
University of Houston
Houston, Texas

Huija Teng
School of Civil and Environmental
        Engineering
Oklahoma State University
Stillwater, Oklahoma

Kirankumar Topudurti
PRC Environmental Management, Inc.
Chicago, Illinois

Richard L. Valentine
Department of Civil and
        Environmental Engineering
University of Iowa
Iowa City, Iowa

C.W. Vaughan
Department of Civil Engineering
University of New Hampshire
Durham, New Hampshire

John N. Veenstra
School of Civil and Environmental
        Engineering
Oklahoma State University
Stillwater, Oklahoma

N. Karpel Vel Leitner
Laboratoire de Chimie de l'Eau et
        des Nuisances
Universite de Poitiers
Poitiers, France

Urs von Gunten
Department of Chemistry
Swiss Federal Institute for
        Environmental Science
        and Technology
EAWAG
Dubendorf, Switzerland

Thomas D. Waite
University of Miami
Coral Gables, Florida

Howard S. Weinberg
Department of Environmental
        Sciences and Engineering
University of North Carolina
Chapel Hill, North Carolina

Paul Westerhoff
Department of Civil and
        Environmental Engineering
Arizona State University
Tempe, Arizona

Yuefeng Xie
Department of Environmental
        Programs
Pennsylvania State University
        at Harrisburg
Middletown, Pennsylvania

Jeyong Yoon
Department of Environmental
        Engineering
Ajou University
Suwon, Korea

Wenyi Zhai
University of Colorado at Boulder
Boulder, Colorado

Preface

Chlorine in the environment has come from savior to suspect over the past several decades. Its incorporation in products gave them desirable properties; thermal stability and resistance to degradation, notably in plastics, plasticizers, solvents, pesticides and heat transfer and dielectric fluids. Early in the century, chlorine was introduced into drinking waters for disinfection against water-borne diseases, a practice that has persisted to date and to which the saving of millions of lives has been attributed.

Concerns about chlorinated compounds began to surface, first in the 1960s with respect to chlorinated pesticides, then with respect to the negative persistence of other chlorinated substances, such as polychlorinated biphenyls (PCBs) and photochemically active solvents like trichloroethylene (TCE). TCE was first associated with atmospheric smog reactions but later became one of the most pervasive contaminants of the nation's groundwaters. Like most other chlorinated compounds, the properties that initially made it a desirable product were the same properties that made it an environmental liability.

Then in 1974, a Dutch Chemist, Johnnes Rook, published results that implicated the use of chlorine in drinking waters as the cause of chloro and bromo trihalomethanes (THMs) found in treated drinking waters. These early disinfection by-products (DBPs), as they are now called, were later joined by several other halogenated organic, non-halogenated organic, and inorganic halogen oxide compounds. In part, these revelations have all come about as the result of improved analytical methods and greater scrutiny directed at the contents of chlorinated drinking waters in the last two decades.

A shift to alternate disinfection processes in order to minimize THM formation has not been without the creation of potential new problems. These alternate processes are not devoid of their own set of DBPs, some of which, like bromate, are not associated with conventional chlorination. While control and minimization of DBPs is necessary and, in fact, coming under wider spread regulation at lower and lower concentration limits, or MCLs (maximum concentration limits) there cannot be compromise on the original goal, proper disinfection of drinking water.

We must continue to understand all aspects of new and substitute processes. This book has resulted from an Americal Chemical Society, Division of Environmental Chemistry symposium on disinfection by-products in water treatment, held in August of 1993. That symposium and this book are intended to continue the tradition of the six "Chlorination Conferences" organized under the leadership of Dr. Robert L. Jolley, now retired from Oak Ridge National Laboratory, and who brought the proceedings of each conference to publication in a classic series.

As Bob Jolley believed, we do also believe that understanding of the disinfection of drinking water is of great importance to enable wise decisions to be made in the balance between potentially harmful products of the disinfection process and the necessary protection from biological disease agents. The chapters contained in this book are intended to continue development of this understanding.

Contents

PART I
General Aspects

CHAPTER 1

Disinfection By-Products: Current Practices and Future Directions

Jeffrey L. Oxenford

INTRODUCTION

Drinking water utilities are preparing for major changes due to the proposed regulations for disinfection by-products (DBPs). The purpose of this chapter is to provide a prospective on the current practices and future research directions.

BACKGROUND AND REGULATORY REQUIREMENTS

Trihalomethanes (THMs) were first reported in drinking water in 1974.[1,2] This was of great concern, since chlorination—which has been used since the early 1900s—had been the major weapon for preventing waterborne disease. In 1979, a maximum contaminant level (MCL) was established for total THMs (TTHMs) at 100 µg/l.[3] The MCL was based on the annual average of four quarterly samples.

In 1981, the U.S. Environmental Protection Agency (EPA) published guidance for controlling THMs. For THM removal, aeration and activated carbon adsorption were discussed as effective technologies. Precursors to THM formation could be controlled by (1) oxidation by ozone or chlorine dioxide, (2) clarification by coagulation, precipitative softening, or direct filtration, or (3) adsorption by activated carbon. Other control technologies included oxidation with potassium permanganate, lowering the pH, or moving the point of chlorination to the end of the plant. Improvement of the source water quality was also mentioned, with the benefits of reducing precursors and chlorine demand.

The guidance document stated that THMs can be controlled by using alternative oxidants such as chloramines, chlorine dioxide, and ozone while maintaining the bacteriological quality of the water. Disadvantages of each alternative were discussed, such as ozone not leaving a residual, chloramines

1-56670-136-8/96/$0.00+$.50

**Table 1   Maximum Contaminant Levels and Maximum Residual Disinfectant
Level Goals to Be Proposed in Stage 1 of the Disinfection/
Disinfection By-Products Rule[4]**

Maximum contaminant levels for DBPs
80 μg/l TTHMs[a]
60 μg/l HAAs[b]
10 μg/l bromate
1.0 mg/l chlorite
Maximum residual disinfectant level goals
4 mg/l chlorine
4 mg/l chloramines
0.8 mg/l chlorine dioxide

[a] Sum of the four trihalomethanes: chloroform, bromodichloromethane, dibromochloromethane, and bromoform.
[b] Sum of five haloacetic acids: monochloracetic acid, dichloroacetic acid, trichloracetic
acid, monobromacetic acid, and dibromoacetic acid.

not being as effective for disinfection as chlorine and having toxicological
properties, and the inorganic contaminants (chlorite and chlorate) associated
with chlorine dioxide. The document also states but does not elaborate that
other by-products may be produced by the alternative oxidants.

Beginning in 1992, water utilities, environmental groups, and EPA began
regulatory negotiations on disinfection and DBPs. Due to the complexity of
the issue and the amount of information still needed, regulations will be
proposed in two stages.

Stage one will be proposed in 1994 and will provide MCLs for four
classes of compounds and maximum residual disinfectant goals for chlorine,
chloramines, and chlorine dioxide.[4] Table 1 lists these requirements. Enhanced
coagulation and granular activated carbon (GAC) were proposed as the best
available technologies for precursor control.

The regulation will be revisited in 1998. It is anticipated that MCLs will
be lowered for the THMs and haloacetic acids (HAAs). MCLs may also be
developed for other by-products.

To provide the information necessary for the second stage, the information
collection rule (ICR) has been proposed.[5] The rule will require utilities serving
a population greater than 10,000 to begin monitoring for microbial contaminants and DBPs. Surface water utilities serving a population greater than
100,000 and possessing raw water total organic carbon (TOC) exceeding 4
mg/l and groundwater utilities serving greater than 50,000 people with a
finished water TOC of greater than 2 mg/l must test GAC and membrane
treatment at the bench or pilot scale.

Another regulation that will play a significant role in the DBP issue is the
surface water treatment rule (SWTR).[4] This rule will require utilities to ensure
adequate safeguards from microbial contamination. Components of the rule
include filtration requirements for surface water facilities and disinfection
capabilities to ensure 99.9% removal of parasites and 99.99% removal of

Table 2   Percentages of Utilities Exceeding Total THM Levels of 80 and 100
           μg/l from the WIDB

	80 μg/l	100 μg/l	Number
Surface water			
Annual average[a]	5.1%	1.0%	583
Maximum sample[b]	33.4%	17.6%	551
Groundwater			
Annual average	1.2%	0.0%	321
Maximum sample	14.2%	5.7%	247

[a] Annual average is the average of four samples taken quarterly over a 1-year period.
[b] Maximum sample is the highest quarterly value.

viruses. This rule, combined with the DBP rule, brings a difficult challenge
to utilities—ensuring disinfection while minimizing the DBPs produced.

CURRENT WATER QUALITY

The statistics provided in this paper were developed from the Water Industry Database (WIDB). The database was developed by the American Water
Works Association Research Foundation (AWWARF) and the American Water
Works Association (AWWA) and contains data from 1128 utilities that serve
populations greater than 10,000. Data included in the database accounts for
systems serving approximately 50% of the U.S. population. However, because
the majority of all systems serve less than 10,000, the WIDB only accounts
for 2% of the total community water systems. Data were collected from 1989
to 1992. New regulations and treatment modifications to meet future rules
may impact the utility practices since the survey. This is important to keep
in mind when reviewing the data presented in this paper.

The percentages of utilities exceeding the 100 μg/l MCL are given in
Table 2. Approximately 1% of the 583 surface water utilities had THM values
exceeding 100 μg/l (the current MCL) based on the annual average of four
quarterly samples. No groundwater utilities exceeded this value. When looking
at the individual quarterly samples, 17.6% of the surface water and 5.7% of
the groundwater facilities had at least one quarterly sample exceeding 100
μg/l. If the THM MCL is lowered to 80 μg/l, as is proposed in the stage-one
regulations, data from the WIDB indicate that approximately 5% of the surface
water and 1.2% of the groundwater facilities would need to reduce their THMs.

In the WIDB only 20% of the surface water and 11% of the groundwater
facilities reported TOC data (Table 3). Of those reporting data in the WIDB,
the average raw water TOC was 4.7 mg/l in surface water and 3.4 mg/l in
groundwater. The median value for groundwater was 0.78 mg/l and for surface
water was 3.9 mg/l. These values are somewhat higher than may be expected.
One possible reason is that the facilities that are monitoring TOC are monitor-

Table 3   Average and Median TOC Values Reported in the WIDB

	Reporting/ Total	Average mg/l	Median mg/l
Groundwater	68/591	3.4	0.78
Surface water	138/703	4.66	3.9

ing because they expect high levels of TOC. Amy et al.,[6] in a recent survey of 100 sites (both groundwater and surface water) found the average TOC to be 2.73 mg/l and the median to be 2.07 mg/l. Estimates during the development of the ICR proposal[7] were that 220 surface water utilities and 33 groundwater utilities would be required to evaluate treatment options to remove TOC.

INDUSTRY PRACTICES

Figure 1 shows the common points for oxidant addition at a water utility. The following sections will identify the common points of oxidant addition and reasons for their use.

PREOXIDATION

Oxidant may be added in transmission lines prior to entering a treatment facility for taste and odor control, to get a head start on disinfection, and to minimize biological growth in the line. Length of transmission lines may vary from relatively short distances to many miles. Biological growth in the line can significantly reduce the carrying capacity. One organism in particular, the zebra mussel, has been affecting transmission lines in the Great Lakes and has reduced flows by as much as 50%.[8] Oxidants, primarily chlorine, have been used for their control.

Oxidant is added during the rapid mix phase of conventional treatment (coagulation, flocculation, filtration) to achieve disinfection, taste and odor control, iron and manganese removal, to improve coagulation and filtration,

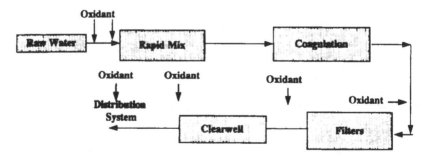

Figure 1   Possible points for addition of oxidant in a water treatment plant.

Table 4   Disinfection Practices of Surface Water Utilities
           From the WIDB

	Population > 50,000 (%)	Population < 50,000 (%)
Preoxidation/disinfection		
Chlorine	64	67
Chlorine dioxide	9	12
Chloramines	15	8
Ozone	0.7	0.7
Potassium permanganate	31	35
Postdisinfection		
Chlorine	71	85
Chlorine dioxide	2	3
Chloramines	25	2
Number reporting	427	276

as well as to keep biological growth from colonizing in treatment basins. New research suggests that oxidation prior to coagulation may be necessary for arsenic removal.[9,10]

Table 4 shows the relative usage of preoxidants from the WIDB. For surface water facilities, approximately 65% of the surface water facilities use chlorine as a preoxidant. There is no significant difference in preoxidant usage between large- and medium-sized systems. The other major preoxidant is potassium permanganate which is used by over 30% of the surface water systems. There is no indication in the WIDB as to whether use of the preoxidant is seasonal or practiced continually. Intermittent oxidation is commonly used to control taste and odor and/or growth in the plant.

INTERMEDIATE OXIDATION

Oxidant may be added prior to filtration to minimize biological growth on the filters, achieve additional disinfection, and oxidize iron and manganese. Oxidant may also be present in the filter backwash water. In the past, utilities in the U.S. have discouraged biological growth in filters. The concern has been over microorganisms passing into the product water. Numerous recent research projects including those by Glaze and Weinberg[11] and Price[12] have documented that biological treatment can be beneficial for reducing the by-products of ozonation and for removal of natural organic matter (NOM).

POSTDISINFECTION

Water leaving the filters is disinfected and usually stored prior to distribution. The concentration of the disinfectant and the amount of time held in storage is related to the disinfection effectiveness. Residual levels may also be boosted up at various points in the distribution system to ensure that a

Table 5   Disinfection Practices of Groundwater Utilities
          From the WIDB

	Population > 50,000 (%)	Population < 50,000 (%)
Any disinfectant	88.4	85.8
Chlorine	84	83
Chlorine dioxide	0.9	0.3
Chloramines	10	5
Ozone	0	0.3
Potassium permanganate	4	6
Number reporting	232	359

disinfectant residual is maintained. In the U.S., utilities are commonly required to maintain a residual to the furthest point in the distribution system.

As shown in Tables 4 and 5, most utilities use chlorine as the postdisinfectant. Chloramines are used in approximately 25% of the large surface water utilities and to a much lesser extent by groundwater utilities and smaller surface water utilities.

BY-PRODUCTS ASSOCIATED WITH OXIDANT USE

BY-PRODUCTS ASSOCIATED WITH CHLORINE

Numerous by-products have been found with chlorine. Classes of compounds include

THMs
HAAs
Haloacetonitriles
Halopicrin
Cyanogen chloride and bromide
Chloral hydrate
MX, EMX

THMs and HAAs are considered to be the major by-products associated with chlorine. MX has been the major mutagenic (as determined by Ames testing) compound identified in drinking water.[13] The list above accounts for only a small percent of the total organic halides (TOX) found in most waters. This indicates that there are still many other halogenated by-products that have escaped detection.

Chlorine can also have contaminants associated with it when sodium hypochlorite is used. Chlorate has been recently reported as being a decomposition product of sodium hypochlorite.[14] An ongoing project is evaluating ways to minimize chlorate formation from sodium hypochlorite.[15]

Table 6   Research Efforts on Oxidation by Utilities
Serving >50,000 People from the WIDB

	Recent/ongoing (%)	Future (%)
Ozone	14	18
Chlorine dioxide	10	9
Chloramines	14	8
Chlorine	22	6
KMnO$_4$	13	6

ALTERNATIVE OXIDANTS

Many oxidants are being considered as alternatives to chlorination to reduce the THMs and HAAs. These include chloramines, chlorine dioxide, ozone, UV, potassium permanganate, and other oxidation technologies. Research efforts on oxidation by utilities serving more than 50,000 is presented in Table 6.

While many of the alternative oxidants will help to minimize the THMs, other by-products will be produced. Table 7 shows some of the common classes of by-products associated with the alternative oxidants. More research is needed to determine other by-products.

For chloramines some of the typical chlorination by-products are often observed, but at lower levels. The presence of chlorination DBPs may be due to the common practice of forming the chloramines by the addition of ammonia

Table 7   Classes of By-Products Associated with
Alternative Oxidants

Chloramines	Haloacetic acids(HAAs)
	Nitrite/nitrate
	Cyanogen chloride/bromide
	1,1-Dichloropropanone
	Trihalomethanes (THMs)*
	?**
Chlorine dioxide	Chlorite
	Chlorate
	?
Ozone	Bromate
	BDOC
	Aldehydes
	Ketoacids
	Bromoform
	Peroxides
	Epoxides
	?
UV	?
KMnO$_4$	?

* If formed by chlorination followed by ammonia addition.

?** By-products may be formed that still have not been identified.

following the addition of chlorine. However, chloramination has been shown to produce some halogenated by-products itself. Singer[16] has shown that dichloroacetic acid (DCAA) may be formed at concentrations similar to and cyanogen at levels greater than chlorination when using preformed monochloramine. Chloramine use may also lead to nitrification in the distribution system.[17,18]

Ozonation has been shown to lead to the formation of aldehydes, ketoacids, bromoform, peroxides, and bromate.[11] Proposed regulations for bromate may limit the application of ozone in high (>0.1 mg/l) bromide waters.[6] Other by-products such as epoxides may be produced, but have evaded current analytical capability. Ozone often increases biodegradable dissolved organic carbon (BDOC) or assimable organic carbon (AOC) content of the water. This can lead to bacterial regrowth problems in the distribution system for certain waters. Also, ozone does not carry a residual into the distribution system and a postdisinfectant will be needed.

For chlorine dioxide, the majority of studies have been concerned with the inorganic by-products chlorite and chlorate. Limited work has been done on the organic by-products associated with chlorine dioxide as well as by-products associated with UV and potassium permanganate.

COMBINATION OF OXIDANTS

Oxidants are commonly used in combination. Table 8 shows some possible combinations. When potassium permanganate, UV, chlorine dioxide, or ozone are used as a preoxidant, chlorine or chloramines must generally be used to provide a disinfectant residual in the distribution system. Some systems that use chloramines may use free chlorine throughout the plant to achieve disinfection, followed by postammonification to produce chloramines in the distribution system.

Combinations of oxidants and their impact on DBP formation are the subject of much current research. Singer[16] has evaluated the impacts of ozonation on chlorination and chloramination by-products. Collins et al.[19] have evaluated the impact of UV on chlorination by-products. By-products not previously identified may also occur with the these combinations. Dietrich and Hoehn[20] found that chlorine dioxide followed by free chlorine can lead to the formation of a "cat urine" or "kerosene" odor at the consumer's tap in homes with new carpeting. They postulated that the chlorine reacts with the

Table 8   Possible Combinations of Oxidants

Ozone/chloramines or chlorine
Prechlorination/ozone/postchlorine or chloramines
UV/chlorine or chloramines
Chlorine dioxide/chlorine or chloramines
Potassium permanganate/chlorine or chloramines

chlorite to produce free chlorine dioxide in the distribution system. The chlorine dioxide then reacts with an airborne substance from new carpeting to produce the odor.

PRECURSOR CONTROL/REMOVAL

The majority of the DBP control strategies focus on removal of the precursors. NOM, often expressed as TOC or dissolved organic carbon (DOC), is considered to be the major precursor to DBP formation. NOM is very site specific and research into its character is providing some important insights for control.[21]

Precursor removal can be grouped into three different categories: control at the source, physical/chemical removal, and oxidation/transformation. Control at the source requires managing the inputs into the watershed. Physical/chemical removal includes coagulation, adsorption, and membrane separation. Oxidation/transformation involves processes that change the form of NOM.

CONTROL AT THE SOURCE

Cooke and Carlson[22] identified parameters that can be used in a water supply management program to reduce precursors. They stated, however, that the effectiveness of these programs on reducing the concentration of THM precursors is unknown.

Cotsaris et al.[23] described a detailed evaluation on one watershed in Australia. They found that the major factor influencing DOC transport into the water is the adsorption capacity of the soils. In areas with low exchange capacity it may be possible to improve the adsorption capacity of the soils by adding an adsorbing material such as lime, gypsum, or possibly alum sludge.

PHYSICAL/CHEMICAL REMOVAL

Three techniques—membranes, enhanced coagulation, and adsorption—can be used for removal of NOM. These techniques vary in effectiveness, cost, operational complexity, and residuals produced.

The greatest removals can be achieved using membranes (up to 95% removal of NOM) but the process is expensive and residuals are a major issue. The residuals include a concentrated brine which can be difficult to dispose. Issues surrounding disposal are discussed in a report by Mickley et al.[24] Reject water from the membrane system can be of concern in some areas, ranging from 10 to 20% of the water entering the facility.[25]

One of the simplest strategies for utilities already using conventional coagulation is enhanced coagulation. Enhanced coagulation may involve an increase in the coagulant dose, lowering of the pH, or changing the coagulant.

There are concerns that the "enhanced" conditions may affect turbidity removal, increase residual disposal, affect corrosion control, and increase contaminants, such as aluminum, in the finished water.[25]

Adsorption of NOM can be achieved using GAC, powdered activated carbon (PAC), or other adsorbing materials.[26] GAC is currently recognized as the most viable adsorption option. Optimization for precursor removal is the subject of a current GAC project.[21] Regeneration and other disposal options were recently reviewed by McTigue and Cornwell.[27]

OXIDATION/TRANSFORMATION

Oxidation can remove NOM by direct oxidation to carbon dioxide, improving coagulation, or by increasing the biodegradability of the NOM. Direct oxidation of the NOM using most oxidants is relatively minor, on the order of 10 to 20%. An innovative photocatalytic oxidation process[28] has achieved up to 90% removal; however, this process is not economically feasible at this time.

Oxidants may improve overall NOM removals by increasing the removals achieved by coagulation. Reckhow et al.[29] showed that removals of NOM using ozone prior to coagulation were quite variable and highly dependent on raw water quality. They concluded that iron, organic matter, and algae were all important on ozone's effect on subsequent coagulation. Systems must conduct pilot testing to determine if their source waters are susceptible to the coagulating effects of ozonation. O'Melia[30] has investigated the mechanism further.

Ozonation coupled with biological treatment seems to hold promise for precursor reduction. It is generally believed that ozone breaks the NOM into smaller pieces, which are more amenable to biodegradation. Biodegradation is then carried out in a biologically active filter. An additional benefit could be a lower demand for postchlorination, thereby minimizing the formation of by-products.

REMOVAL OF DBPs

After the DBPs have formed it may be possible to remove them by subsequent treatment processes. Air stripping and GAC were proposed by the EPA[31] as techniques for removal of the THMs. However, the discovery of HAAs makes air stripping less attractive, and GAC has low capacities for the THMs, especially chloroform. Removal of chlorite, associated with chlorine dioxide, was investigated by Hoehn[32] and Knocke and Iatrou.[33] Research indicated that the use of ferrous iron is the most effective technique for reduction of chlorite. The majority of ozonation DBPs are biodegradable; however, bromate is not. Bromate removal is the subject of current research.[34]

While options for removing DBPs after they are formed are needed, the goal should be to minimize the formation of the DBPs in the first place through precursor removal, manipulation of water quality parameters, and minimizing the use of oxidants while still achieving adequate disinfection.

SUMMARY AND CONCLUSIONS

The DBP issue continues to be one of the most challenging issues facing water utilities today. Utilities want to minimize the potential long-term health impacts associated with the DBPs while maintaining adequate disinfection. Effective disinfection must remain the primary concern of water utilities.

Utilities must also consider the cost, reliability, and impacts on other treatment processes. By solving one problem, they must be sure not to create another. Preoxidation is such a case. Even though preoxidation leads to increased formation of DBPs, many utilities will continue to use it. Preoxidation is needed for many treatment objectives and it will require delicate balancing between these objectives and DBP control.

Before changing oxidants, it should be assumed that all oxidants will produce by-products. By changing oxidants to reduce THMs and HAAs, potential by-products with the alternative oxidant should be considered. It should be remembered that chlorination DBPs have been studied since the 1970s and to date only a small percentage of the halogenated DBPs have been determined. Additional research is still needed on the by-products produced by the alternative oxidants.

Removal of the precursors should be the preferred method for controlling DBPs. Enhanced coagulation, GAC, membranes, and ozone/biological treatment are currently the best options available. The performance of these options will vary based on water quality and other treatment objectives. Cost, operational complexity, and environmental factors will need to be evaluated for each option. Removing the DBPs after formation while technically possible should be a secondary consideration.

Research continues to be needed. The AWWA Research Foundation has sponsored 72 projects on disinfection and DBPs, with an investment of almost $20 million. At a recent workshop[35] it was estimated that an additional $30 to 50 million would be needed in the next 5 years. Cooperative efforts among the water suppliers, research community, and EPA will be needed to address these issues.

DISCLAIMER

The comments and views expressed in this paper are those of the author and not the AWWA Research Foundation.

ACKNOWLEDGMENT

The author would like to thank Vern Achtermann, the Water Industry Database Manager at the American Water Works Association, for his assistance in developing the data on water industry practices.

REFERENCES

1. Rook, J. J., Formation of haloforms during chlorination of natural water, *Water Treat. Exam.*, 23, 234, 1974.
2. Beller, T. A., Lichtenberg, J. J., and Kroner, R. C., The occurrence of organohalides in chlorinated drinking water, *J. AWWA*, 66(12), 703, 1974.
3. *Fed. Reg.*, 44(231), 68624, 1979.
4. Pontius, F. W., D-DBP rule to set tight standards, *J. AWWA*, 85(11), 22, 1993a; *Fed. Reg.*, 54(124), 27486, 1989.
5. *Fed. Reg.*, 59(28), 6332, 1994.
6. Amy, G., Siddiqui, M., Zhai, W., DeBroux, J., and Odem, W., Survey of Bromide Ion Concentrations in Drinking Water Sources and Impact on DBP Formation, AWWA Research Foundation, Denver, 1994.
7. DBP Technologies Workgroup, unpublished communication, May 12, 1993.
8. Van Cott, W., Controlling Zebra Mussels at Water Treatment Plant Intakes, AWWA Research Foundation, Denver, 1996.
9. Waer, M. A., Arsenic Removal Using Potassium Permanganate, presentation at the 1993 AWWA Water Quality Technology Conference, Miami, November 8, 1993.
10. Cheng, R. C., Arsenic Removal through Enhanced Coagulation, presentation at the 1993 AWWA Water Quality Technology Conference, Miami, November 8, 1993.
11. Glaze, W. H. and Weinberg, H., Identification and Occurrence of Ozonation By-Products in Drinking Water, Order No. 90625, AWWA Research Foundation and American Water Works Association, Denver, 1993.
12. Price, M. P., Ozone and Biological Treatment for DBP Control and Biological Stability, AWWA Research Foundation, Denver, Order #90649, 1994.
13. Christman, R. F., Kronkerg, L., Sing, R., Ball, L. M., and Johnson, J. D., Identification of Mutagenic By-Products from Aquatic Humic Chlorination, Order No. 90568, AWWA Research Foundation and American Water Works Association, Denver, 1990.
14. Bolyard, M., Fair, P. S., and Hautman, D. P., Sources of chlorate ion in US drinking water, *J. AWWA*, 85(9), 81, 1993.
15. Gordon, G. et al. Minimizing Chlorate Ion Formation in Drinking Water When Hypochlorite Ion is the Chlorinating Agent, AWWA Research Foundation, Denver, Order #90675, 1994.
16. Singer, P. C., Impacts of Ozonation on the Formation of Chlorination and Chloramination By-Products, AWWA Research Foundation, Denver, 1996.
17. Kirmeyer, G. J., Foust, G. W., Pierson, G. C., and Simmler, J. J., Optimizing Chloramine Treatment, Order No. 90620, AWWA Research Foundation and American Water Works Association, Denver, 1993.

18. Kirmeyer, G. J., Foust, G. W., Pierson, G. C., and Simmler, J. J., Occurrence of Nitrification in Chloraminated Water Systems, AWWA Research Foundation, Denver, 1995.

19. Collins et al., Evaluation of the By-products Produced by the Treatment of Groundwaters with Ultraviolet Radiation, AWWA Research Foundation, Denver, 1995.

20. Dietrich, A. M. and Hoehn, R. C., Taste-and Odor Problems Associated with Chlorine Dioxide, Order No. 90589, AWWA Research Foundation and American Water Works Association, Denver, 1991.

21. Owen, D. M., Amy, G. L., and Chowdhury, Z. K., Characterization of Natural Organic Matter and its Relationship to Treatability, Order No. 90631, AWWA Research Foundation and American Water Works Association, Denver, 1993.

22. Cooke, G. D. and Carlson, R. E., Reservoir Management for Water Quality and THM Precursor Control, Order No. 90569, AWWA Research Foundation and American Water Works Association, Denver, 1989.

23. Cotsaris, E., Bursill, D. B., Nelson, P. N., and Oades, J. M., The influences of soil properties on dissolved organic matter in stream water, in Natural Organic Matter in Drinking Water: Origin, Characterization, and Removal, Order No. 90671, AWWA Research Foundation and American Water Works Association, Denver, 1994.

24. Mickley, M., Hamilton, R., Gallegos, L., and Truesdell, J., Membrane Concentrate Disposal, Order No. 90637, AWWA Research Foundation and American Water Works Association, Denver, 1993.

25. AWWA Research Foundation, Disinfection By-Product Precursor Control, Video Report, #DII-1, 1994a.

26. Benjamin, M. M., Chang, V. J., Li, C. W., and Korshin, G., NOM Adsorption Onto Iron-Oxide-Coated Sand, Order No. 90632, AWWA Research Foundation and American Water Works Association, Denver, 1993.

27. McTigue, N. E. and Cornwell, D., The Hazardous Potential of Activated Carbons Used in Water Treatment, AWWA Research Foundation and American Water Works Association, Order No. 90637, Denver, 1994.

28. Hand, D. W., Crittenden, J. C., and Periam, D. L., Destruction of DBP Precursors Using Photoassisted Heterogeneous Catalytic Oxidation, Order No. 90622, AWWA Research Foundation and American Water Works Association, Denver, 1993.

29. Reckhow, D. A., Edzwald, J. K., and Tobiason, J. E., Ozone as an Aid to Coagulation and Filtration, Order No. 90643, AWWA Research Foundation and American Water Works Association, Denver, 1993.

30. O'Melia, C. R., Optimizing Ozonation for Turbidity and Organics Removal by Coagulation and Filtration, AWWA Research Foundation, Denver, 1996.

31. U.S. Environmental Protection Agency, Treatment Techniques for Controlling Trihalomethanes in Drinking Water, EPA-600/2-81-156, Cincinnati, 1981.

32. Hoehn, R. C., Control of Chlorine Dioxide, Chlorite, and Chlorate Residual, AWWA Research Foundation and American Water Works Association, Denver, Order #90656, 1994.

33. Knocke, W. R. and Iatrou, A., Chlorite Ion Reduction by Ferrous Iron Addition, Order No. 90627, AWWA Research Foundation and American Water Works Association, Denver, 1993.

34. Amy, G. and Siddiqui, M., Strategies to Control Bromide and Bromate, 1994b.
35. Bellamy, W. D. and McGuire, M. J., Report from Expert Workshop on Microbial and Disinfection By-Product Research Needs, AWWA Research Foundation and AWWA Water Utility Council, Denver, 1994.
36. Collins et al., Evaluation of the By-products Produced by the Treatment of Groundwaters with Ultraviolet Radiation, AWWA Research Foundation, Denver, 1995.

CHAPTER 2

Minimizing Disinfection By-Product Formation while Ensuring *Giardia* Control

Robert C. Andrews and M. J. Ferguson

ABSTRACT

An investigation was conducted to examine conventional disinfection options that would ensure control of *Giardia* and viruses yet minimize disinfection by-products (DBPs) and conform to disinfectant-disinfection by-product regulations. Primary issues were the types of disinfectants applied (chlorine or chlorine dioxide in combination with chloramines) and their point of application in a large Canadian drinking water treatment plant. Each application point was associated with a different chemical environment that affected disinfectant effectiveness and DBP formation. The effects of implementing different disinfection options on both inorganic and organic DBP formation were evaluated at bench and pilot scale for various raw water conditions. Examination of extremely cold (winter) and high organic content (spring) waters, where ensuring adequate disinfection has typically been difficult, were of special interest. Analyses were conducted for disinfectant residuals, inorganic by-products (chlorite, chlorate), organic by-products (adsorbable organic halides, trihalomethanes, haloacetonitriles, haloketones, haloacetic acids, chloropicrin, and cyanogen chloride), and odors. Promising disinfection strategies were also investigated using a simulated distribution system methodology that allowed the effect of distribution system conditions on DBP formation to be observed. Results have shown that the use of free chlorine as a primary disinfectant followed by chloramination to achieve a calculated overall 4.0 log *Giardia* reduction (following filtration) would not be compromised by the production of organic DBPs.

INTRODUCTION

The surface water treatment rule (SWTR) introduced by the EPA[1] was designed to protect the public from adverse health effects associated with the waterborne occurrence of *Giardia lamblia*, viruses, *Legionella*, coliforms, and heterotrophic bacteria. A *CT* (concentration of residual disinfectant × time of contact) requirement is defined in the rule to ensure that water quality goals for *Giardia* and virus reduction are achieved. An equally important issue when considering changes to disinfection practice is to assess the impact of increasing *CT* for microbial control on the production of DBPs. Lykins et al.[2] examined the application of four disinfectants at pilot scale in terms of both by-product production and *CT*. These authors concluded that no single disinfectant was applicable in terms of controlling DBPs. Problems in maintaining trihalomethane (THM) concentrations within compliance levels while satisfying a required *CT* for *Giardia* inactivation using free chlorine were reported by Hess et al.[3] It is anticipated that the effects of major variables such as precursor type and concentration; disinfectant type, concentration, and contact time; and temperature and pH will have a major impact on the formation of subsequent DBPs for a given microbial control strategy.

Although prior studies have to a limited extent described disinfectant-disinfection by-product (D-DBP) production associated with *CT* requirements for *Giardia* inactivation, there has been no reported study that examined DBP formation when disinfection requirements for *Cryptosporidium* inactivation are considered. The need for reliable DBP data associated with disinfection requirements to minimize microbial hazards has been expressed by Sobsey et al.[4] and Glaze et al.[5] such that appropriate risk assessments may be conducted and subsequently allow informed comparisons of microbiological vs. chemical risks to be completed.

A study was conducted at the E.L. Smith Water Treatment Plant (WTP), Edmonton, Alberta, Canada, to examine alternative disinfection options in terms of their potential to produce both organic and inorganic DBPs in both finished water and simulated distribution system (SDS) samples. Although use of more powerful disinfectants such as ozone are gaining increased popularity in the U.S., and to limited extent in Canada, the scope of this study is limited to the evaluation of disinfectants currently in use (chlorine dioxide and chloramines) or that could easily be introduced by the City of Edmonton (free chlorine). A flow diagram of the existing E.L. Smith Plant illustrating individual processes and chemical addition points is shown in Figure 1a. Chlorine dioxide has been applied at the plant since 1982, primarily for control of taste and odor events that occur in the spring as a result of surface runoff.

Properly designed treatment plants employing filtration and disinfection routinely achieve 99.9% reduction (3 log) in viable *Giardia* cysts and a 99.99% reduction (4 log) in viruses.[6] On the basis of this criteria, the SWTR Guidance

Manual[1] stipulates that water treatment plants employing both filtration and disinfection must achieve a minimum 99.9% and 99.99% reduction in *Giardia* cysts and viruses, respectively. The Manual[1] provides a credit of 2.5 log *Giardia* reduction (2.0 log virus reduction) to conventional water treatment plants that are properly operated and are capable of achieving filtered water turbidities of less than or equal to 0.5 NTU in 95% of monthly measurements. Reductions of up to 3.0 log for *Giardia* and viruses in well-operated conventional treatment plants have been reported by LeChevallier and Norton,[7] Logsdon et al.,[8] and Roebick et al.[9] The "*CT*" concept as described in the manual requires that an adequate disinfection concentration (C) is maintained in the treatment process for a period of time (T) which is long enough to inactivate the most disease-causing organisms. The T value used in calculations is actually a T_{10} value that represents the time it takes for 10% of the water to pass through a given unit process. *CT* values required to obtain a given level of microbial inactivation have been calculated in the SWTR Guidance Manual for individual disinfectants over a wide range of water temperatures and pH values.

The actual level of *Giardia* and virus reduction required for a given treatment process remains a function of the raw water quality.[1] In the absence of actual *Giardia* monitoring data, the recommended level of overall treatment that should be provided is based on source water characteristics. The Manual suggests that the locations of discharges or other possible polluting activities upstream of a water supply intake should be considered when determining the actual amount of reduction that is required. For water supplies where sewage or agricultural discharges occur upstream, a required treatment level that would attain an overall 5.0 log reduction of *Giardia* cysts should be maintained, whereas the minimum required 3.0 log reduction would be sufficient in cases where no significant microbiological contamination from human sources is evident. A 4.0 log reduction requirement would generally be applicable to water sources that fall between these two extremes. At the time that the study in Edmonton was undertaken it was known that agricultural discharges from feedlots did exist upstream of the treatment facility. As well, preliminary raw water *Giardia* monitoring data suggested that the influent cyst concentration would likely be in the range of 1 to 10 cysts per 100 liters. The SWTR suggests that for this range of cyst concentrations a minimum 4.0 log reduction be provided. To address both of these criteria a 4.0 log *Giardia* reduction target was selected for all disinfection investigations. A credit of 2.5 log *Giardia* reduction is allocated by the SWTR to physical removal by filtration, thereby leaving disinfection to account for a 1.5 log inactivation, in cases where a 4.0 log overall reduction/inactivation is required.

The primary objective of this study was to determine optimal disinfectant types and feed locations to satisfy disinfection requirements for year-round water quality conditions while minimizing the formation of DBPs. Due to the

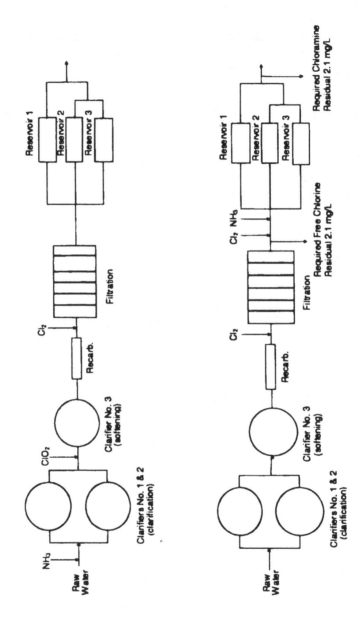

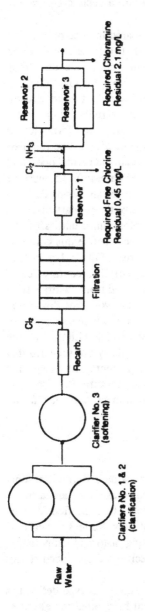

Figure 1 Comparison of pilot-scale chlorine residuals required for phase I and II to existing E.L. Smith WTP practice—utilizing chlorine dioxide. a (Top): Existing E.L. Smith WTP (Estimated overall *Giardia* Reduction of 2.9 log); b (middle): Parallel Reservoir Operation Alternative (High Free Chlorine Residual, HR-CL₂); c (bottom): Series Reservoir Operation Alternative (Low Free Chlorine Residual, LR-Cl₂).

low temperatures experienced during winter months, these conditions have been shown to be the most difficult in which to obtain a desired *Giardia* or virus reduction.

On the basis of bench-scale studies, which compared the use of chlorine dioxide to free chlorine, two alternative scenarios that utilized free chlorine as the primary disinfectant were selected for evaluation at pilot scale. Pilot trials pursued only disinfection alternatives that could achieve a 4.0 log reduction/ inactivation in *Giardia* cysts and an associated 5.0 log reduction in viruses. Results obtained from prior bench experiments tentatively ruled out the possible use of chlorine dioxide as a primary disinfectant due to the formation of inorganic by-products (chlorite, chlorate) which when added to the required residual would exceed the upper limit defined in the treatment plants' operating license (0.5 mg/l as chlorine dioxide + chlorite + chlorate). To increase contact time and correspondingly lower required disinfectant residuals for a desired *CT*, one pilot scenario was designed to maximize the primary disinfectant contact period by postponing the addition of chloramines until after the contact time provided by a 45-ML reservoir. In future practice this could be accomplished by operating on-site reservoirs in-series (Figure 1c) instead of using the existing parallel configuration (Figure 1b). The contact period associated with the primary disinfectant would then extend from recarbonation through filtration to the effluent of the first series reservoir. For discussion purposes, this scenario in which the free chlorine contact is associated with filtration and reservoir storage, thus requiring a low residual (LR) for a high contact period, is defined as LR-Cl$_2$ (Figure 1c). Similarly the scenario that utilizes parallel-reservoir operation and provides a lower contact time requiring a high chlorine residual (HR) is defined as HR-Cl$_2$ (Figure 1b). Chlorine residuals required to provide a 4.0 log overall reduction/inactivation in *Giardia* are shown in Figures 1b and c.

SCHEDULING OF PILOT TRIALS

As raw water conditions were known to vary widely during seasonal climatic changes, it was necessary to define typical seasonal characteristics such that appropriate disinfection residual values could be identified for each water quality type and to allow scheduling of pilot runs. Ranges of seasonal water quality parameters were defined (Table 1) using historical temperature, turbidity, and color data, such that the extent of specific seasonal occurrences could be estimated.

Four pilot runs were scheduled for the predominant water type defined as "winter normal" such that a statistical "t-test" could be applied to determine if alternative disinfectant strategies produced DBPs that were different from each other and the status quo. This approach was previously employed by Miltner et al.[11] in a study that evaluated the formation and control of DBPs. Prior to beginning each pilot run an analysis of the raw water was conducted

Table 1 Parameter Ranges for Seasonal Raw Water Classification

		Parameter ranges		
Type	Classification	Temperature (°C)	Turbidity (NTU)	Color (TCU)
I	Winter normal	<2	<10	<10
II	Spring runoff	<2	<10	>10
III	Spring breakup	<7	>10	>10
IV	Summer rains	7–17	>15	>5
V	Summer flood	>17	>5	>5
VI	Summer normal	7–17	<15	>5
VII	Fall normal	2–7	<10	<15

to verify that parameter values were representative of the seasonal water classification scheduled for evaluation.

Upon completion of the winter normal comparison runs, the typical water type that represented 37% of the days expected during a given year had been accounted for. This period, referred to as "phase I" also addressed conditions during which the highest required CT (coldest temperature) values occurred (Figure 2). Additional pilot runs during phase II comprised of "spring runoff" (Figure 3) and "spring breakup" conditions allowed disinfectant and DBP data to be collected under the most severe treatment conditions. Maximum by-product production was expected to occur during these periods when organic loadings are at a maximum.

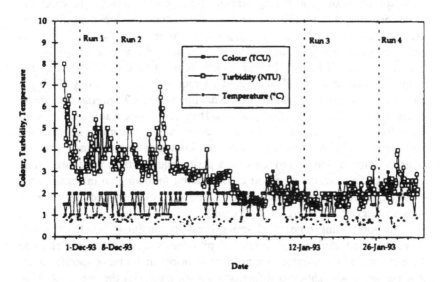

Figure 2 Water quality conditions for phase I (winter conditions) runs 1 to 4.

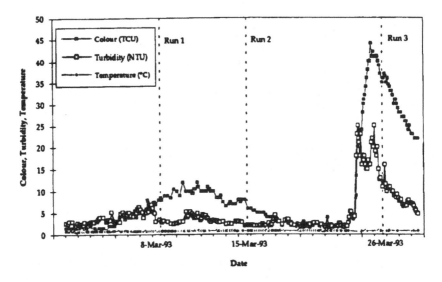

Figure 3   Water quality conditions for phase II (spring runoff conditions) runs 1 to 3.

EXPERIMENTAL DESIGN

Pilot studies were designed to evaluate two alternative chlorination scenarios that employed chlorine as the primary disinfectant under winter normal, spring runoff, and spring breakup treatment conditions. To calculate reservoir contact times (T) and required disinfectant residual concentrations (C), a ratio of the T_{10} to $T_{theoretical}$ contact times (T_{10}/T_{theor}) of 0.5 was assumed to represent future improved hydraulic characteristics, following the installation of baffles. Earlier bench studies had shown that excessive primary disinfectant residuals would be required if the existing T_{10}/T_{theor} ratio of 0.17 was not increased prior to disinfecting to *CT* criteria for *Giardia* inactivation. In scenario HR-Cl_2 (Figure 1b) free chlorine was added immediately following pH adjustment (recarbonation) with chloramines being added to a sample of pilot plant water collected immediately following filtration. This sample was allowed to react for a period equivalent to the contact time of the existing parallel reservoir configuration following an increase in the T_{10}/T_{theor} ratio to 0.5. Scenario LR-Cl_2 (Figure 1c) examined the use of a series reservoir operating configuration.

For both spring runoff and spring breakup conditions bench-scale runs were also conducted that coincided with pilot runs such that the use of chlorine dioxide could be evaluated as a primary disinfectant for these specific water conditions. It was anticipated that odors associated with the use of chlorine could potentially occur during these periods. Process water used in all bench-scale experiments was obtained from the pilot plant following recarbonation. Required chlorine dioxide residuals were calculated in a similar manner as described previously for free chlorine, to provide a 1.5 log *Giardia* inactivation (4.0 log overall) and an associated 3.0 log (5.0 log overall) virus inactivation. Required primary and secondary disinfectant residuals for bench-scale experiments are shown in Figures 4a and b. To differentiate between the runs conducted for free chlorine (pilot scale) and for chlorine dioxide (bench scale) the designation "ClO_2" was added to the HR or LR scenario which evaluated the alternative use of chlorine dioxide in the bench studies. Although bench-scale scenarios HR-ClO_2 and LR-ClO_2 were designed in terms of contact times to represent the pilot plant scenarios HR-Cl_2 and LR-Cl_2, it was recognized that any impacts of filtration could not be addressed. All bench-scale experiments were conducted under refrigerated conditions such that temperatures would represent those encountered at the full-scale treatment facilities.

MONITORING PARAMETERS

Routine analyses included monitoring of free chlorine and chloramine residuals, pH, temperature, nonpurgeable total organic carbon (NPTOC), and flavor profile analyses (FPAs). DBP analyses included the following classes of compounds that are extractable using pentane (PE-DBPs): trihalomethanes, haloacetonitriles, haloketones, and chloropicrin. Other classes of DBP analyses included haloacetic acids and aldehydes. Cyanogen chloride and adsorbable organic halide (AOX) analyses were conducted on a contractual basis by the University of Alberta, Edmonton, Canada. AOX provided an estimate of halogenated compounds that were not accounted for in analyses designed to target specific by-products.

RESULTS AND DISCUSSION

PHASE I—WINTER NORMAL WATER QUALITY CONDITIONS

Confirmation of Water Quality Conditions

Raw water quality conditions for the four pilot runs conducted during winter normal conditions are shown in Figure 2. For all four runs, the color (<10 TCU), turbidity (<10 NTU), and temperature (<2°C) values were within the parameter limits previously defined for this water type (Table 1).

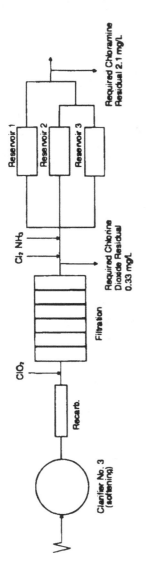
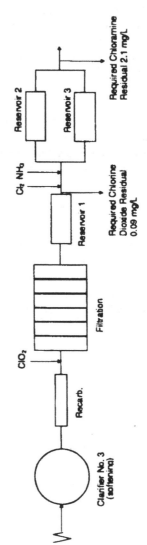

Figure 4　Comparison of bench-scale chlorine dioxide residuals required for phase II — spring runoff and spring breakup water quality conditions. a (Top): Bench-Scale Simulated Parallel Reservoir Operation Alternative (High Chlorine Dioxide Residual, HR-ClO₂); b (bottom): Bench-Scale Simulated Parallel Reservoir Operation Alternative (Low Chlorine Dioxide Residual, LR-ClO₂).

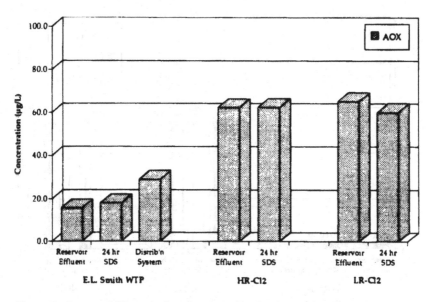

Figure 5 Average AOX results for pilot phase I—winter normal.

Halogenated Disinfection By-Product Results

Results presented for phase I pilot runs 1 to 4 (winter normal) are an average of data obtained from four individual runs that were conducted at similar water quality conditions. These runs compared the use of free chlorine in parallel pilot runs at calculated residual concentrations that would be required to obtain 1.5 log inactivation (4.0 log overall reduction) of *Giardia* cysts and a corresponding 3.0 log inactivation (5.0 log overall reduction) in viruses.

AOX results were examined as a possible surrogate for individual halogenated DBPs. A comparison of AOX formation for the two pilot scenarios and the existing E.L. Smith WTP is shown in Figure 5. In the E.L. Smith WTP, AOX concentrations increased from an influent raw water value of 6.7 µg Cl per liter to a reservoir effluent (finished water) value of 15.7 µg Cl per liter following the addition of both chlorine dioxide and chloramines. These existing disinfection conditions, however, would only provide an estimated 2.9 log overall *Giardia* reduction/inactivation. Following a 24-h contact period under simulated distribution system (SDS) temperature conditions, the AOX concentration increased to 18.2 µg Cl/l. However, this value represented only 64% of the AOX measured in actual distribution system samples. In contrast large increases were observed in both pilot trains where chlorine was employed as a primary disinfectant followed by chloramines. For the pilot plant scenario HR-Cl$_2$ an average finished water AOX concentration of 62.4 µg Cl per liter was observed following chloramination. Similarly an AOX concentration of

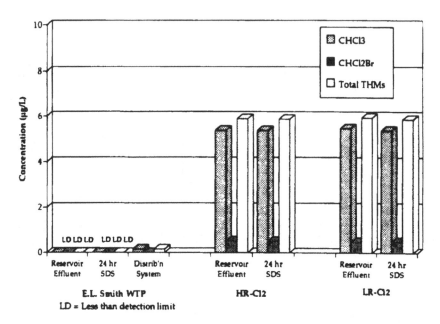

Figure 6 Average trihalomethane results for pilot phase I—winter normal.

65.4 µg Cl per liter was formed in scenario LR-Cl$_2$ following chloramination. Of the AOX present in pilot finished water 85 to 100% was present following chlorination, prior to the addition of chloramines. AOX SDS concentrations in pilot plant water either remained constant or decreased slightly, when compared to finished water values (Figure 5), suggesting that further increases in the distribution system should not be a problem.

Total trihalomethane (TTHM) concentrations associated with all E.L. Smith WTP samples were always less than 1 µg/l and typically at or below detection limits (Figure 6). Chloroform, the major component of TTHMs, was observed to increase in either pilot train following the addition of chlorine. The average chloroform value following chlorination in the LR-Cl$_2$ pilot train of 5.0 µg/l was slightly higher than the corresponding sample point in the HR-Cl$_2$ train of 4.2 µg/l. These data suggest that, for the water matrix under consideration, kinetics for chloroform formation in cold weather temperatures are slow. Almost identical values of chloroform and corresponding TTHMs were observed for finished waters from both pilot trains, suggesting that in terms of THM production little would be gained from using a lower chlorine residual in combination with a longer contact time. TTHMs for the winter period averaged 5.9 µg/l for finished water from both pilot trains, a value that is much lower than the proposed 100 µg/l proposed Canadian limit.[11] Finished water TTHM values from either pilot train were essentially identical to their corresponding SDS values. These data show that, as reported earlier

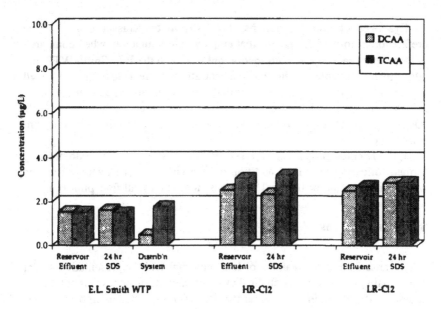

Figure 7 Average haloacetic acid results for pilot phase I—winter normal.

for AOX, once formed, TTHMs would not continue to increase through the distribution system for winter normal-type water quality conditions.

Analyses were conducted for dichloroacetic acid (DCAA) and trichloroacetic acid (TCAA) since these were shown by Krasner et al.[12] to be the most prevalent of the haloacetic acid (HAA) compounds observed in a U.S. nationwide survey. The formation trend for both DCAA and TCAA generally followed that observed for TTHMs, except at lower concentrations. Lykins et al.[2] made a similar observation when comparing individual pilot streams that were disinfected with chlorine dioxide and free chlorine. Unlike THMs, however, both DCAA and TCAA were present in all E.L. Smith WTP samples. Formation of trichloroacetic acid (TCAA) was consistently higher for the pilot train which maintained a high chlorine residual for a short contact time (HR-Cl$_2$; Figure 7). The average observed DCAA concentration for both pilot trains' reservoir effluent samples was the same at 2.5 μg/l. For both DCAA and TCAA only small increases were noted following an SDS contact time of 24 h. In the E.L. Smith WTP, TCAA formation was similar to that observed in pilot studies, whereas the concentration of DCAA was observed to decrease from 1.5 μg/l in the reservoir effluent to below the detection limits in an actual sample obtained from the distribution system following a similar contact time. The same decrease was not observed in the 24-h SDS suggesting that for DCAA actual distribution system behavior may not be easily modeled in the laboratory (Figure 7).

Cyanogen chloride (CNCl) has been shown by Krasner et al.[13] to be preferentially formed in systems that employ chloramination, when compared to those that practice free chlorination only. In both the E.L. Smith WTP and pilot streams, cyanogen chloride concentrations were below the 0.14 μg/l detection limit for all samples. Similarly, chloropicrin concentrations were less than the detection limit of 0.17 μg/l in all samples. Dichloroacetonitrile (DCAN) concentrations were the same at 0.6 μg/l for both pilot train finished waters.

1,1,1-Trichloropropanone (1,1,1-TCP) was observed to be predominantly formed following the initial addition of free chlorine. The average finished water concentration of 0.5 μg/l was consistent among all four pilot runs.

Summary of Phase I DBPs

Because data were obtained for four pilot runs under representative winter water quality conditions, a statistical evaluation of mean values was conducted to assess differences in DBP formation. Data from selected monitoring points were compared using a t-test at the 95% confidence level in a similar manner as that described by Miltner[14] to assess differences between pilot scenarios employing high vs. low chlorine residuals.

There was no observed difference in the formation of monitored DBPs when comparing a high chlorine residual, low contact time to a low chlorine residual, high contact time (Table 2). Therefore, for winter water quality no preference could be given to either chlorination scenario in terms of DBP formation. Similar results were obtained when comparing 24-h SDS results for the same scenarios (Table 3).

Table 2 DBP t-Test between High (Pilot Train 1) and Low (Pilot Train 2) Chlorine Residuals for Phase I Reservoir Waters

| | Concentration μg/l | | t-Test (between high and low Cl₂) | | |
Compound	HR-Cl₂	LR-Cl₂	Better with high residual	Same	Better with low residual
Total THMs	5.9	5.9		X	
CHCl₃	5.4	5.5		X	
CHCl₂Br	0.5	0.5		X	
Chloropicrin	LD	LD		X	
DCAN	0.6	0.6		X	
1,1,1-TCP	0.4	0.4		X	
Total HAAs	5.6	5.2		X	
DCAA	2.5	2.5		X	
TCAA	3.1	2.7		X	
CNCl	LD	LD		X	
AOX (μg Cl/l)	62.4	65.4		X	

Note: LD = less than detection limit; high chlorine residual = 2.10 mg/l; low chlorine residual = 0.45 mg/l.

Table 3 DBP t-Test between High and Low Chlorine Residuals for Phase I Simulated Distribution Waters

Compound	Concentration (μg/l)		t-Test (between high and low Cl₂)		
	HR-Cl₂	LR-Cl₂	Better with high residual	Same	Better with low residual
Total THMs	6.0	5.9		X	
CHCl₃	5.5	5.4		X	
CHCl₂Br	0.5	0.5		X	
Chloropicrin	LD	LD		X	
DCAN	0.6	0.5		X	
1,1,1-TCP	0.4	0.3		X	
Total HAAs	5.6	5.8		X	
DCAA	2.4	2.9		X	
TCAA	3.1	2.9		X	
CNCl	LD	LD		X	
AOX (μg Cl/l)	61.6	60.2		X	

Note: LD = less than detection limit; high chlorine residual = 2.10 mg/l; low chlorine residual = 0.45 mg/l.

Flavor Profile Analyses

Results of flavor profile analyses (FPAs) are summarized in Figure 8. Descriptors of "chlorinous" or "chlorine" were most frequently observed by the three- or four-member FPA panels to describe finished water samples from either the E.L. Smith WTP or the pilot plant. Average intensity values for the

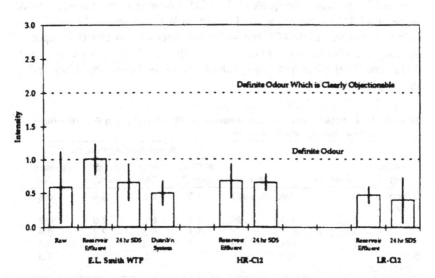

Vertical bars indicate standard deviation

Figure 8 FPA results for pilot phase I—winter normal.

two pilot trains were lower than those observed for the E.L. Smith WTP effluent (Figure 8). SDS results for E.L. Smith were slightly higher than those obtained from the distribution system. The LR-Cl$_2$ scenario was shown to have an SDS FPA intensity value that was lower than that observed for either the E.L. Smith WTP SDS or distribution system sample. In contrast, the HR-Cl$_2$ scenario produced intensity values that were similar to those obtained from the E.L. Smith WTP.

PHASE II—SPRING RUNOFF AND SPRING BREAKUP WATER QUALITY CONDITIONS

Phase II studies that included an evaluation of both spring runoff and spring breakup water quality conditions commenced on March 8, 1993. Pilot- and bench-scale runs conducted on March 8 and 15 represented spring runoff conditions exhibiting low color values of 8.5 and 8.0 TCU, respectively (Table 4, Figure 3). The first two runs were conducted despite the fact that the color values had not attained the minimum value of 10 TCU which is used to define typical spring runoff conditions. At the time that these runs were conducted, it was anticipated that a major color event would not occur in 1993. However, during the next 10 d the maximum spring 1993 color peaked at 42 TCU. Pilot- and bench-scale runs were conducted on March 26, 1993, at a color of 35 TCU (Figure 3). Because the first two runs were conducted at similar color values of 8.5 and 8.0 TCU, results were averaged and compared to those obtained at the high color peak of 35 TCU. The maximum raw water color value of 42 TCU observed at the E.L. Smith WTP on March 25, 1993, was below the average of 96 TCU but within the range of 5 to 240 TCU reported for the years 1981 to 1991 by Gammie.[15] Of these, only the years 1984, 1987, 1988, and 1991 displayed color values that were below the 1993 spring runoff peak.

Table 4   Raw Water Quality Characteristics for Phase II (Spring Runoff—Runs 1 to 3; Spring Breakup—Run 4)

Water type	Run no.	Date	NPTOC (mg/l)	Color (TCU)	Turbidity (NTU)	Temperature (°C)
Spring runoff	1	March 8, 1993	1.8	8.5	3.4	0.8
	2	March 15, 1993	1.7	8.0	2.0	0.8
	Avg. 1–2	—	1.7	8.3	2.7	0.8
	3	March 26, 1993	5.3	35.0	11.0	0.8
Spring breakup	4	April 26, 1993	1.9	5.0	110.0	1.3

Note: Spring runoff characterized as Color >10 TCU, Turbidity <10 NTU, Temperature <2°C. Spring breakup characterized as Color >10 TCU, Turbidity >10 NTU, Temperature <7°C.

The spring breakup period, defined by the raw water characteristics of temperature <7°C, turbidity > 10 NTU, and color > 10 TCU, was evaluated in pilot- and bench-scale studies as "run 4". The characteristic spring breakup turbidity peak occurred April 25, 1993. Pilot- and bench-scale studies were conducted near this peak at a turbidity value of 110 NTU.

Halogenated Disinfection By-Product Results

AOX concentrations in pilot plant finished water for phase II runs 1 and 2 (spring runoff, low color conditions) as shown in Figure 9a were approximately a factor of 2 higher than the observed average for phase I winter normal water quality conditions. This increase was most likely due to changes in the type of precursor material that was present since the average NPTOC concentrations for phase I runs 1 to 4, and phase II runs 1 and 2 at the point of chlorination, following recarbonation, was the same at a value of 1.0 mg/l. Raw water NPTOC concentrations however increased from 1.2 to 1.7 mg/l. The finished water AOX concentration was slightly higher (100 μg Cl per liter) for the LR-Cl$_2$ scenario when compared to the HR-Cl$_2$ scenario (90 μg Cl per liter). For both HR-Cl$_2$ and LR-Cl$_2$ scenarios a decrease was observed in AOX following chloramination. The highest observed AOX concentration of 180 μg Cl per liter (average of runs 1 and 2) immediately following chlorination for the HR-Cl$_2$ scenario decreased by almost 50% following chloramination.

In contrast to runs 1 and 2, AOX concentrations continued to increase for both HR-Cl$_2$ and HR-Cl$_2$ scenarios following chloramination during the spring runoff color peak (run 3). For this water type the highest AOX concentration recorded during the study of 267 μg Cl per liter was reported in the HR-Cl$_2$ finished water (Figure 9a). The NPTOC of the process water at the point of chlorination was 2.4 mg/l. AOX concentrations in finished water for both pilot disinfection scenarios decreased dramatically for the spring breakup water (run 4), coinciding with a large decease in raw water color from 35 to 5 TCU, and NPTOC concentration at the point of chlorination from 2.4 to 1.0 mg/l. AOX concentrations were lower than for phase II runs 1 and 2 (Figure 9a) despite similar NPTOC concentrations of 1.0 mg/l at the point of chlorination, again suggesting a change in precursor characteristics. AOX SDS results showed no appreciable increase following an additional 24-h contact time (Figure 9a) for any of the phase II runs. Finished water AOX concentrations for bench-scale experiments conducted using chlorine dioxide in run 3 and run 4 were between 18 and 75% lower, when compared to similar scenarios using free chlorine (Figure 9a and b).

With the exception of run 3 (spring runoff, high color) 80 to 90% of the TTHMs formed in pilot studies by either the HR-Cl$_2$ or LR-Cl$_2$ scenarios were present prior to the addition of ammonia to form chloramines. In run 3, however, only 19.7 μg/l of TTHMs (54%) were formed following the addition

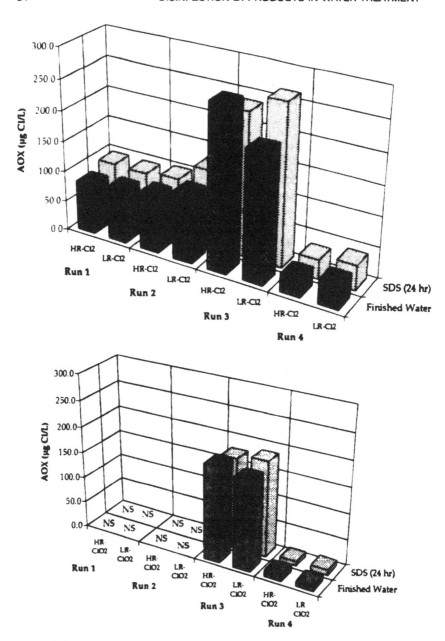

Figure 9 Comparison of finished water to SDS AOX for phase II runs 1 to 4. NS = not sampled. a (Top): pilot plant runs using free chlorine + chloramines; b (bottom): bench-scale runs using chlorine dioxide + chloramines.

of free chlorine in the HR-Cl$_2$ scenario. The highest TTHM finished water concentration obtained in this study of 36.4 µg/l was present in this run following chloramination.

As reported for phase I winter water quality conditions, chloroform was the major component of TTHMs, typically accounting for over 90% of the TTHMs. With the exception of run 3, TTHM values in the pilot LR-Cl$_2$ scenario were observed to be higher than those observed in the HR-Cl$_2$ scenario. These data are consistent with that of phase I (winter conditions), suggesting that the kinetics for TTHM formation in cold water temperatures (<2°C) are slow. The exception to this case would be during occurrences of high color and NPTOC as was the case at the peak of spring runoff in run 3.

SDS TTHM values from either pilot train showed only minor increases when compared to their corresponding finished water values (Figure 10a). The maximum SDS TTHM value observed in this study of 40.4 µg/l (run 3, HR-Cl$_2$) was only slightly higher than the value of 36.4 µg/l reported for the finished water. As previously discussed for the winter conditions of phase I, these results suggest that once formed in the treatment process THMs should not increase appreciably in the distribution system. In all cases finished water TTHM concentrations were well below the proposed Canadian limit of 100 µg/l.

As reported for AOX, finished water TTHM values displayed the same dependence on raw water NPTOC concentration, with the exception of run 4 (high turbidity) where TTHMs were observed to increase following a decrease in NPTOC in both the raw water and subsequently at the point of chlorination. This relationship suggests that more THM-specific precursors as opposed to AOX precursors were present in the raw water during this period.

THMs were only examined in bench-scale chlorine dioxide studies for run 3 (high color) and run 4 (high turbidity). Samples analyzed for THMs immediately following contact with chlorine dioxide LR-ClO$_2$ showed concentrations to be less than the detection limit for both runs 3 and 4. To minimize analyses, samples were not analyzed for the HR-ClO$_2$ scenario prior to the addition of chloramines, as it was anticipated that this scenario would produce unacceptable levels of chlorite and chlorate. TTHM concentrations following chloramination for the HR-ClO$_2$ and LR-ClO$_2$ scenarios of run 3 were 13.0 and 17.2 µg/l, respectively (Figure 10b). To achieve a *CT* credit for chloramination the SWTR requires that free chlorine be added prior to ammonia. It is anticipated that THMs were the result of the approximate 10-s exposure to free chlorine. The THM values for chlorine dioxide were much lower than the values of 36.4 and 25.1 µg/l reported for the same run when using free chlorine as the primary disinfectant (Figure 10a and b). For run 4 finished water THM values were less than the detection limits for both bench chlorine dioxide scenarios.

Finished water total HAAs (TCAA + DCAA) were similar for both low color pilot chlorination scenarios (runs 1 and 2). As previously observed for

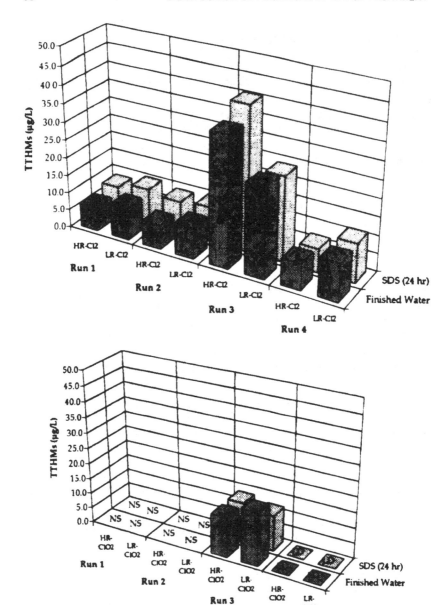

Figure 10 Comparison of finished water to SDS TTHMs for phase II runs 1 to 4 NS = not sampled; LD = less than detection limit. a (Top): pilot plant runs using free chlorine + chloramines; b (bottom): bench-scale runs using chlorine dioxide + chloramines.

AOX, concentrations of total HAAs increased by a factor of approximately 2.5 for run 3 pilot scenarios, with concentrations of 26.0 and 25.6 µg/l, respectively, being obtained for the HR-Cl$_2$ and LR-Cl$_2$ scenarios, respectively. For run 4 (spring breakup) conditions, all observed total HAA concentrations were less than 10 µg/l. With the exception of the HR-Cl$_2$ scenario for run 3, no significant increases in HAAs were observed in SDS samples (Figure 11a). HAAs formed in bench-scale experiments using chlorine dioxide were similar for both HR-ClO$_2$ and LR-ClO$_2$ scenarios (Figure 11b). For the high color peak of spring runoff (run 3) total HAA concentrations were observed at 50% of those reported in pilot studies when using free chlorine. However, similar SDS concentrations were observed as previously reported for free chlorine following 24 h of contact time.

Although analyses were conducted for only two of the five HAA components scheduled for monitoring under the D-DBP negotiated regulations (*Fed. Reg.*, 1994), TCAA and DCAA likely represent the most prevalent components.[12] At a maximum observed concentration of 26.0 µg/l (TCAA + DCAA) it is unlikely that the proposed 1997 EPA limit of 60 µg/l would be exceeded.

Dichloroacetonitrile (DCAN) concentrations in the four runs of phase II were always equal to or higher than the average concentration of 0.5 µg/l reported for the four pilot runs employing free chlorine in phase I. The formation trend of DCAN was parallel to that described for TTHMs in that, with the exception of run 3, concentration values were higher for the chlorination scenario that utilized an LR with high contact time. The highest value of 5.2 µg/l was obtained for the peak spring runoff color conditions (run 3) HR-Cl$_2$ scenario. For the same run, DCAN concentrations were always below detection limits in all E.L. Smith WTP samples. DCAN concentrations were only reported above the detection limit in bench-scale experiments using chlorine dioxide for run 3. The value of 0.9 µg/l obtained using the high chlorine dioxide residual, low contact time represented only 16% of the value reported when chlorine was used in a similar pilot scenario. No significant changes in DCAN concentrations were observed in SDS samples when considering the use of either free chlorine or chlorine dioxide.

The highest concentrations of 1,1,1-TCP occurred during peak spring runoff color conditions, with a finished water value of 3.0 µg/l for the HR-Cl$_2$ scenario. In bench-scale studies using chlorine dioxide 1,1,1-TCP concentrations were only observed above the detection limits in run 3. The highest observed value of 0.9 µg/l (HR-ClO$_2$ scenario, finished water) represented 28% of the comparable value obtained when using free chlorine. For all chlorine dioxide scenarios 1,1,1-TCP concentrations remained below the detection limit until after the addition of chloramines.

Chloropicrin (CHP) was observed in concentrations above the detection limit of 0.17 µg/l for all except one of the pilot runs (finished water) which employed free chlorine as a primary disinfectant. The maximum concentration of 0.5 µg/l was observed for the HR-Cl$_2$ scenario of run 3. For the same

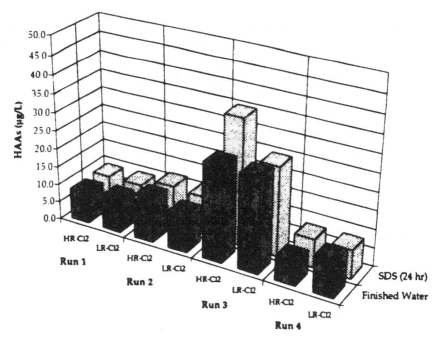

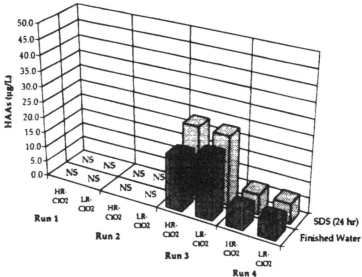

Figure 11 Comparison of finished water to SDS HAAs for phase II runs 1 to 4. NS =
not sampled; LD = less than detection limit. a (Top): pilot plant runs using
free chlorine + chloramines; b (bottom): bench-scale runs using chlorine
dioxide + chloramines.

scenario a maximum value of 0.4 µg/l was observed for finished water bench-scale experiments utilizing chlorine dioxide.

Cyanogen chloride (CNCl) was present in concentrations above the detection limit of 0.14 µg/l in all except one finished water sample for pilot plant trials using free chlorine during the three spring runoff runs. CNCl analyses were not conducted for run 4. The highest concentration of cyanogen chloride observed in phase II studies using free chlorine was for run 1 (spring runoff—low color) where concentrations of 3.3 and 2.7 µg/l were observed, respectively, for the pilot HR-Cl$_2$ and LR-Cl$_2$ scenario finished waters. With the exception of run 1 HR-Cl$_2$ scenario, CNCl concentrations continued to increase following the addition of chloramines. The high concentrations of CNCl reported for run 1 (spring runoff, low color) appeared to be independent of either the NPTOC concentration or color. In general, CNCl was observed to increase slightly in SDS samples.

For bench-scale studies conducted using chlorine dioxide, the maximum observed CNCl concentration of CNCl of 1.3 µg/l for the HR-ClO$_2$ scenario prior to chloramination (run 3) was higher than the 0.4 µg/l observed for the same scenario when using free chlorine. As for pilot studies conducted with free chlorine, an increase in CNCl formation was noted following the addition of chloramines. These observations are consistent with those reported by Krasner et al.,[13] in which CNCl values of 0.4 and 1.6 µg/l were observed in a nationwide study of finished waters disinfected with free chlorine, and chloramines, respectively. It was suggested that CNCl was formed by the reaction of humic acid with hypochlorous acid in the presence of ammonium ions. Results obtained in phase II for spring runoff and spring breakup water quality conditions were in direct contrast to those reported in phase I for winter conditions where CNCl concentrations in finished waters were always less than the detection limit.

Contribution of Known DBPs to AOX

To provide a better understanding of the percentage of AOX that could be attributed to known DBPs or conversely to identify the fraction of unknown origin, concentrations of monitored DBPs were converted from micrograms per liter to micrograms of chloride per liter to allow quantification of their contribution to AOX. For spring runoff runs 1 and 2 (low color), DBP concentration values expressed in micrograms per liter were averaged prior to being converted to micrograms of chloride per liter since both had been conducted under similar conditions. TTHMs and HAAs were the major contributors to AOX, accounting for approximately 13% of the observed AOX in finished waters. An example comparison is provided in Figure 12; complete data for all runs are shown in Tables 5 and 6. Although the concentration of AOX was 2.2 times as high for the LR-Cl$_2$ scenario when compared to the HR-Cl$_2$ scenario, the percentage of AOX that could be accounted for by identified

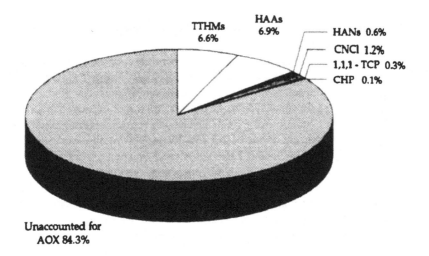

Total AOX = 89.5 µg Cl/L

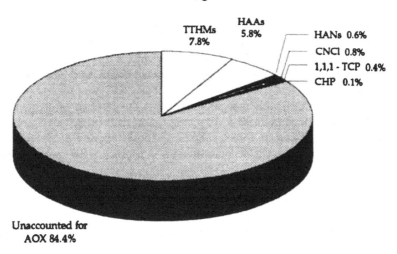

Total AOX = 201.2 µg Cl/L

Figure 12 Average contribution of known DBPs to finished water AOX for phase II pilot runs 1 to 2 (spring runoff—low color conditions). a (Top): pilot train 1 - HR-Cl₂ scenario; b (bottom): pilot train 2 - LR-Cl₂ scenario.

Table 5 Comparison of Percentage of AOX Contribution by Individual DBP Components for Phase II Pilot Runs (HR-Cl$_2$ and LR-Cl$_2$)

Class	Component	Contribution to AOX (%)					
		Avg. Run 1 and 2		Run 3		Run 4	
		HR-Cl$_2$	LR-Cl$_2$	HR-Cl$_2$	LR-Cl$_2$	HR-Cl$_2$	LR-Cl$_2$
THMs	CHCl$_3$	6.3	7.5	11.8	10.4	21.2	23.2
	CHCl$_2$Br	0.3	0.3	0.2	0.2	0.8	0.8
	CHClBr$_2$	LD	LD	LD	LD	LD	LD
	Total THMs	6.6	7.8	12.1	10.6	21.9	24.0
HAAs	DCAA	3.1	2.5	2.8	3.7	6.7	6.6
	TCAA	3.9	3.3	3.1	3.6	5.9	8.5
	Total HAAs	6.9	5.8	5.8	7.3	12.6	15.1
HANs	DCAN	0.6	0.6	1.3	1.0	1.1	1.3
Misc.	CNCl	1.2	0.8	0.1	0.3	NS	NS
	1,1,1-TCP	0.3	0.4	0.8	0.6	1.0	1.0
	CHP	0.1	0.1	0.1	0.1	0.4	0.4
Total % accounted for		15.7	15.6	20.1	19.8	37.0	41.7
Total % unaccounted for		84.3	84.4	79.9	80.2	63.0	58.3
AOX concentration (µg Cl/l)		89.5	201.2	267.1	210.3	30.0	39.2

Note: LD = concentration was less than detection limit.

halogenated DBPs (16%) was essentially the same. Miltner[14] conducted a similar pilot study where chlorine was added to filtered water and DBPs were expressed on a weight basis as a fraction of AOX. In his studies THMs and HAAs were reported to be the major contributors to AOX with respective contributions of 26.4 and 16.6%. Of the known DBPs, these two groups of compounds contributed more than 87% of the AOX that was accounted for. In pilot studies conducted at the E.L. Smith WTP a similar 87% of the AOX

Table 6 Comparison of Percentage of AOX Contribution by Individual DBP Components for Phase II Bench Runs (HR-ClO$_2$ and LR-ClO$_2$)

Class	Component	Contribution to AOX (%)			
		Run 3		Run 4	
		HR-ClO$_2$	LR-ClO$_2$	HR-ClO$_2$	LR-ClO$_2$
THMs	CHCl$_3$	6.1	8.7	LD	LD
	CHCl$_2$Br	0.1	0.1	LD	LD
	CHClBr$_2$	LD	0	LD	LD
	Total THMs	6.3	8.8	LD	LD
HAAs	DCAA	3.5	4.4	15.6	16.6
	TCAA	2.0	2.3	13.5	14.0
	Total HAAs	5.5	6.7	29.0	30.6
HANs	DCAN	0.3	0.3	LD	LD
Misc.	CNCl	0.4	0.3	NS	NS
	1,1,1-TCP	0.3	0.2	LD	LD
	CHP	0.1	LD	LD	1.8
Total % accounted for		13.0	16.2	29.0	32.4
Total % unaccounted for		87.0	83.8	71.0	67.6
AOX concentration (µg Cl/l)		183.9	173.2	12.5	10.9

Note: LD = concentration was less than detection limit.

accounted for was attributable to the combined total of THMs and HAAs for the low color spring runoff conditions (runs 1 and 2). In pilot run 3 (spring runoff, high color conditions) only 20% of the AOX could be accounted for after taking into account all monitored by-products (Table 5). As well, in this run the HR-Cl$_2$ scenario produced a higher AOX value. In run 4 (spring breakup, low color, high turbidity water) approximately 38% of the AOX could be accounted for by the known DBPs (Table 5). However, at the same time AOX concentrations decreased to approximately 15% of those observed during the high color conditions associated with run 3.

Similar results are shown for the bench-scale chlorine dioxide data for spring runoff (run 3) and spring breakup (run 4) in Table 6. For spring runoff high color conditions, chlorine dioxide was observed to produce less AOX when compared to free chlorine (183 vs. 267 μg Cl per liter); however, on a percentage basis chlorine dioxide produced slightly more unaccountable AOX. Use of chlorine dioxide during the spring breakup (high turbidity conditions) period results in a much lower production of THMs and therefore displays a larger apparent contribution of HAAs to overall AOX. However, it must be recognized that, for similar HAA concentrations, free chlorine for the same run produced two to three times as much AOX. Should free chlorine be selected as a primary disinfectant at the E.L. Smith WTP, efforts should be made to identify individual components that contribute to the large fraction of unidentified AOX.

Inorganic Disinfectant By-Products

Bench-scale disinfection studies were conducted using chlorine dioxide as the primary disinfectant for three spring runoff runs and one spring breakup run. Chlorine dioxide was added to water obtained from the pilot plant following recarbonation, immediately upstream of the free chlorine addition point. Dosing trials were conducted with at least three initial concentrations such that an appropriate dosage of chlorine dioxide could be selected to nominally obtain the desired residuals of 0.33 mg/l (HR-ClO$_2$) and 0.09 mg/l (LR-ClO$_2$), associated with the low and high contact times, respectively (Figure 4).

For the first two runs (spring runoff, low color conditions), increases were observed in both the concentration of chlorite and chlorate following the contact time associated with chloramination. This result was as expected due to the conversion of chlorine dioxide to its inorganic by-products chlorite and chlorate; however, the large increase in chlorate concentration could not be readily explained at the time. To address this effect, further bench-scale experiments were designed and conducted in an attempt to define the origin of the chlorate increase. Samples were analyzed by ion chromatography following the individual addition of chlorine dioxide, chlorine, and ammonia to reverse osmosis (RO) water and pilot plant water following recarbonation. Approximately 0.2 mg/l of chlorate was present following the addition of chlorine as sodium hypochlorite prior to the addition of ammonia to form chloramines. The presence of chlorate in solutions of hypochlorite used for disinfection in water treatment has been reported by Bolyard and Snyder-Fair.[16] Therefore, when interpreting bench-scale chlorate results following the addition of chloramines, it was assumed that values would be approximately 0.2 mg/l lower if gaseous chlorine as opposed to hypochlorite was applied in full-scale treatment.

The highest demands for chlorine dioxide in HR and LR scenarios were observed in spring runoff run 3. When comparing finished water chlorite and chlorate concentrations, large increases were noted in chlorate concentrations following chloramination whereas slight decreases were observed in the concentration of chlorite. The total concentrations of chlorate in the finished water were 0.58 and 0.54 mg/l, respectively, for the HR-ClO_2 and LR-ClO_2 scenarios. Assuming that 0.2 mg/l of chlorate was attributable to the hypochlorite solution, corrected values would be 0.38 and 0.34 mg/l, respectively. Considering that spring runoff conditions present a high chlorine dioxide demand water matrix, the observed combined residual concentrations of chlorite + chlorate + chlorine dioxide for the high and low chlorine dioxide residuals of 0.55 and 0.53 mg/l, respectively, were only slightly above the license to operate limit of 0.5 mg/l.

Similar results to those reported above were also observed for run 4 representing the spring breakup turbidity peak. For both bench-scale scenarios, a finished water chlorite + chlorate + chlorine dioxide total of 0.54 mg/l was observed, assuming as before that 0.2 mg/l of chlorate could be attributed to the hypochlorite solution.

Summary of Phase II DBPs

A summary of the DBPs observed for phase II runs 1 to 4 is shown in Figures 13 and 14 for pilot studies involving the use of free chlorine and bench-scale studies utilizing chlorine dioxide. When considering the use of free chlorine, maintaining a high chlorine residual in combination with low contact time prior to addition of chloramines would be the preferred alternative to minimize THMs, HAAs, HANs, and AOX for typical spring runoff and spring breakup conditions, except during the spring runoff color peak. For this period combination, a low chlorine residual, high contact time should minimize DBP formation for cold water temperatures ($< 2°C$) in which color exceeds 35 TCUs. In general, formation of DBPs when using chlorine as a primary disinfectant appears to be more dependent on the type of precursors associated with raw water color than the measurement of NPTOC.

A direct comparison of the use of chlorine vs. chlorine dioxide as a primary disinfectant during severe raw water conditions of high color and high turbidity are shown in Figures 15 and 16. When considering spring runoff peak color conditions (>35 TCU), chlorine dioxide applied as a primary disinfectant would be the preferred alternative to minimize DBPs. The optimal method of

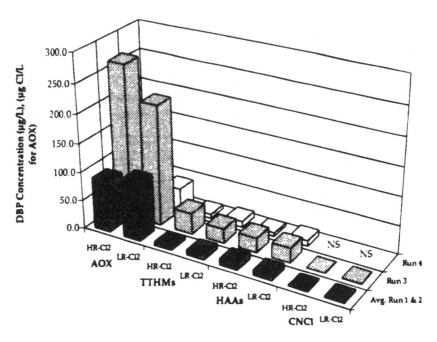

Figure 13 Comparison of DBP concentrations for phase II pilot plant finished water (free chlorine + chloramines).

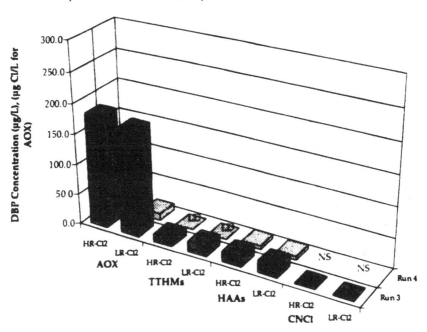

Figure 14 Comparison of DBP concentrations for phase II bench-scale finished water (chlorine dioxide + chloramines).

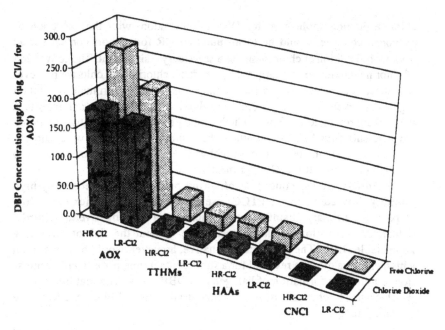

Figure 15 Comparison of finished water DBP concentration for free chlorine and chlorine dioxide—phase II run 3 (spring runoff—high color).

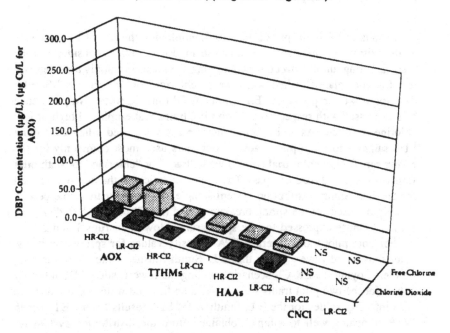

Figure 16 Comparison of finished water DBP concentration for free chlorine and chlorine dioxide—phase II run 4 (spring breakup—high turbidity, low color).

chlorine dioxide application for DBP minimization, with the exception of cyanogen chloride, would be to maintain an HR for a short contact period prior to the addition of chloramines as a secondary disinfectant. The conversion of chlorine dioxide to its inorganic by-products chlorite and chlorate however would require an increase in the existing provincial government maximum regulatory levels of 0.5 mg/l (chlorine dioxide + chlorite + chlorate) to a value of approximately 1.0 mg/l which would provide a conservative operating range. Results from bench-scale studies have shown that finished water concentrations of approximately 0.6 mg/l (chlorite + chlorate) may be required to ensure an overall 4.0 log *Giardia* inactivation.

When considering conditions of spring breakup, characterized by high turbidity, low color, low NPTOC, use of either chlorine dioxide or free chlorine as primary disinfectants to achieve a 4.0 log overall *Giardia* reduction/inactivation would be acceptable without the risk of producing appreciable DBPs. The preferred alternative, however, would be to maintain a high chlorine dioxide residual for a short contact time prior to chloramination to minimize formation of halogenated DBPs providing that the chlorite + chlorate + chlorine dioxide concentration is within an acceptable operating range.

Flavor Profile Analyses

Results of FPAs for pilot runs 1 to 4, employing the use of free chlorine as the primary disinfectant, are shown in Figure 17. When considering runs 1 and 2, spring runoff (low color) conditions, the most commonly used descriptors for pilot plant finished water were "chlorine" or "chlorinous." Similar descriptors were reported for E.L. Smith WTP finished water. No preference was expressed with respect to the two pilot trains that employed high or low chlorine residuals. As well, similar descriptors were used when describing SDS samples for both pilot trains. Descriptors used most commonly for the spring runoff high color trial (run 3) were "trace" or "chlorine" for both pilot finished water and SDS samples. These were identical to the descriptors used for the E.L. Smith distribution system sample collected for the same period. Descriptors used for the spring breakup period (run 3) of high turbidity were similar to those expressed for the spring runoff low color runs 1 and 2.

For both pilot trains, average FPA intensity values for spring runoff low color runs 1 and 2 were lower that E.L. Smith finished water, with the pilot train employing the LR-Cl_2 scenario yielding the lowest value. SDS intensity values for the two pilot trains were lower than finished water values, as were SDS intensity values for the E.L. Smith WTP. SDS results for the E.L. Smith WTP compared well to samples obtained from the distribution system for each of the four runs, confirming that the SDS methodology provided an adequate model.

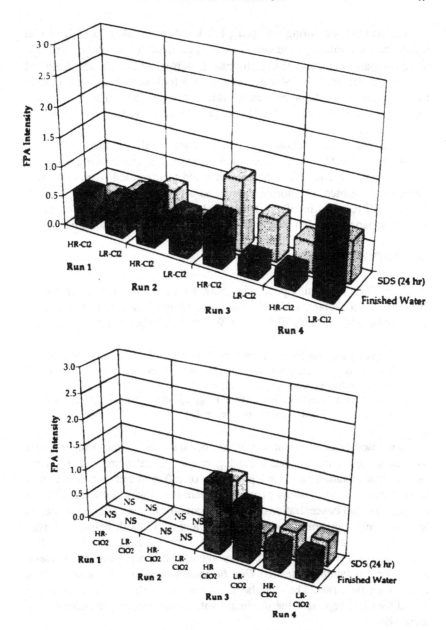

Figure 17 Comparison of finished water to SDS FPA for phase II runs 1 to 4. NS = not sampled. a (Top): pilot plant runs using free chlorine + chloramines; b (bottom): bench-scale runs using chlorine dioxide + chloramines.

Results obtained during the spring breakup bench-scale runs suggest that higher intensity values would be obtained when chlorine dioxide was applied, when compared to the use of free chlorine. Descriptors of bench-scale finished water of "chlorinous" or "chlorine" were similar for both high and low residual trials for runs 3 and 4 and compared well to descriptors used for E.L. Smith finished water. Descriptors of chlorine or chlorinous were also reported for SDS samples.

In summary, disinfection with free chlorine would produce lower FPA intensities if low chlorine residuals were maintained in combination with long contact times. No conclusive differences could be identified when using free chlorine as opposed to chlorine dioxide as a primary disinfectant. As well, no major differences in FPA intensity values or descriptors were observed when comparing the high and low chlorine dioxide residual scenarios.

OPERATIONAL STRATEGIES

To conform to existing provincial inorganic by-product regulations, minimization of disinfectant and DBPs while maximizing disinfectant effectiveness may best be achieved at the E.L. Smith WTP through the combined actions of:

1. Changing the primary disinfectant from chlorine dioxide to free chlorine.
2. Relocating the disinfectant addition locations and revising dosages to obtain residual concentrations shown in Figure 1b.
3. Increasing disinfectant contact time (T_{10}) by improving the hydraulics of three existing 45-ML reservoirs (Figure 1b).

For raw water temperatures of less than or equal to 0.5°C these changes would provide a minimum calculated 4.0 log overall *Giardia* and 5.0 log virus reduction/inactivation based on a plant flow of 180 ml/d and an 80% reservoir level. Figures 18 to 21 show plant flow operating ranges and corresponding chlorine residual concentrations required to attain a desired *Giardia* reduction for raw water temperatures of 0.5, 5, 10, and 15°C, both before and after reservoir modifications which would increase the T_{10}/T_{theor} from 0.17 to 0.5. Temperatures of 0.5, 5, and 10°C correspond to the minimum seasonal temperatures where obtaining a desired *Giardia* inactivation may be difficult for a given plant flow rate. The calculated overall *Giardia* reduction/inactivation would therefore represent the minimum values that could be obtained during these periods.

A review of Figure 18 shows that, by changing from the existing disinfection practice using chlorine dioxide to free chlorine as defined in Figure 1b, the *Giardia* reduction/inactivation at 0.5°C could be increased from 2.9 to 3.3 log at a flow of 180 ML/d without reservoir upgrades and to 4.0 log following upgrades. For a temperature of 5°C, representing minimum summer normal conditions as shown in Figure 19, a 4.0 log overall reduction/inactivation

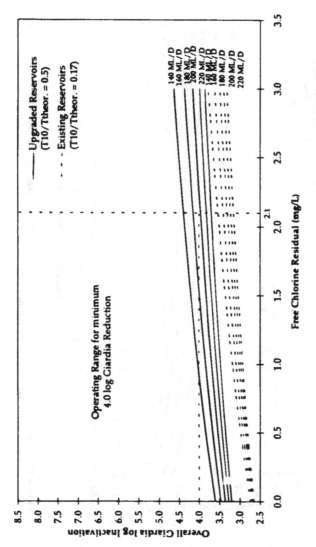

Figure 18 Impact of reservoir upgrading on free chlorine residual required to obtain a desired *Giardia* inactivation at a raw water temperature of 0.5°C.

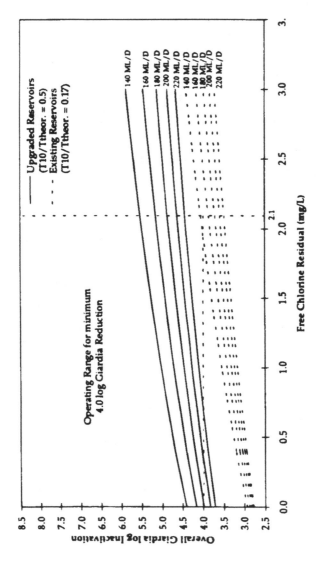

Figure 19 Impact of reservoir upgrading on free chlorine residual required to obtain a desired *Giardia* inactivation at a raw water temperature of 5°C.

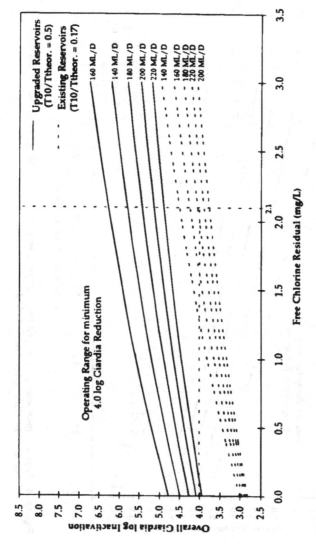

Figure 20 Impact of reservoir upgrading on free chlorine residual required to obtain a desired *Giardia* inactivation at a raw water temperature of 10°C.

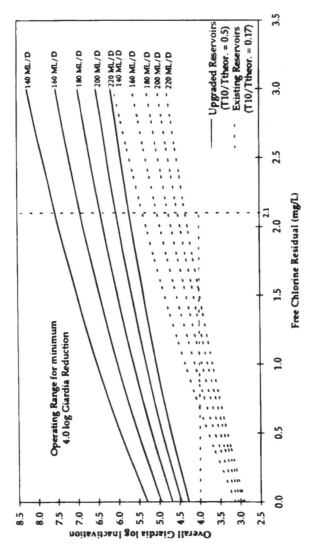

Figure 21 Impact of reservoir upgrading on free chlorine residual required to obtain a desired *Giardia* inactivation at a raw water temperature of 15°C.

could only be obtained by limiting flow to 150 ML/d prior to the completion of upgrades, and assuming a maximum chlorine residual of 2.1 mg/l. However, following upgrades the E.L. Smith WTP could obtain the same disinfection efficiency at a flow of 220 ml/d, and at a chlorine residual of only 0.7 mg/l. The *Giardia* reductions/inactivations shown at a free chlorine residual of 0 mg/l were obtained by adding the basic 2.5 log credit for coagulation/filtration to the inactivation attributable to chloramination. Figure 20 shows that, during summer rains or summer flood conditions, over 4.5 log reduction/inactivation of *Giardia* could be obtained, even at plant flows as high as 220 ml/d, following implementation of reservoir improvements.

In lieu of obtaining these high *Giardia* reductions following reservoir hydraulic improvements a 4.0 log *Giardia* reduction/inactivation could be obtained at a plant flow of 180 ml/d for chlorine residual concentrations of much less than 2.1 mg/l, whenever raw water temperatures exceed 10°C (Figure 21). In all cases a chloramine residual of 2.1 mg/l is assumed to be maintained prior to finished water entering the distribution system.

CONCLUSIONS

1. The use of either free chlorine or chlorine dioxide as primary disinfectants followed by chloramination to achieve an overall 4.0 log *Giardia* reduction/ inactivation calculated on the basis of the SWTR would not be compromised in terms of existing or proposed regulatory levels by the production of organic DBPs. However, both total trihalomethane and total haloacetic acid concentrations would exceed those produced by the existing (chlorine dioxide) disinfection practice at the E.L. Smith WTP which provides only a 2.9 log overall *Giardia* reduction/inactivation. The inorganic by-products of chlorine dioxide, chlorite, and chlorate would likely exceed the maximum 0.5 mg/l outlined in the current operating license, should the chlorine dioxide alternative be implemented.
2. Minimization of organic DBPs could best be achieved by introducing chlorine dioxide immediately prior to filtration, with free chlorine and ammonia addition prior to on-site reservoir storage. Free chlorine when introduced at the same location would produce a finished water with higher concentrations of TTHMs, HAAs, HANs, and AOX.
3. Raw water color is more important as an indicator of potential halogenated DBP production than NPTOC for the production of organic DBPs when using either chlorine or chlorine dioxide as primary disinfectants for color values in the range of 8.0 to 35.0 TCU.
4. Of the DBPs identified in this study, the major contributors to AOX for the spring runoff period are THMs and HAAs. Together these accounted for up to 87% of the identified contributors to AOX.
5. No obvious differences in flavor profile analysis descriptors or intensities are expected in E.L. Smith WTP finished water should free chlorine be applied in lieu of chlorine dioxide for typical winter or spring water quality

conditions. Chlorine dioxide residuals required to provide a 4.0 log overall *Giardia* reduction/inactivation produced higher FPA intensities when compared to free chlorine for the spring runoff color peak.

REFERENCES

1. U.S. Environmental Protection Agency, *Guidance Manual for Compliance with the Filtration and Disinfection Requirements for Public Water Systems Using Surface Water Sources*, U.S. EPA Office of Drinking Water, Washington, DC, October 1990.
2. Lykins, B. W., Goodrich, J. A., Koffskey W., and Griese, M. H., Controlling disinfection by-products with alternative disinfectants, in Proc. AWWA Annual Conf., American Water Works Association, Orlando, FL, 1991.
3. Hess, A. F., Dunn, H. J., Huntley, G. M., Affinito, J. R., Smith, D. B., and Grant, N. B., Operational and Water Quality Problems Associated with CT Compliance, in Proc. AWWA Annu. Conf., Vancouver, British Columbia, 1992.
4. Sobsey, M. D., Dufour, A. P., Greba, C. P., LeChevallier, M., and Payment, P., Using a conceptual framework for assessing risks to health from microbes in drinking water, *J. AWWA*, 85, 3, 44, 1993.
5. Glaze, W. H., Andelman, J. B., Bull, R. J., Conoly, R. B., Hertz, C. D., Hood, R. D., and Pegram, R. A., Determining health risks associated with disinfectants and disinfection by-products: research needs, *J. AWWA*, 85, 3, 53, 1993.
6. Von Huben, H., Sunface Water Treatment: The New Rules, American Water Works Association, Denver, 1991.
7. LeChevallier, M. W. and Norton, W. D., Examining relationships between particle counts and *Giardia, Cryptosporidium* and turbidity, *J. AWWA*, 84, 12, 1992.
8. Logsdon, G. S., Symons, J. M., Hoyle, R. L., and Arozarena, M. M., Alternative filtration methods for the removal of cysts and cyst models, *J. AWWA*, 73, 2, 1981.
9. Roebick, G. G., Clarke, N. A., and Dostal, K. A., Effectiveness of water treatment processes in virus removal, *J. AWWA*, 54, 10, 1275, 1962.
10. Miltner, R. J., Rice, E. W., and Stevens, A. A., Pilot-Scale Investigation of the Formation and Control of Disinfection By-products, in Proc. AWWA Annual Conf., American Water Works Association, Cincinnati, 1990.
11. Canadian Water and Wastewater Association, Memorandum Re: USA Reg/Neg for DBP Rule, CWWA, Ontario, January, 1993.
12. Krasner, S. W., Reagan, K. M., Jacangelo, J. G., Patania, N. L., Aieta, E. M., and Gramith K. M., Relationships between Disinfection By-products and Water Quality Parameters: Implications for Formation and Control, in Proc. AWWA Annual Conf., American Water Works Association, Cincinnati, 1990.
13. Krasner, S. W., McGuire, M. J., Jacangelo, J. G., Patania, N. L., Reagan, K. M., and Aieta, E. M., The occurrence of disinfection by-products in US drinking water, *J. AWWA*, 81, 8, 41, 1989.
14. Miltner, R. J., Treatment for Control of Disinfection Byproducts. Seminar on Current Research Activities in Support of USEPA's Regulatory Agenda, in Proc. 1990 AWWA Annual Conf., Cincinnati, 1990.

15. Gammie, L., City of Edmonton internal correspondence, 1992.
16. Bolyard, M. and Snyder-Fair, P., Occurrence of Chlorate in Hypochlorite Solutions Used for Drinking Water Disinfection, in Proc. AWWA Annu. Conf., Vancouver, British Columbia, 1992.
17. Federal Register, Vol. 59, No. 145, Proposed Rules: Parts II & III, EPA, July 29, 1994, pp. 33668–38858.

PART II
Chlorine By-Products

CHAPTER 3

The Impact of TOC and Bromide on Chlorination By-Product Formation

S. W. Krasner, M. J. Sclimenti, R. Chinn, Z. K. Chowdhury,
and D. M. Owen

ABSTRACT

The EPA has proposed the Disinfectants/Disinfection By-Products (D/DBP) Rule, which includes a treatment requirement to remove DBP precursors, as measured by total organic carbon (TOC), through enhanced coagulation or softening. Because this technology does not remove bromide, bromide-to-TOC ratios in the treated waters are higher. A combination of two water sources, yielding a matrix of 25 waters (five TOC levels and five bromide levels), was evaluated for the formation of trihalomethanes (THMs), haloacetic acids (HAAs), and haloacetonitriles (HANs). These and other data are being used to assess not only the effects of the precursor levels, but also the ratio of inorganic to organic precursors on DBP formation.

INTRODUCTION

REGULATORY BACKGROUND

The EPA has proposed the D/DBP Rule, under which modified and new standards have been established for trihalomethanes, other DBPs, DBP precursor levels, and disinfectants themselves.[1] Stage 1 of the proposed rule will include maximum contaminant levels (MCLs) for total THMs (TTHMs) and certain haloacetic acids; the TTHM MCL will be set at 80 μg/l, and the MCL for the sum of the five regulated HAAs (HAA5) will be set at 60 μg/l. In addition, there will be a treatment requirement for surface waters to remove DBP precursors through enhanced coagulation or softening. The removal of TOC will be used as a treatment performance indicator for compliance with

Table 1   Theoretical Health Effects of Individual THMs

THM	MCLG (µg/l)	Theoretical excess cancer risk level (µg/l)		
		10^{-4}	10^{-5}	10^{-6}
CHCl$_3$	0	600	60	6
CHCl$_2$Br	0	60	6	0.6
CHClBr$_2$	60	NA	NA	NA
CHBr$_3$	0	400	40	4

Note: NA = not available.

the precursor removal criteria. Stage 1 of the proposed rule will control DBPs through the MCLs for TTHMs and HAA5, as well as the DBP precursor control requirement.

Stage 2 of the proposed rule will remain a proposal (as a "backstop" option) until after further research can be completed and evaluated over the next 4 years. Stage 2 would lower the TTHM and HAA5 MCLs (for utilities that treat surface water and serve >10,000 people) to 40 and 30 µg/l, respectively. In addition, the use of granular activated carbon (GAC) would be considered "best available technology" for the control of DBP precursors.

Because bromodichloromethane (CHCl$_2$Br) poses a higher potential cancer risk than chloroform (CHCl$_3$) (Table 1),[1,2] the EPA had considered regulating individual THMs in stage 1. The EPA sets MCLs as close to the MCL goals (MCLGs) as is technically feasible, with cost as an additional consideration. For suspected carcinogens, the MCLG is zero. In such instances, the MCL is typically set between the 10^{-4} and 10^{-6} levels of theoretical excess cancer risk. Because precursor removal technologies do not remove bromide, bromide-to-TOC ratios in the treated waters from these processes are higher. Consequently, there is a shift to brominated by-products,[3-5] some of which pose higher health risks. Health effects data, however, are currently not available for the brominated HAAs. The EPA has the objective of developing a more complete health effects database so that the regulation of individual DBPs can be revisited in future rulemaking efforts.

During the development of the proposed D/DBP rule, there was concern over the control of other chlorination DBPs. In addition, a commercial standard for a sixth HAA, bromochloroacetic acid (BCAA), recently became available, but there were insufficient data to perform a regulatory impact analysis on the occurrence and control of BCAA. To resolve these concerns, the EPA has proposed an information collection rule (ICR)[6] which will require systems serving >100,000 people to monitor for THMs, six HAAs (HAA6), haloacetonitriles, and several other DBPs. In addition, systems with high TOC (e.g., surface waters with >4.0 mg/l) will have to conduct bench- or pilot-scale studies on the use of GAC or membranes to remove DBP precursors. These data will be used in developing a revised stage 2 proposal.

SAN FRANCISCO BAY/SACRAMENTO-SAN JOAQUIN RIVER DELTA

The San Francisco Bay/Sacramento-San Joaquin River Delta (Delta) is a source of drinking water for 20 million Californians. Sacramento River water above the Delta typically contains 1 to 2 mg/l TOC and ≤ 0.02 mg/l bromide. However, water pumped from the Delta to utilities in northern and southern California typically contains 3 to 7 mg/l TOC and 0.1 to 0.5 mg/l bromide. This degradation in water quality has presented users of Delta water with a number of challenges in meeting disinfection and DBP regulations.[7,8]

A significant source of TOC in Delta water is from agricultural tracts of land in the Delta composed of peat soil. Agricultural drainage waters from these tracts are high in both TOC (up to 30 to 50 mg/l) and THM formation potential (THMFP). Studies have indicated that the agricultural drainage in the Delta during the summer irrigation months (as well as in the winter, when channel waters are used to leach out salts that have built up in the soil during the year) can account for up to approximately one half of the THMFP of exported water.[9,10]

During drought, high-tide, and/or low-river-flow conditions, saltwater intrusion from the San Francisco Bay has resulted in higher bromide levels in water exported from the Delta (rising to as high as 0.8 mg/l). This has led to an increase in TTHMs as well as a shift to more brominated species.

OBJECTIVES

This paper presents the results of investigations into (1) the effects of TOC, bromide, and the bromide-to-TOC ratio on the formation and speciation of DBPs of health and regulatory interest and (2) their implications for compliance with pending DBP regulations. In addition, a primary objective in conducting these experiments was to generate a robust database on THM formation in Delta waters for a modeling effort.[11] THM production is a function of TOC and bromide levels, chlorine dose and contact time, temperature, etc. By varying these parameters and measuring the individual and total THMs produced, data would be available to develop empirical equations to predict THM formation. Furthermore, the Delta study results were compared to precursor removal studies of some other surface-water supplies.

EXPERIMENTAL METHODS

EVALUATION OF BROMIDE AND TOC MATRIX

To assess the possible combinations of TOC and bromide that could emanate from existing or possible future Delta transfer facilities, a five-by-five

(5 × 5) matrix of waters spanning a wide range of TOC and bromide levels (as well as bromide-to-TOC ratios) was evaluated. Sacramento River water upstream of the Delta was used as the baseline water (1.1 mg/l TOC and ≤0.01 mg/l bromide). A TOC spike was prepared from agricultural drainage water (having a TOC of 35 mg/l) sampled from a peat-soil tract of land. Thus, the TOC spike represented the actual type of natural organic matter present in the Delta. Sodium bromide was used to augment bromide levels.

Five TOC levels were evaluated, from ambient to 4 mg/l (the latter level represents the maximum TOC that has been measured in Delta water exported to Metropolitan). Five bromide levels were tested, from ambient to 0.8 mg/l. These 25 unique water qualities were chlorinated under conditions similar to those used in simulated-distribution-system (SDS) testing.[12] To assess the amount and speciation of DBPs that could be formed in actual full-scale operations, SDS-type testing was performed.[12,13]

These samples, instead of being jar-treated as is done in the SDS protocol, were filtered through a 0.45-μm filter to remove turbidity but not dissolved organic carbon (DOC). Thus, these experiments were actually based on the DOC concentrations. Measurements of both TOC and DOC in Delta waters, however, have shown that these parameters typically agree to within 10%. Filtered samples were then chlorinated under an array of temperatures and reaction times and analyzed for THMs, HAA6, and HANs. Chlorine dosages were adjusted to achieve a target chlorine residual of 0.5 to 1.5 mg/l, a residual that would be typical in Metropolitan's full-scale operations. SDS tests were conducted at 10 and 25°C at pH 8.2 for 1, 3, 8, 24, and 48 h. These temperatures were chosen to represent the impact of winter and summer water temperatures, when formation can be the lowest and highest, respectively. The incubation pH is a typical pH level that Metropolitan uses for distributing water to meet the requirements of the Lead and Copper Rule. Samples chlorinated for 1 and 3 h represent the approximate free-chlorine contact time in Metropolitan's treatment plants, whereas samples incubated for 8, 24, and 48 h allow for an evaluation of the formation of DBPs in a chlorinated distribution system.

PRECURSOR REMOVAL STUDIES

In a demonstration-scale test of the use of GAC for DBP precursor removal in Ohio River water (ORW), samples were collected at different points in the TOC breakthrough curve.[14] For SDS testing of the GAC influent, a 4.5-mg/l chlorine dose was required for the 3-d, pH 8.2, 25°C incubation (corresponding to a terminal location in that utility's distribution system with a summer water temperature). For the SDS testing of the GAC effluent, only a 2.0-mg/l chlorine dose was needed.

In a bench-scale study using a rapid small-scale column test (RSSCT), Mississippi River water was evaluated.[4] The RSSCT was run until there was a 1-mg/l TOC breakthrough; there was 0.1 mg/l TOC in the nonadsorbable

Table 2 Minimum Reporting
Levels (MRLs) of DBPs

DBP	MRL (μg/l)
THMs	
CHCl$_3$	0.1
CHCl$_2$Br	0.1
CHClBr$_2$	0.1
CHBr$_3$	0.1
HAAs	
MCAA	1.0
DCAA	0.6
TCAA	0.6
BCAA	0.8
MBAA	0.5
DBAA	0.6
HANs	
TCAN	0.03
DCAN	0.03
BCAN	0.04
DBAN	0.05

fraction. These samples were chlorinated with a 1-mg/l dose at pH 8.2 and incubated for 5 d at 25°C.

A demonstration-scale evaluation of enhanced coagulation for DBP precursor control in the California Delta water supply was performed.[15] In this study, SDS chlorination for 1 h at 25°C and pH 8.2 was performed to simulate chlorination through a clearwell (assuming postchloramination).

CHEMICAL ANALYSES

HAA analysis was typically performed for six species: monochloro-, dichloro-, trichloro-, bromochloro-, monobromo-, and dibromoacetic acid (MCAA, DCAA, TCAA, BCAA, MBAA, and DBAA).[16] Briefly, the HAA method involved a microextraction with *t*-butyl methyl ether, derivatization with diazomethane, and gas chromatography (GC) analysis. The four THMs and the four HANs (trichloro-, dichloro-, bromochloro-, and dibromoacetonitrile, or TCAN, DCAN, BCAN, and DBAN) were analyzed together in a separate extraction with pentane, followed by GC.[17]

Bromide was analyzed by ion chromatography.[16] TOC/DOC samples were analyzed by the ultraviolet (UV)/persulfate oxidation method.[16] Table 2 lists the MRLs for each DBP. These values are equal to or greater than the calculated method detection limits. To determine the accuracy of the methods, at least one out of every 10 samples was matrix spiked. For determination of the precision of the methods, at least one out of every 10 samples was analyzed in duplicate. Accuracy for THM samples was usually between 83 and 126%, and replicate samples differed by ≤11%. For HAA samples, accuracy was between 80 and 125%, and replicate samples differed by

Table 3   Evaluation of Synthetic Delta Sample

Parameter	Delta outflow	Synthetic sample[a]
DOC (mg/l)	3.65	3.53
UV (cm^{-1})	0.122	0.126
Br$^-$ (mg/l)	0.48	0.48
3-h SDS THM (μg/l)		
CHCl$_3$	12	13
CHCl$_2$Br	34	36
CHClBr$_2$	67	70
CHBr$_3$	37	38
TTHMs	150	157
24-h SDS THM (μg/l)		
CHCl$_3$	34	34
CHCl$_2$Br	65	73
CHClBr$_2$	102	117
CHBr$_3$	36	40
TTHMs	237	263

[a] Synthetic water = 90% Sacramento River water + 10% agricultural drainage + bromide spike.

≤19%. Accuracy for HANs was usually between 81 and 132%, and replicate samples differed by ≤16%.

RESULTS AND DISCUSSION

IMPACT OF BROMIDE AND DOC ON DBP FORMATION AND SPECIATION

Impact on THMs

The 5 × 5 matrix sought to address all possible combinations of DOC and bromide that might be experienced with alternative Delta transfer facilities. To ensure that an agricultural drain sample could be used as a DOC spike, a preliminary test compared a sample of Delta outflow to a synthetic sample consisting of 90% Sacramento River water (from upstream of the Delta) and 10% agricultural drainage, with an appropriate bromide spike (Table 3). The synthetic sample matched the actual Delta outflow in DOC, UV absorbance at 254 nm, and bromide levels, and similar amounts of individual and total THMs were produced.

Figure 1 shows the impact of bromide and TOC/DOC on THM speciation for the 3-h, 25°C tests in the 5 × 5 matrix. As DOC increased or bromide decreased, CHCl$_3$ formation increased. As bromide increased, bromoform (CHBr$_3$) increased. However, as DOC increased, CHBr$_3$ formation increased, reached a maximum, and then decreased. The maximum was at a different DOC level for each bromide concentration. As DOC increased, the required chlorine dosage increased. Therefore, the chlorine-to-bromide ratio increased

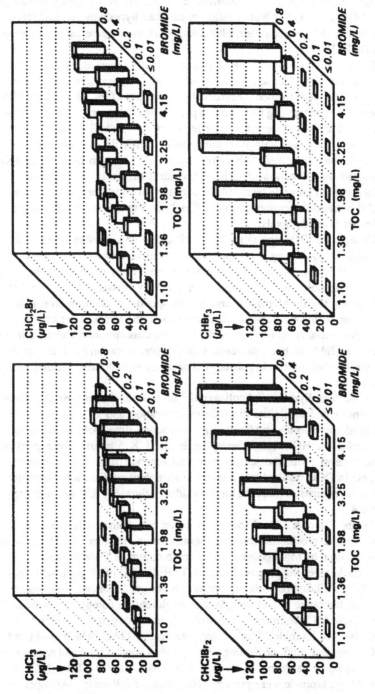

Figure 1 Impact of TOC and bromide on THM speciation (temp = 25°C, pH = 8.2, incubation time = 3 h). (Source: Krasner, S.W., Sclimenti, M.J., and Means, E.G. Quality degradation: Implications for DBP formation. Jour. AWWA, 86 [6], 34, 1994.)

as DOC increased for a fixed chlorine residual goal. Although hypobromous acid (HOBr) is a better halogenation agent than hypochlorous acid (HOCl), if sufficient chlorine is added, HOCl can start to compete more effectively with HOBr.[13]

The mixed chlorobromo THM species initially increased with increasing bromide but then decreased in concentration as more brominated species (e.g., $CHBr_3$) were formed. Dibromochloromethane ($CHClBr_2$), like $CHBr_3$, increased with increasing DOC up to a point. Then its concentration decreased, except at the highest bromide concentration at which $CHClBr_2$ formation continued to increase with increasing DOC. $CHCl_2Br$, at least under these test conditions, increased with increasing DOC. Although minimizing DOC minimized the formation of $CHCl_2Br$, control of the latter THM was best realized at high bromide levels (≥ 0.2 mg/l). At 0.1 mg/l bromide, decreasing DOC decreased the concentration of $CHCl_2Br$, but not as significantly as at the higher bromide levels.

Changes in THM speciation are attributable to a number of parameters. One means of evaluating THM speciation is by using the bromine incorporation factor n, the molar amount of bromine in the THMs divided by the molar TTHM concentration.[18] The value of n can vary between 0 and 3, where 0 corresponds to all chloroform and 3 corresponds to all bromoform. In the 5 × 5 matrix, for a particular DOC level (with the temperature and reaction time held constant), an increase in the bromide concentration yielded a higher value of n (Table 4). For a particular bromide concentration (with the temperature and reaction time held constant), a decrease in DOC also yielded a larger n (Table 4).

In addition, Symons and colleagues[13] utilized the fraction Br^-/Cl^+, where Br^- represented the initial bromide concentration and Cl^+ was the average free available chlorine (FAC) during the reaction time (both in molar units), to examine the HOBr/HOCl interplay. In the 5 × 5 matrix, for a particular bromide level (with the temperature and reaction time held constant), an increase in DOC resulted in a higher chlorine demand (i.e., a higher chlorine dose was required to meet the target residual); thus, Br^-/Cl^+ and n decreased (Table 4).

Figure 2 shows TTHM formation in the 5 × 5 matrix for the 3-h, 25°C tests. With a prechlorination/postchloramination treatment of Delta waters (which is used by Metropolitan and others to minimize THM formation), Delta water could meet the proposed stage 1 TTHM MCL of 80 μg/l, with up to 4 mg/l DOC, if bromide were not present. As bromide concentration increases, however, the range of DOC levels that would enable compliance with an 80-μg/l TTHM MCL decreases, even with enhanced coagulation (which removes DOC, but not bromide). Furthermore, the use of GAC (which could reduce DOC down to 1 mg/l, depending on empty-bed contact time and regeneration rate[4]) would be hampered in meeting a proposed stage 2 TTHM MCL of 40 μg/l if high bromide were present. (Note that, in full-scale GAC operations,

Table 4 Effects of DOC, Br⁻, and FAC on n in 5 × 5 Matrix

DOC (mg/l)	Br⁻ᵃ (mg/l)	Temp (°C)	Time (h)	Cl₂ dose (mg/l)	Cl₂ residual (mg/l)	Cl₂ demand (mg/l)	Average FAC (mg/l)	Br⁻/Cl⁺ (μmol/ μmol)	n
1.1	0.005	25	1	1.5	0.91	0.59	1.21	0.004	0.16
1.1	0.2	25	1	1.75	0.97	0.78	1.36	0.130	1.88
1.1	0.8	25	1	2	0.75	1.25	1.38	0.516	2.55
1.1	0.005	25	3	2	1.25	0.75	1.63	0.003	0.17
1.1	0.1	25	3	2.25	1.39	0.86	1.82	0.049	1.22
1.1	0.2	25	3	2.25	1.29	0.96	1.77	0.100	1.80
1.1	0.4	25	3	2.4	1.25	1.15	1.83	0.194	2.26
1.1	0.8	25	3	2.5	1.29	1.21	1.9	0.375	2.56
1.1	0.005	25	8	2.5	1.4	1.1	1.95	0.002	0.16
1.1	0.2	25	8	2.75	1.37	1.38	2.06	0.086	1.77
1.1	0.8	25	8	3	1.45	1.55	2.23	0.319	2.57
1.1	0.005	25	24	2.5	1.35	1.15	1.93	0.002	0.14
1.1	0.2	25	24	3	1.35	1.65	2.18	0.082	1.69
1.1	0.8	25	24	3.5	1.6	1.9	2.55	0.278	2.53
1.1	0.005	25	48	3	1.35	1.65	2.18	0.002	0.12
1.1	0.2	25	48	3.5	1.45	2.05	2.48	0.072	1.61
1.1	0.8	25	48	4	1.6	2.4	2.8	0.254	2.52
1.36	0.005	10	1	1.7	0.79	0.91	1.25	0.004	0.14
1.36	0.22	10	1	1.9	0.95	0.95	1.43	0.137	1.67
1.36	0.79	10	1	2.2	1.08	1.12	1.64	0.427	2.44
1.36	0.005	10	3	2	1.15	0.85	1.58	0.003	0.15
1.36	0.22	10	3	2.25	1.07	1.18	1.66	0.118	1.62
1.36	0.79	10	3	2.5	1.07	1.43	1.79	0.393	2.50
1.36	0.005	10	8	2.5	1.25	1.25	1.88	0.002	0.14
1.36	0.22	10	8	2.75	1.28	1.47	2.02	0.097	1.56
1.36	0.79	10	8	3	1.2	1.8	2.1	0.334	2.51
1.36	0.005	10	24	2.5	1.1	1.4	1.8	0.002	0.13
1.36	0.22	10	24	2.75	1.08	1.67	1.92	0.102	1.48
1.36	0.79	10	24	3	0.9	2.1	1.95	0.359	2.56
1.36	0.005	10	48	3	1.32	1.68	2.16	0.002	0.12
1.36	0.22	10	48	3.25	1.2	2.05	2.23	0.088	1.42
1.36	0.79	10	48	3.5	0.95	2.55	2.23	0.315	2.53
1.36	0.005	25	1	2	1.15	0.85	1.58	0.003	0.14
1.36	0.22	25	1	2.25	1.1	1.15	1.68	0.117	1.55
1.36	0.79	25	1	2.5	1.4	1.1	1.95	0.359	2.50
1.36	0.005	25	3	2.5	1.3	1.2	1.90	0.002	0.12
1.36	0.11	25	3	2.75	1.3	1.45	2.03	0.048	0.93
1.36	0.22	25	3	2.75	1.3	1.45	2.03	0.096	1.52
1.36	0.4	25	3	2.9	1.3	1.6	2.10	0.169	2.13
1.36	0.79	25	3	3	1.2	1.8	2.10	0.334	2.50
1.36	0.005	25	8	3	1.65	1.35	2.33	0.002	0.11
1.36	0.22	25	8	3.25	1.4	1.85	2.33	0.084	1.45
1.36	0.79	25	8	3.5	1.3	2.2	2.40	0.292	2.52
1.36	0.005	25	24	3	1.18	1.82	2.09	0.002	0.09
1.36	0.22	25	24	3.5	1.2	2.3	2.35	0.083	1.35
1.36	0.79	25	24	4	1.15	2.85	2.58	0.272	2.51
1.36	0.005	25	48	3.5	1.5	2	2.50	0.002	0.08
1.36	0.22	25	48	4	1.38	2.62	2.69	0.073	1.26
1.36	0.79	25	48	4.5	1.18	3.32	2.84	0.247	2.48
1.98	0.005	10	1	1.75	0.81	0.94	1.28	0.003	0.14
1.98	0.2	10	1	2.25	1.05	1.2	1.65	0.108	1.42
1.98	0.79	10	1	2.75	1.2	1.55	1.98	0.355	2.42
1.98	0.005	10	3	2	0.88	1.12	1.44	0.003	0.13
1.98	0.2	10	3	2.5	1.12	1.38	1.81	0.098	1.40

Table 4 (continued)

DOC (mg/l)	Br⁻ˢ (mg/l)	Temp (°C)	Time (h)	Cl₂ dose (mg/l)	Cl₂ residual (mg/l)	Cl₂ demand (mg/l)	Average FAC (mg/l)	Br⁻/Cl⁺ (μmol/ μmol)	n
1.98	0.79	10	3	3	1.19	1.81	2.10	0.335	2.45
1.98	0.005	10	8	2.75	1.25	1.5	2.00	0.002	0.12
1.98	0.2	10	8	3.25	1.5	1.75	2.38	0.075	1.32
1.98	0.79	10	8	3.75	1.5	2.25	2.63	0.267	2.45
1.98	0.005	10	24	3	1.15	1.85	2.08	0.002	0.11
1.98	0.2	10	24	3.5	1.25	2.25	2.38	0.075	1.19
1.98	0.79	10	24	4	1.25	2.75	2.63	0.267	2.44
1.98	0.005	10	48	3.5	1.3	2.2	2.40	0.002	0.10
1.98	0.2	10	48	4	1.4	2.6	2.70	0.066	1.16
1.98	0.79	10	48	4.5	1.1	3.4	2.80	0.250	2.38
1.98	0.005	25	1	2.25	1.3	0.95	1.78	0.002	0.13
1.98	0.2	25	1	2.75	1.4	1.35	2.08	0.086	1.33
1.98	0.79	25	1	3.25	1.3	1.95	2.28	0.308	2.40
1.98	0.005	25	3	2.5	1.15	1.35	1.83	0.002	0.12
1.98	0.11	25	3	2.75	1.3	1.45	2.03	0.048	0.76
1.98	0.2	25	3	3.25	1.6	1.65	2.43	0.073	1.27
1.98	0.39	25	3	3.5	1.55	1.95	2.53	0.137	1.89
1.98	0.79	25	3	3.75	1.4	2.35	2.58	0.272	2.41
1.98	0.005	25	8	3.25	1.5	1.75	2.38	0.002	0.10
1.98	0.2	25	8	3.75	1.7	2.05	2.73	0.065	1.17
1.98	0.79	25	8	4.5	1.65	2.85	3.08	0.228	2.37
1.98	0.005	25	24	3.5	1.05	2.45	2.28	0.002	0.08
1.98	0.2	25	24	4.25	1.3	2.95	2.78	0.064	1.09
1.98	0.79	25	24	5	1.15	3.85	3.08	0.228	2.34
1.98	0.005	25	48	4	1	3	2.50	0.002	0.08
1.98	0.2	25	48	4.75	1.2	3.55	2.98	0.060	1.00
1.98	0.79	25	48	5.5	1.15	4.35	3.33	0.211	2.32
3.25	0.005	25	1	3	1.25	1.75	2.13	0.002	0.10
3.25	0.23	25	1	3.75	1.5	2.25	2.63	0.078	0.94
3.25	0.86	25	1	4.25	1.35	2.9	2.80	0.273	2.20
3.25	0.005	25	3	3.25	1.05	2.2	2.15	0.002	0.08
3.25	0.12	25	3	3.6	1.2	2.4	2.40	0.044	0.47
3.25	0.23	25	3	4	1.35	2.65	2.68	0.076	0.85
3.25	0.44	25	3	4.25	1.3	2.95	2.78	0.141	1.44
3.25	0.86	25	3	4.5	1.15	3.35	2.83	0.270	2.17
3.25	0.005	25	8	4	1.05	2.95	2.53	0.002	0.08
3.25	0.23	25	8	4.5	1.35	3.15	2.93	0.070	0.79
3.25	0.86	25	8	5.25	1.15	4.1	3.20	0.238	2.08
3.25	0.005	25	24	5	1	4	3.00	0.001	0.05
3.25	0.23	25	24	5.75	1.4	4.35	3.58	0.057	0.66
3.25	0.86	25	24	6.5	1.18	5.32	3.84	0.199	1.93
3.25	0.005	25	48	5.5	0.81	4.69	3.16	0.001	0.06
3.25	0.23	25	48	6.25	1.1	5.15	3.68	0.056	0.59
3.25	0.86	25	48	7	0.85	6.15	3.93	0.194	1.83
3.25	0.005	10	3	2.75	1.05	1.7	1.90	0.002	0.10
3.25	0.23	10	3	3.25	1.3	1.95	2.28	0.090	0.96
3.25	0.86	10	3	3.75	1.1	2.65	2.43	0.315	2.24
3.25	0.005	10	8	3.5	1.35	2.15	2.43	0.002	0.11
3.25	0.23	10	8	4	1.55	2.45	2.78	0.074	0.91
3.25	0.86	10	8	4.5	1.35	3.15	2.93	0.261	2.22
3.25	0.005	10	24	3.75	0.95	2.8	2.35	0.002	0.09
3.25	0.23	10	24	4.25	1.22	3.03	2.74	0.075	0.83
3.25	0.86	10	24	4.75	1.25	3.5	3.00	0.254	2.17
3.25	0.005	10	48	4.25	1.13	3.12	2.69	0.002	0.08

Table 4   (continued)

DOC (mg/l)	Br⁻ᵃ (mg/l)	Temp (°C)	Time (hr)	Cl_2 Dose (mg/l)	Cl_2 Residual (mg/l)	Cl_2 Demand (mg/l)	Average FAC (mg/l)	Br⁻/Cl⁺ (μmol/ gmmol)	n
3.25	0.23	10	48	4.75	1.23	3.52	2.99	0.068	0.78
3.25	0.86	10	48	5.25	0.95	4.3	3.10	0.246	2.09
4.15	0.005	25	1	3.75	1.35	2.4	2.55	0.002	0.09
4.15	0.26	25	1	4.25	1.5	2.75	2.88	0.080	0.61
4.15	0.88	25	1	4.75	1.31	3.44	3.03	0.258	1.97
4.15	0.005	25	3	4	0.93	3.07	2.47	0.002	0.08
4.15	0.12	25	3	4.25	1.03	3.22	2.64	0.040	0.38
4.15	0.26	25	3	4.5	1.25	3.25	2.88	0.080	0.57
4.15	0.43	25	3	4.75	1.21	3.54	2.98	0.128	1.13
4.15	0.88	25	3	5	0.9	4.1	2.95	0.265	1.92
4.15	0.005	25	8	5	0.95	4.05	2.98	0.001	0.06
4.15	0.26	25	8	5.5	1.15	4.35	3.33	0.069	0.63
4.15	0.88	25	8	6	0.95	5.05	3.48	0.225	1.80
4.15	0.005	25	24	6	0.7	5.3	3.35	0.001	0.06
4.15	0.26	25	24	6.75	0.98	5.77	3.87	0.060	0.49
4.15	0.88	25	24	7.5	0.85	6.65	4.18	0.187	1.60
4.15	0.005	25	48	7	0.6	6.4	3.80	0.001	0.05
4.15	0.26	25	48	7.75	0.79	6.96	4.27	0.054	0.48
4.15	0.88	25	48	8.5	0.73	7.77	4.62	0.169	1.49
4.15	0.005	10	1	3.25	1.5	1.75	2.38	0.002	0.10
4.15	0.26	10	1	3.75	1.55	2.2	2.65	0.087	0.77
4.15	0.88	10	1	4.25	1.6	2.65	2.93	0.267	2.03
4.15	0.005	10	3	3.5	1.3	2.2	2.40	0.002	0.09
4.15	0.26	10	3	4	1.5	2.5	2.75	0.084	0.75
4.15	0.88	10	3	4.5	1.35	3.15	2.93	0.267	2.04
4.15	0.005	10	8	4.25	1.25	3	2.75	0.002	0.09
4.15	0.26	10	8	4.75	1.55	3.2	3.15	0.073	0.70
4.15	0.88	10	8	5	1.18	3.82	3.09	0.253	1.90
4.15	0.005	10	24	5	1.25	3.75	3.13	0.001	0.07
4.15	0.26	10	24	5.5	1.5	4	3.50	0.066	0.66
4.15	0.88	10	24	5.75	1.15	4.6	3.45	0.226	1.78
4.15	0.005	10	48	5.5	1.15	4.35	3.33	0.001	0.07
4.15	0.26	10	48	6	1.4	4.6	3.70	0.062	0.61
4.15	0.88	10	48	6.5	1.17	5.33	3.84	0.204	1.72

ᵃ The ambient bromide level was <0.01 mg/l; however, a value of 0.005 mg/l was used in estimating the bromide level in the Br⁻/Cl⁺ ratio in this table.

one could take advantage of "staged" carbon contactors, as well as less THM production, when the water temperature is colder. However, the effects of temperature on HOBr are not quite the same as for HOCl; see below.) Thus, to control the potential carcinogenic risk from THMs, both DOC and bromide must be controlled.

Impact on HAAs

Figure 3 shows the impact of bromide and DOC on HAA speciation for four of the six measured HAAs for the 3-h, 25°C tests in the 5 × 5 matrix. (Note that HAA analyses were not done for two components of the matrix:

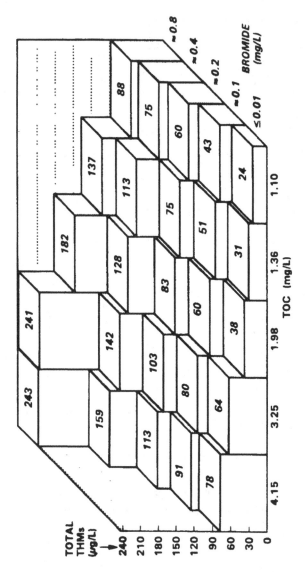

Figure 2   Impact of TOC and bromide on TTHM formation (temp = 25°C, pH = 8.2, incubation time = 3 h). (Source: Krasner, S.W., Sclimenti, M.J., and Means, E.G. Quality degradation: Implications for DBP formation. Jour. AWWA, 86 [6], 34, 1994.)

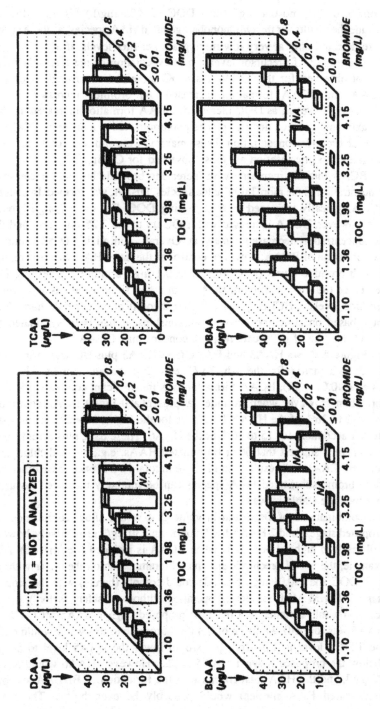

Figure 3 Impact of TOC and bromide on HAA speciation (temp = 25°C, pH = 8.2, incubation time = 3 h).

bromide of 0.1 and 0.4 mg/l for a DOC of 3.25 mg/l.) Figure 3 does not include the results for the two monohalogenated HAA species, as these species were detected only at very low microgram-per-liter levels, if at all.

The impact of precursor levels on the speciation of the HAAs is similar to that observed for the THMs. As DOC increased or bromide decreased, DCAA and TCAA formation increased. As bromide increased, DBAA increased. However, as DOC increased, DBAA formation increased, reached a maximum, and then decreased. The maximum was at a different DOC level for each bromide concentration. These maxima were similar to (but not always at the exact same location as) those observed for $CHBr_3$.

BCAA initially increased with increasing bromide but then decreased in concentration as the more brominated species, DBAA, was formed. In addition, BCAA typically increased with increasing DOC. For the 0.2-mg/l bromide test, however, BCAA increased with DOC up to the 3.25-mg/l test, and then its concentration decreased somewhat. Overall, minimizing DOC minimized the formation of BCAA; however, control of the latter HAA was best realized at high bromide levels (≥ 0.2 mg/l). At 0.1 mg/l bromide, decreasing DOC decreased the concentration of BCAA, but not as significantly as at the higher bromide levels. $CHCl_2Br$ poses a higher theoretical cancer risk than $CHCl_3$; thus, based on structural activity relationships, it has been speculated that BCAA may be of more carcinogenic concern than DCAA.[19]

Figure 4 shows HAA5 and HAA6 (i.e., HAA5 plus BCAA) formation in the 5×5 matrix for the 3-h, 25°C tests. For the ambient bromide level tested (≤ 0.01 mg/l), BCAA formation was insignificant compared to HAA5 formation. However, when bromide was present at 0.1 mg/l, BCAA contributed from 15 to 26% of the HAA6. Similarly, for the higher bromide levels tested, BCAA accounted for 15 to 30% of the HAA6. These data, plus recent studies elsewhere,[20] suggest that other brominated HAAs, e.g., bromodichloroacetic acid (BDCAA), will result in an increase in total HAAs once these other three more bromo-substituted HAA species can be routinely measured (assuming that they are stable in water).

Combinations of DOC and bromide that would allow compliance with proposed stage 1 and 2 HAA5 MCLs of 60 and 30 μg/l, respectively, however, will result in higher HAA sums when BCAA is included (i.e., HAA6). As an example, the use of prechlorination/postchloramination on the water with 4.15 mg/l DOC and ≤ 0.01 mg/l bromide could comply (barely) with the proposed stage 1 TTHM MCL (assuming operation without a "safety factor"), but this scenario would definitely not meet the proposed stage 1 HAA5 MCL (76 μg/l HAA5 was formed). Addition of 0.4 mg/l bromide to this water would double the TTHM level to 159 μg/l, but would lower the HAA5 level to 54 μg/l (below the proposed Stage 1 HAA5 MCL). However, HAA6 would now be 77 μg/l, and the HAA total (because of the formation of BDCAA and other unmeasured HAA species) would probably be even higher. Thus, more

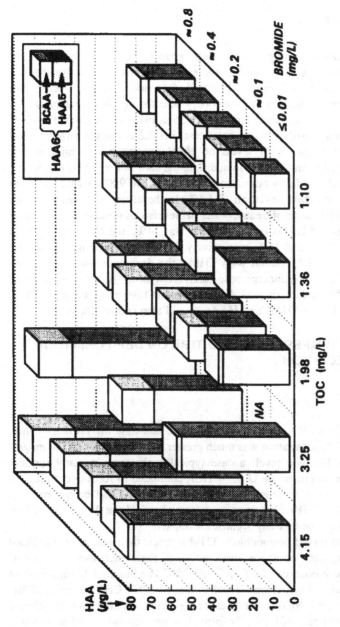

Figure 4 Impact of TOC and bromide on formation of HAA5 and HAA6 (temp = 25°C, pH = 8.2, incubation time = 3 h).

research into the occurrence and control of all nine HAAs is required to fully regulate this class of DBPs.

Impact on HANs

Figure 5 shows the impact of bromide and DOC on HAN speciation for three of the four HANs, as well as the sum of all four HANs, for the 3-h, 25°C tests in the 5 × 5 matrix. Figure 5 does not include the results for TCAN, as formation of this HAN was only detected in a few samples at 0.03 to 0.04 μg/l. TCAN was rarely detected in a 35-utility nationwide DBP study.[14] TCAN hydrolyzes easily,[21] so its absence at pH 8.2 is not surprising.

The impact of precursor levels on HAN speciation is similar to that observed for THMs and HAAs. In addition, the sum of the HANs divided by the TTHM level was fairly constant: a range of 6.1 to 9.8%, with an average of 7.4% and a standard deviation of 1.1%. In the aforementioned 35-utility survey (in which a range of treated-water pH levels was observed), the median ratio of the sum of HANs to the TTHMs was 11%, and the 25th and 75th percentiles were 7.3 and 15%, respectively.[14] Although there appears to be a correlation of HAN formation with THMs, the decomposition mechanisms are not alike. HANs can undergo base-catalyzed hydrolysis, whereas increased pH favors THM production.[22] TCAA, however, is favored at low pH.[22] Thus, the use of MCLs for THMs and HAAs should also control for HANs.

IMPACT OF TEMPERATURE AND TIME ON DBP FORMATION AND SPECIATION

Impact on THMs

Figure 6 shows the impact of temperature and time on THM formation when the DOC concentration was much greater (more than an order of magnitude) than the bromide level, a case typically observed in many surface-water treatment plants in the U.S.[23] $CHCl_3$ continued to form over time, and approximately twice as much was formed at 25°C as had been formed at 10°C (Figure 6). $CHBr_3$, however, was formed quickly at 10°C, and neither temperature nor time had any significant impact.

For the two mixed bromochloro THM species, their formation increased with increasing time or temperature, although these increases were not as dramatic as those observed for $CHCl_3$. At 25°C, the initial (1-h) formation of $CHCl_3$ (on a weight basis) was less than that of either $CHCl_2Br$ or $CHClBr_2$ in this sample, yet $CHCl_3$ was highest in concentration after 24 h of chlorine contact time. The rate of $CHCl_2Br$ formation was somewhat higher than that observed for $CHClBr_2$ in this sample. These data point out the difference in the kinetics of halogenation between HOBr and HOCl; that is, halogenation

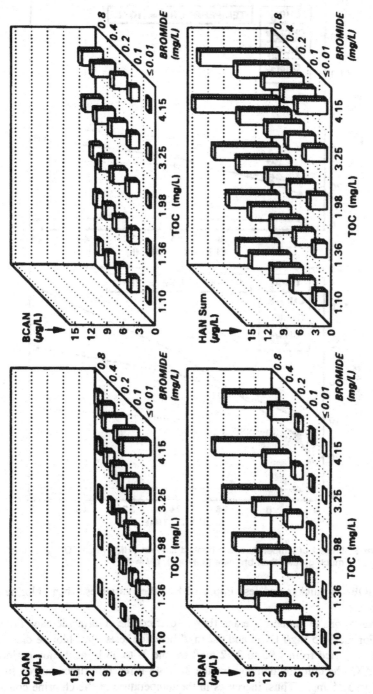

Figure 5 Impact of TOC and bromide on HAN speciation and sum (temp = 25°C, pH = 8.2, incubation time = 3 h).

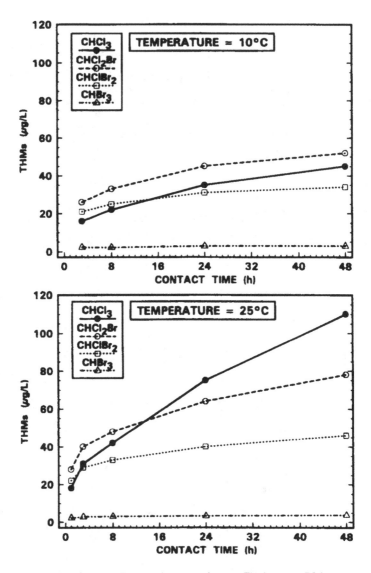

Figure 6 Impact of temperature and contact time on THMs when TOC (3.25 mg/l) is much greater than bromide (0.2 mg/l); pH = 8.2.

by HOBr is quicker and less impacted by other variables when chlorine is in excess.

Variations in temperature and time also affect the chlorine demand (Table 4). For the 3.25 mg/l DOC and 0.2 mg/l bromide tests, the chlorine demand at 10°C varied from 1.95 to 3.52 mg/l for the 3- to 48-h incubation periods. At 25°C, the chlorine demand for the 3- to 48-h reaction times ranged from 2.65 to 5.15 mg/l. Thus, increases in the temperature or the chlorine contact

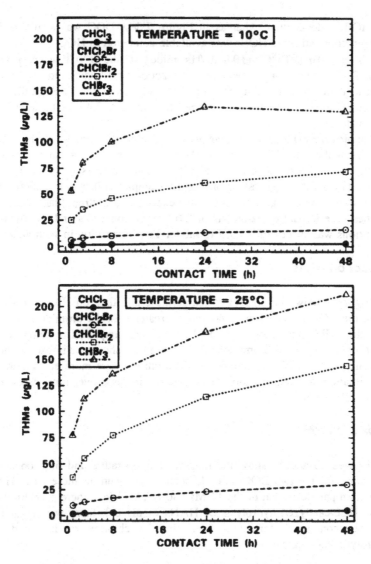

Figure 7 Impact of temperature and contact time on THMs when TOC (2.0 mg/l) and bromide (0.8 mg/l) have same-range values; pH = 8.2.

time resulted in a higher chlorine demand and, in turn, a lower Br$^-$/Cl$^+$ and n (Table 4).

In Figure 7, when DOC and bromide concentrations were more equivalent (a case that can be observed in coastal surface or groundwater heavily impacted by saltwater intrusion), CHBr$_3$ formation appears to be strongly impacted by reaction time and temperature. In Figure 6, DOC was not a limiting factor for HOBr halogenation, whereas in Figure 7 less DOC was available for halogenation and more HOBr was available compared to HOCl.

For the 2.0 mg/l DOC and 0.8 mg/l bromide tests, the Br$^-$/Cl$^+$ was relatively high. Thus, changes in temperature and reaction time, which did impact the chlorine demand, had a relatively minor impact on n (Table 4). Regardless of the temperature or time tested, n ranged from 2.32 to 2.45 for these precursor conditions.

Empirical equations were subsequently developed to predict the formation of individual and total THMs during the chlorination of Delta waters.[11] These equations utilized the data from the 5×5 matrix discussed above, as well as results from the SDS testing of monthly samples collected in Sacramento River water, the Delta, and Delta water exported to Metropolitan and other utilities. The impact of variations in DBP precursor levels and chlorination conditions was successfully used to model THM formation in such waters.

Impact on HAAs

Figure 8 shows the impact of temperature and time on HAA formation when the DOC concentration was much greater than the bromide level. As with the THMs, the chlorinated HAAs were more impacted by temperature and time than were the brominated HAAs. Figure 9 shows the impact of these variables when DOC and bromide concentrations were more equivalent. In this situation, the concentration of DBAA did increase over time, especially in the 25°C tests.

Impact on HANs

Figures 10 and 11 show the impact of temperature and time on HAN formation for the two DOC/bromide scenarios. Again, in these tests, HAN formation paralleled that of the THMs and HAAs. While there probably was some base-catalyzed hydrolysis of HANs at pH 8.2, there was a balance between continuous formation from the free-chlorine residual and the hydrolytic degradation.

COMPARISONS OF THESE RESULTS TO SOME PRECURSOR REMOVAL STUDIES

Impact of GAC on DBP Reduction and Speciation

Table 5 shows the data for the ORW GAC study at a point where there was 0.85 mg/l TOC breakthrough (2.3 mg/l TOC in the GAC influent). The removal of 63% of the TOC by the GAC enabled a 56% reduction in the chlorine dosage (and a 72% lower chlorine demand).

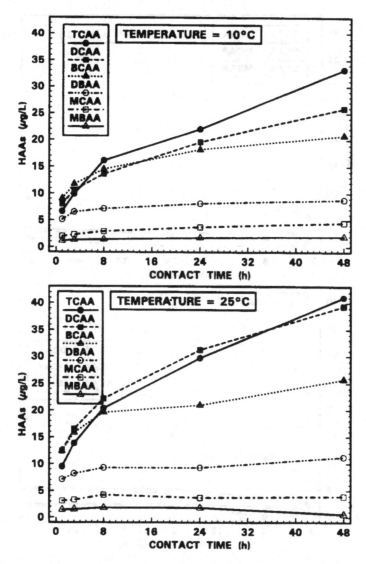

Figure 8 Impact of temperature and contact time on HAAs when TOC (3.25 mg/l) is much greater than bromide (0.2 mg/l); pH = 8.2.

This combined reduction in TOC and chlorine dosage resulted in 72 and 85% decreases in TTHMs and HAA5, respectively. Figure 12, however, shows that the impact of GAC on individual DBPs was somewhat different. The reductions in $CHCl_3$, DCAA, and TCAA were greater than or equal to those observed for the respective DBP class sum. On the other hand, $CHCl_2Br$ had a lower reduction (58%), whereas the concentrations of $CHClBr_2$, $CHBr_3$, and DBAA were higher in the chlorinated GAC effluent than in the chlorinated

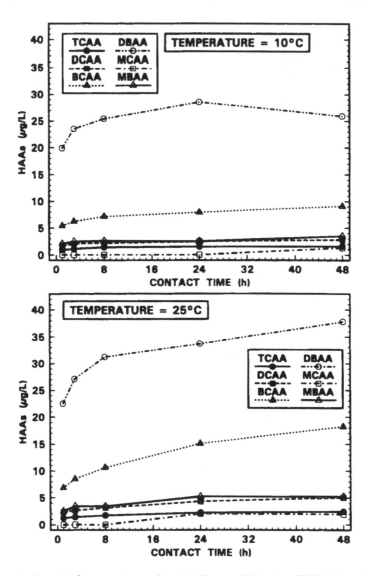

Figure 9   Impact of temperature and contact time on HAAs when TOC (2.0 mg/l) and bromide (0.8 mg/l) have same-range values; pH = 8.2.

GAC influent. The latter DBPs were from 1 to 5 μg/l higher as a result of GAC removing TOC but not bromide, thus shifting the DBP formation to the more brominated species. Similar speciation results for GAC treatment of ORW have been reported by Summers and co-workers.[5]

Chlorination of the GAC influent and effluent yielded n values of 0.27

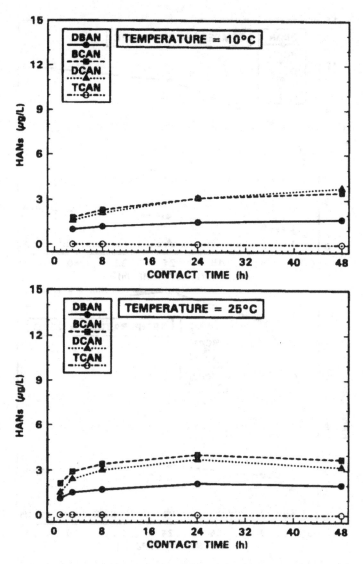

Figure 10 Impact of temperature and contact time on HANs when TOC (3.25 mg/l) is much greater than bromide (0.2 mg/l); pH = 8.2.

and 1.05, respectively. As was similarly observed in the 5 × 5 matrix for the 1.1 mg/l DOC tests, the ORW GAC effluent had a limited amount of organic precursors. Because the kinetics of THM formation favor bromine incorporation, HOBr consumed most of the active sites, leaving few active sites for chlorine substitution.

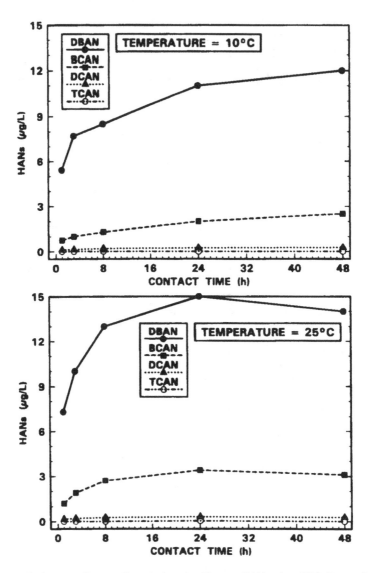

Figure 11   Impact of temperature and contact time on HANs when TOC (2.0 mg/l) and bromide (0.8 mg/l) have same-range values; pH = 8.2.

Impact of TOC Breakthrough (during GAC Treatment) on DBP Speciation

The influent Mississippi River water had 2.7 mg/l TOC and was low in bromide at ~0.03 mg/l (however, reliable bromide measurements were not

Table 5 Effects of GAC on DBP Reduction and Speciation

Parameter	Influent	Effluent
TOC (mg/l)	2.3	0.85
Br⁻ (mg/l)	0.06	0.06
Cl_2 dose (mg/l)	4.5	2.0
Cl_2 residual[a] (mg/l)	0.95	1.0
Cl_2 demand[a] (mg/l)	3.55	1.0
Br⁻/Cl⁺ ($\mu mol/\mu mol$)	0.020	0.035
DBPs[a] ($\mu g/l$)		
$CHCl_3$	94	8.8
$CHCl_2Br$	31	13
$CHClBr_2$	8.6	14
$CHBr_3$	<0.96	2.2
TTHMs	134	38
n (dimensionless)	0.27	1.05
MCAA	2.3	<1.0
DCAA	35	4.9
TCAA	28	1.6
MBAA	0.7	0.6
DBAA	1.8	3.4
HAA5	68	10

Note: ORW demonstration-scale study.
[a] Three-day, 25°C SDS testing.

available at that time). Over the course of the breakthrough curve, TTHM concentrations increased from 2 to 46 $\mu g/l$. Figure 13 shows the impact of the variation in bromide-to-TOC ratio, with a constant ratio of chlorine dose to bromide concentration. All of these samples had a positive chlorine residual at the end of the 5-d incubation period, but the Br⁻/Cl⁺ increased somewhat over the course of the breakthrough curve. Because the influent water did not maintain a positive chlorine residual over 5 d, the influent SDS data are not included for comparison. Scaling relationships used in sizing the RSSCT indicated that the breakthrough curve (up to RSSCT code number 27 in Figure 13) corresponded to approximately 50 d.[4]

For most of this curve, $CHBr_3$ (followed in concentration by $CHClBr_2$) dominated the THM formation. Toward the end of the breakthrough, however, the concentrations of $CHCl_3$ and $CHCl_2Br$ rose, whereas the $CHBr_3$ concentration dropped. The shift in speciation of the THMs in this GAC study parallels that observed in the 5 × 5 matrix. Furthermore, the data in this GAC study extrapolate the results of the 5 × 5 matrix below a 1 mg/l TOC organics loading.

Historically, GAC usage for DBP precursor removal has operationally focused on the TOC breakthrough level and the TTHMs produced. As health-effects data are developed on individual DBPs (e.g., the brominated HAAs), the control of theoretical cancer risk will be more complicated because GAC produces water with a variable bromide-to-TOC ratio; thus, there will be a time period in which brominated DBP formation will predominate.

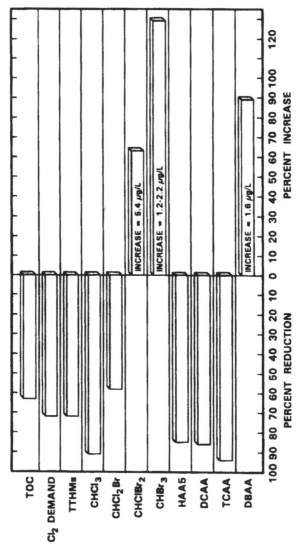

Figure 12 Impact of GAC on DBP reduction and speciation: ORW.

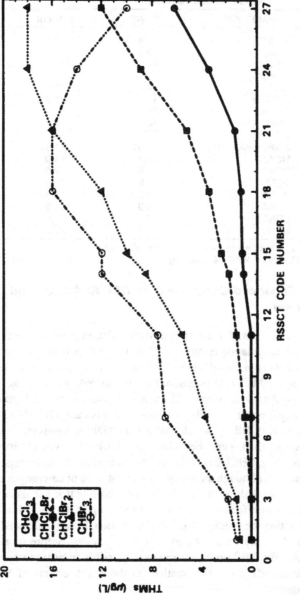

Figure 13 Impact of TOC breakthrough (during GAC treatment) on DBP speciation: Mississippi River water.

Table 6   Effects of Enhanced Coagulation on DBP
           Reduction and Speciation

Parameter	Influent	Effluent
TOC (mg/l)	3.4	2.3
Br⁻ (mg/l)	0.34	0.34
THMFP (μg/l)	351	232
SDS DBPs[a] (μg/l)		
CHCl₃	6.2	2.1
CHCl₂Br	19	8.9
CHClBr₂	35	21
CHBr₃	16	21
TTHMs	76	53
n (dimensionless)	1.61	1.97
MCAA	1.7	<1.0
DCAA	4.9	2.5
TCAA	2.9	1.5
BCAA	9.9	4.8
MBAA	2.3	1.8
DBAA	13	9.5
HAA6	34	20

Note: Delta water demonstration-scale study.
[a] One-hour, 25°C SDS testing.

Impact of Enhanced Coagulation on DBP Reduction and Speciation

Table 6 shows data for the enhanced coagulation study of Delta water. In this test, alum coagulation removed 32% of the TOC and 34% of the THMFP. SDS TTHMs were reduced by 30% (from 76 to 53 μg/l, safely below the proposed stage 1 MCL). In addition, HAA6 was reduced by 41%.

Figure 14 shows the impact of enhanced coagulation, in high-bromide-level (0.34 mg/l) water, on DBP speciation. In this case, all DBP levels were reduced except for CHBr₃. The other brominated DBPs, however, were reduced to lesser extents than were the chlorinated DBPs. As was observed in the 5 × 5 matrix study, CHCl₂Br can be reduced in concentration for high-bromide waters (Figure 1), although an amount sufficient to be of carcinogenic concern may remain. Nonetheless, these types of data indicate that enhanced coagulation, which is less effective than GAC in reducing TOC, has a less dramatic impact on shifting the speciation of DBP formation. In fact, enhanced coagulation increased n only from 1.61 to 1.97 in this water. Before more systems move to GAC for DBP precursor removal (possibly in order to comply with a stage 2 standard), more information on the health effects of brominated DBPs is required.

CONCLUSIONS

Both TOC level and bromide level affect DBP formation, but it is the chlorine-to-bromide ratio (as well as the bromide-to-TOC ratio) that affects

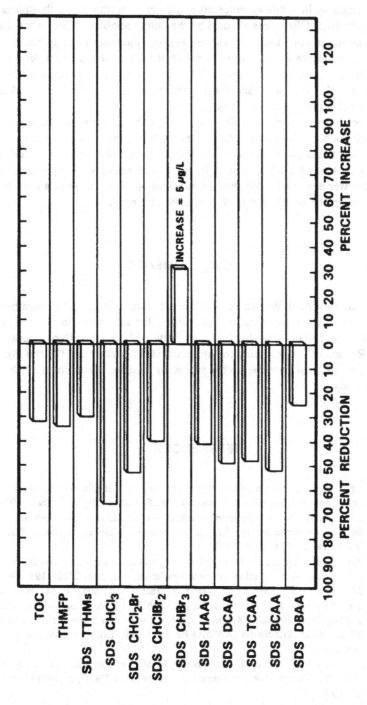

Figure 14 Impact of enhanced coagulation on DBP reduction and speciation: California Delta water.

DBP speciation. In addition, temperature and reaction time affect the speciation. Typically, HOBr is less affected by the latter variables, except when the TOC concentration is limited. However, because increases in temperature or reaction time increase the chlorine demand, there is a concomitant decrease in the Br^-/Cl^+ ratio and the bromine incorporation factor n. Because of their effects on THM formation and speciation, these factors have been used to develop THM predictive equations.

Precursor removal technologies (e.g., GAC, enhanced coagulation) remove TOC, but not bromide. These technologies result in a shift to more brominated DBPs. In particular, when the organic precursor levels are limiting (e.g., during early GAC breakthrough), HOBr will react with active sites more readily, leaving fewer active sites for chlorine substitution. Thus, there is a need for more data on the health effects of brominated by-products.

ACKNOWLEDGMENTS

The authors thank Warren Schimpff and Suzanne Teague, staff members at Metropolitan's Water Quality Laboratory, for assistance in the sample analysis. The authors also thank (1) the State of California Department of Water Resources for providing the agricultural drainwater and (2) Jim West of the City of West Sacramento for providing baseline water for the 5 × 5 matrix study.

REFERENCES

1. EPA, Drinking water; national primary drinking water regulations: disinfectants and disinfection byproducts; proposed rule, *Fed. Reg.*, 59(145), 38668, 1994.
2. EPA, Status report on the development of draft MCLGs for disinfectants and by-products, memorandum, EPA Office of Groundwater and Drinking Water, Washington, DC, October 1992.
3. Amy, G. L., Tan, L., and Davis, M. K., The effects of ozonation and activated carbon adsorption on trihalomethane speciation, *Wat. Res.*, 25(2), 191, 1991.
4. McGuire, M. J., Davis, M. K., Liang, S., Tate, C. H., Aieta, E. M., Wallace, I. E., Wilkes, D. R., Crittenden, J. C., and Vaith, K., *Optimization and Economic Evaluation of Granular Activated Carbon for Organic Removal.* American Water Works Association (AWWA) Research Foundation and AWWA, Denver, 1989.
5. Summers, R. S., Benz, M. A., Shukairy, H. M., and Cummings, L., Effect of separation processes on the formation of brominated THMs, *J. AWWA*, 85(1), 88, 1993.

6. EPA, National primary drinking water regulations: monitoring requirements for public drinking water supplies: Cryptosporidium, Giardia, viruses, disinfection byproducts, water treatment plant data and other information requirements; proposed rule, *Fed. Reg.*, 59(28), 6332, 1994.

7. State of California, Delta Water Quality: A Report to the Legislature on Trihalomethanes and the Quality of Drinking Water Available from the Sacramento-San Joaquin Delta, State Water Resources Control Board, Department of Health Services, and Department of Water Resources (DWR), Sacramento, 1991.

8. Krasner, S. W., Sclimenti, M. J., and Means, E. G., Quality degradation: implications for DBP formation, *J. AWWA*, 86(6), 34, 1994.

9. State of California, Delta Island Drainage Investigation Report, DWR Division of Local Assistance, Sacramento, 1990.

10. Amy, G. L., Thompson, J. M., Tan, L., Davis, M. K., and Krasner, S. W., Evaluation of THM precursor contributions from agricultural drains, *J. AWWA*, 82(1), 57, 1990.

11. Malcolm Pirnie, Inc., Bay-Delta Water Quality Modeling, final report for Metropolitan Water District of Southern California (MWDSC), MWDSC Water Quality Division, La Verne, CA, 1993.

12. Koch, B., Krasner, S. W., Sclimenti, M. J., and Schimpff, W. K., Predicting the formation of DBPs by the simulated distribution system, *J. AWWA*, 83(10), 62, 1991.

13. Symons, J. M., Krasner, S. W., Simms, L. A., and Sclimenti, M. J., Measurement of THM and precursor concentrations revisited: the efffect of bromide ion, *J. AWWA*, 85(1), 51, 1993.

14. James, M. Montgomery Consulting Engineers, Inc., and MWDSC, Disinfection By-Products in United States Drinking Water, final report for EPA and Association of Metropolitan Water Agencies, MWDSC Water Quality Division, La Verne, CA, 1989.

15. Cheng, R. C., Krasner, S. W., Green, J. F., and Wattier, K. L., Demonstration-scale enhanced coagulation as a disinfection by-product control strategy, Proc. 1993 AWWA Annu. Conf. (Water Quality), AWWA, Denver, 1993.

16. American Public Health Association (APHA), *Standard Methods for the Examination of Water and Wastewater*, 18th ed., APHA, AWWA, and Water Environment Federation, Washington, DC, 1992.

17. Koch, B., Crofts, E. W., Schimpff, W. K., and Davis, M. K., Analysis of halogenated disinfection by-products by capillary chromatography, Proc. 1988 AWWA WQTC, AWWA, Denver, 1989.

18. Gould, J. P., Fitchorn, L. E., and Urheim, E., Formation of brominated trihalomethanes: extent and kinetics, in *Water Chlorination: Environmental Impact and Health Effects*, Vol. 4, Jolley, R. L. et al., Eds., Ann Arbor Science, Ann Arbor, MI, 1983.

19. McKinney, J., Overview of metabolism studies, presented at Science Advisory Board Drinking Water Committee meeting, EPA Environmental Research Center, Research Triangle Park, NC, December 18, 1991.

20. Smith, M. E., Cowman, G. A., and Singer, P. C., The impact of ozonation and coagulation on disinfection by-product formation, Proc. 1993 AWWA Annu. Conf. (Water Research), AWWA, Denver, 1993.

21. Croué, J.-P. and Reckhow, D. A., Destruction of chlorination byproducts with sulfite, *Environ. Sci. Technol.*, 23(11), 1412, 1989.
22. Stevens, A. A., Moore, L. A., and Miltner, R. J., Formation and control of non-trihalomethane disinfection by-products, *J. AWWA*, 81(8), 54, 1989.
23. Amy, G., Siddiqui, M., Zhai, W., DeBroux, J., and Odem, W., Nation-wide survey of bromide ion concentrations in drinking water sources, Proc. 1993 AWWA Annu. Conf. (Water Research), AWWA, Denver, 1993.

CHAPTER 4

Influence of Bromide Ion on Trihalomethane and Haloacetic Acid Formation

James M. Symons, Stuart W. Krasner, Michael J. Sclimenti, Louis A. Simms, Harvey Wayne Sorensen, Jr., Gerald E. Speitel Jr., and Alicia C. Diehl

INTRODUCTION AND OBJECTIVES

BACKGROUND

Since the publication of the paper[1] outlining a technique for the measurement of trihalomethane (THM) precursors, the technique has been used and discussed by many authors and was included as a tentative method (No. 5710) in the 17th edition of *Standard Methods for the Examination of Water and Waste Water.*[2] This technique to measure total trihalomethane (TTHM) *precursor* called for measuring the difference between the TTHM concentration at the beginning and the end of a test in which a sample is first dosed with enough free chlorine to maintain a free chlorine residual throughout the test and then stored under selected conditions of temperature and pH for a selected time (commonly called the incubation time). The choice of incubation conditions was site specific. The resulting parameter was called the trihalomethane formation potential (THMFP).

Any sample will contain a variety of precursors that could potentially be reactive with free chlorine. Which precursors actually do react with free chlorine from this milieu depends on the sample and the reaction conditions. Thus, nature of the organic compounds, pH, temperature, incubation time, free chlorine concentration, and bromide ion concentration all will influence the final TTHM concentration at the end of a precursor determination.[3] Thus, THMFP must have an operational definition.

Subsequent to the publication of the precursor measurement paper,[1] a related test was developed with a different specific purpose, that of *predicting*

the TTHM concentration at some selected point in a given distribution system.[4]
The original description of this determination was to store a sample under
conditions representative of the distribution system under study for a time
equal to the flow time to the selected point in the distribution system and to
measure the TTHM concentration in the stored sample at the end of the
incubation time. The resulting parameter was called the simulated distribution
system THM (SDS THM). This test was included in the 17th edition of
Standard Methods for the Examination of Water and Waste Water[2] as
method 5710E.

Note that the SDS THM test usually does not measure THM precursor
because the chlorination conditions are often "milder" than would occur in a
precursor test. A properly run SDS test will contain the *same* disinfectant
residual type and concentration at the end of the incubation period as occurs
at the chosen point in the distribution system. Specifically, an SDS sample
might *not* contain a free available chlorine (FAC) residual or it might contain
a combined chlorine residual at the end of incubation and the test would still
be valid, if this was representative of distribution system conditions. On the
other hand, a THMFP test *must* contain an FAC residual at the end of incubation
to be valid.

In summary, the THM concentration at the end of incubation [$TTHM_t$] in
a finished water precursor test could be equal to or greater than (the usual
case) the SDS THM in the same water, but would only be less in the unlikely
situation where the FAC levels in the actual distribution system were greater
that those chosen for the THMFP test. Correlations between the THMFP and
SDS THM tests are possible if water quality and treatment conditions are
fairly constant, but any correlations are site specific.

Because the incubation conditions of the THMFP test may be varied to
suit local conditions, standardizing more than an approach is not possible.
The 19th edition of *Standard Methods*[5] will include a "standard" THMFP test
(like 5710B, but with the FAC residual fixed) to allow comparisons of precursors from location to location, followed by a general outline of the "variable"
THMFP test.

OBJECTIVES

Although all of the environmental conditions mentioned in the previous
section influence the reaction of precursors and free chlorine, bromide ion
has a major impact on both THM formation and speciation,[6-8] and on both
haloacetic acid (HAA) formation and speciation.[9] Thus, the purpose of this
paper is to comprehensively review and provide mechanistic interpretations
of the impact of bromide ion on both the formation potential (FP) and SDS
tests, particularly on their ability to predict halogen substitution into the THM
and HAA species.

Collecting data from the literature, the following questions on the reactions were explored:

1. What is the impact of the initial [Br⁻]/average [Cl⁺] molar ratio?
2. What is the impact of chlorine residual?
3. What is the impact of incubation conditions?
4. What is the impact of the initial [Br⁻]/[DOC] ratio?

METHODOLOGY

EXPERIMENTAL DESIGN AND DEFINITIONS

The data reviewed for this paper[7,10-13] were in two parts: one part conducted in the laboratory, where variables could easily be changed over a wide range, and the second conducted in the field, using samples from the field to confirm the laboratory tests. The experimental procedures have been documented in the literature cited above and will not be repeated here.

In analyzing the data, two methods were used to assess bromine substitution during DBP formation. The first technique[14] was to calculate the bromine incorporation factor "n" in the THMs formed using Equation 1 for data collected at the same incubation time,

$$n = \mu mol/l \ TTHMBr/\mu mol/l \ TTHM \qquad (1)$$

where $TTHMBr = \sum_{i=1}^{3} i \times CHCl_{3-i}Br_i$ in micromoles per liter. Depending on the degree of bromine substitution, n can vary between 0 and 3. Thus, n = 0 corresponds to the formation of only chloroform, whereas n = 3 corresponds to THM formation exclusively as bromoform. In typical waters, n is between 0 and 3, as a mixture of THMs are formed.

The bromine substitution in the HAAs is somewhat more difficult to quantify because only five HAAs were measured: monochloroacetic acid (MCAA), dichloroacetic acid (DCAA), trichloroacetic acid (TCAA), monobromoacetic acid (MBAA), and dibromoacetic acid (DBAA), i.e., three chlorine substituted species and two bromine substituted species. At the time of the work reviewed for this paper, commercial standards for these five of the nine possible HAAs were the only ones available. Thus, the total micromoles per liter of chlorine substitution in the six theoretical possible chlorine-substituted HAAs (HAACl6), the total μmol/l of bromine substitution in the theoretical six possible bromine-substituted HAAs (HAABr6), and the total micromoles per liter of HAA (HAA9) could not be calculated. The [HAACl3], [HAABr2], and [HAA5] were calculated, however, with the concentrations of each being the sum of the species measured.

The HAA analog to n for the THMs in this study therefore is

$$n'(2/5) = HAABr2/HAA5, \; \mu mol/\mu mol \qquad (2)$$

Note that, because of the multiple halogens in the THMs,

$$[TTHMBr] + [TTHMCl] \neq [TTHM] \qquad (3a)$$

but

$$[TTHMBr] + [TTHMCl] = [TTHMX], \qquad (3b)$$

where X = either chlorine or bromine.

For the same reasons, in the measured HAAs,

$$[HAACl3] + [HAABr2] \neq [HAA5] \qquad (4a)$$

but

$$[HAACl3] + [HAABr2] = [HAAX5] \qquad (4b)$$

Of course, if all nine HAAs could be measured, then

$$[HAACl6] + [HAABr6] = [HAAX9] \qquad (4c)$$

[TTHM] is used in the denominator in Equation 1 rather than [TTHMX] as [TTHM] is the regulated parameter. [TTHM] represents the moles of carbon (DOC in the water) that were converted to THMs, whereas [TTHMX] equals the moles of halogens substituted into the THMs. Because only THMs are formed, [TTHMX] in micromoles per liter = 3 × [TTHM].

As an example of these calculations, if one had the following analysis for THMs, $CHCl_3$ = 1.8 µg/l, $CHBrCl_2$ = 4.1 µg/l, $CHBr_2Cl$ = 12.6 µg/l, and $CHBr_3$ = 25.1 µg/l, converting each concentration to micromoles per liter and summing the THMBrs would yield TTHMBr = 0.44 µmol/l. The sum of the THMs, [TTHM], would yield 0.20 µmol/l; thus, n = 0.44/0.20 or 2.22. For the HAAs, if one had the following analysis; MCAA = 0 µg/l, DCAA = 25 µg/l, TCAA = 40 µg/l, MBAA = 0 µg/l, and DBAA = 18 µg/l, converting each concentration to micromoles per liter and summing would yield 0.17 µmol/l of HAABr2, and 0.52 µmol/l of HAA5. Thus, n' (2/5) would be 0.17/0.52 or 0.32.

MECHANISTIC METAPHOR

Some early work on bromine substitution in THMs by USEPA[6] led to three questions: (1) Why does a higher molar ratio of bromine to chlorine occur in the THMs formed when compared to the bromide ion to FAC molar ratio in the chlorinated water? (2) How does mixed-halogen substitution occur? and (3) Why does more bromine substitution occur in THMs as the concentration of the bromide ion in water increases? To explain the mechanisms involved, Stevens[15] developed the following analogy. HOBr is a stronger halogenating agent, analogous to a "bully," while HOCl is a weaker halogenating agent, but more of it is present, analogous to a "mob." Thus, metaphorically speaking, a few bullies are competing with a mob for active sites on precursors.

As an example of the difference in halogenating power of HOBr and HOCl, in one study reported by Symons et al.,[7] in the presence of 0.67 mg/l (8.4 μmol/l) of bromide ion, 30.9% was substituted into THMs, while only 1.2% of the 14.3 mg/l (201 μmol/l) of average free chlorine [(FAC dose + FAC residual)/2] was substituted into the THMs. In this same sample, 11.5 mg/l of FAC was consumed from the initial dose of 20 mg/l; thus, 56.3% of the FAC was used to satisfy the oxidative demand.

As noted in the example above, relative to their numbers in the chlorinated water, the bullies outcompete the mob for precursor reaction sites, in spite of their relatively low numbers. This increases the [TTHMBr] with respect to the [TTHMCl] and produces a higher molar ratio of bromine to chlorine in the THMs ([TTHMBr]/[TTHMCl] = 1.09 in the example above) than in the water ([Br$^-$]/[average FAC] = 0.04), the answer to question one. Mixed halogen substitution occurs because sometimes, as chlorine is attempting to form chloroform, a bromine atom or two "pushes a chlorine atom aside" as a bully might, to attach to a specific site. This produces a mixed halogen THM, the answer to question two. Finally, as the bromide ion increases, the number of the bullies increase and the TTHMBr concentration would increase, as now even more of the strong halogenators are present, the answer to question three.

In summary, the type and degree of halogen substitution depends on the relative numbers of the mob and the bullies. This metaphor will be used in interpreting the data reviewed in this paper.

ANALYTIC METHODS

THMs were determined by the liquid-liquid extraction method.[16,17] THMs were extracted from water with 2 ml of pentane in 42-ml vials or with 1.5 ml of pentane in 65-ml bottles. After pentane addition, the vials or bottles were placed on shaker table for 1 h to complete the extraction. Trichloroethene was used as an internal standard. The extract was analyzed on a gas chromatograph equipped with a DB-624 megabore column (J&W Scientific) and an electron capture detector. The reported THM concentrations are the average of three replicate injections.

The microextraction procedure[18,19] was used to analyze for the HAAs. Twelve grams of Na_2SO_4, 3 g of $CuSO_4$, 1.5 ml of concentrated H_2SO_4, and 3 ml of standard methyl-*tert*-butyl ether (MTBE) were added to 30 ml of water. This mixture was shaken at high speed for 9 min. Exactly 2 ml of the organic layer was removed using a 5-ml glass syringe and placed in a reaction vial. After cooling, the HAAs were esterified with 0.250 ml of freshly generated diazomethane. The esterified extract was analyzed on a gas chromatograph equipped with a DB-5 capillary column (J&W Scientific) and an electron capture detector. The reported HAA concentrations are the average of three replicate injections. Five haloacetic acids were measured: monochloroacetic acid (MCAA), dichloroacetic acid (DCAA), trichloroacetic acid (TCAA), monobromoacetic acid (MBAA), and dibromoacetic acid (DBAA). Preliminary experiments with standards indicated that a concentration of 2.5 μg/l of monochloroacetic acid or 0.2 μg/l of monobromoacetic acid produced a very slight response on the gas chromatogram. No response at these retention times was obtained from any of the samples in this study: therefore, the concentrations of these two HAAs were taken as zero in the calculation HAA5. Thus, the HAA5 concentration might be low by a maximum of 3 μg/l because of this assumption.

Bromide and bromate ion concentrations were measured by a Dionex® ion chromatograph (IC) 16 with a Dionex® IONPAC® AS9 analytical column in accordance with EPA test method 300.0 A and B.[20] Bromide ion analysis (method A) required two reagents, the eluent solutions 1.7 mM sodium carbonate and 1.8 mM sodium bicarbonate in deionized water. The regeneration solution for the micromembrane suppresser was 0.025 N sulfuric acid. Untreated samples were injected into the IC and results were recorded on a Hewlett Packard (HP) 3390A integrator. Bromate ion analysis (method B) requires the following reagents. The eluent was a solution of the two reagents, 1.7 mM sodium carbonate and 1.8 mM sodium bicarbonate in deionized water. The regeneration solution for the micromembrane suppresser was 0.025 N sulfuric acid. Prior to injection, all bromate samples were passed through a Dionex OnGuard-Ag® cartridge.[21] The cartridge contains a silver form, high-capacity, strong acid cation exchange resin that removes interfering ions (e.g., chloride) by precipitation. The cleaned sample was then injected and results were recorded on the HP integrator. Concentrations were calculated from a four point calibration curve constructed from prepared standards. The calibration standards were injected in duplicate before each analytical session and verified every ten injections.

Toward the end of the study, the system was upgraded with an IONPAC® AS9-SC analytical column.[22] A borate eluent (22 mM H_3BO_3 plus 22 mM $Na_2B_4O_7$) was used in place of the traditional carbonate eluate because of better resolution. Both bromide and bromate can be analyzed with a single injection and filtering of samples was not required. All samples were injected in duplicate and concentrations were taken from the average of the two injections. Concentrations were calculated from a four-point calibration curve constructed from prepared standards. The calibration standards were injected in duplicate before each analytical session and verified every ten injections. This provided similar results to the method described above, but more easily.

Dissolved organic carbon was measured on a Dohrmann DC-180 or DC-80 total organic carbon analyzer. The instruments use the UV-persulfate technique for converting organic carbon to inorganic carbon for subsequent analysis by an infrared carbon dioxide analyzer, according to method 5310 C.[19] The DC-180 TOC analyzer was equipped with a multisampler and was totally automated in operation. A 2-ml sample loop was used with five replicate injections. The DOC concentration in the first injection was consistently higher than that of the remaining four replicates; consequently, only the last four replicates were used in calculating the DOC concentration. For the DC-80 instrument, the DOC concentration was taken as the average of three replicate syringe injections.

DATA HANDLING

The primary independent variable in this study was the molar ratio of the bromide ion to the FAC concentration as previous data[6,23] showed that this strongly impacted bromine substitution during THM and HAA formation. The concentrations of both the bromide ion and FAC change during the THM formation reaction. The bromide ion is oxidized by the FAC and the resulting hypobromous acid causes the substitution of bromine into THMs as well as HAAs and other disinfection by-products (DBPs). The hypobromous acid also acts as an oxidant, resulting in some recycling of the bromide ion. In this study, the calculations of bromine uptake only focused on the regulated THMs and HAAs proposed to be regulated. Similar studies in the future that include the other DBPs would be quite interesting.

The FAC concentration ($[Cl^+]$) is also changing as it oxidizes organic matter and reduced inorganic ions such as sulfide, reacts with ammonia, and causes the substitution of chlorine into THMs, HAAs, and other DBPs. In this study, in the absence of any inorganic chlorine demand, the substitution of chlorine into the THMs and HAAs was typically less than about 10% of the total chlorine demand.

For treating the data in this study, the following approach was used. Because the key variable, Br^-/Cl^+ molar ratio, could not be calculated at every moment during the incubation, the initial bromide concentration (initial $[Br^-]$) was chosen as the numerator of the ratio. This concentration relates to the potential for bromine substitution into the THMs and HAAs.

The samples tested both in the laboratory and in the field contained neither ammonia, nor reduced inorganic materials such as sulfide, ferrous iron, and

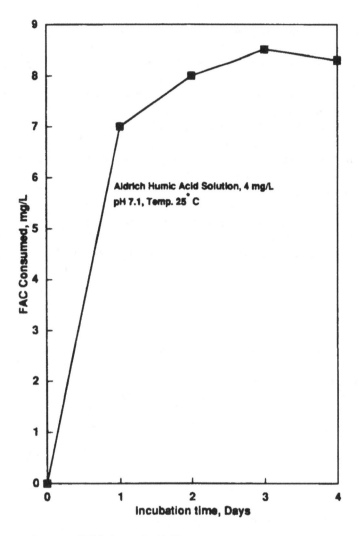

Figure 1 Exertion of FAC demand with time.

so forth. Thus, the change in FAC concentration (chlorine demand exerted) after 1 d of contact time was fairly slow and gradual throughout the remainder of the incubation period, about 18% of the total after day one, for the example shown in Figure 1. Therefore, using the average FAC concentration during the incubation, ([initial Cl^+] + [final Cl^+])/2 = (average [Cl^+]) was representative of the FAC concentration "driving" the THM and HAA formation reaction during incubation. The primary independent variable thus becomes, "initial [Br^-]/average [Cl^+] molar ratio." Using the chlorine demand for the denominator of this ratio was considered, but it did not seem an appropriate parameter to indicate what was driving the THM and HAA formation reaction as so many additional oxidation reaction consume free chlorine residual. In waters with a rapid chlorine demand, such as those containing ammonia, the chlorine dose minus the instantaneous demand should be used as the [initial Cl^+]. This did not occur in the data reviewed for this paper.

Although molar concentrations were used almost exclusively throughout this paper, because the TTHM regulation is based on the arithmetic sum of the weights per unit volume of the four THM species, changes in molar concentrations will not be the same as changes in weight concentrations. For example, if free chlorination of a bromide-free sample produced 0.5 μmol/l of chloroform (n = 0), the TTHM would be 0.06 mg/l, less than the current maximum contaminant level (MCL) of 0.10 mg/l. If this same sample were spiked with an excess of bromide ion such that chlorine substitution was completely suppressed and 0.5 μmol/l of only bromoform were produced (n = 3), the TTHM would now be 0.13 mg/l, well over the MCL, although the molar concentration would be the same in both samples. All of the data for individual THM and HAA species in this paper can easily be converted to a weight per unit volume basis by multiplying the molar concentration by the appropriate molecular weight. Of course, the actual MCL is based on a 12-month running average of four quarterly samples.

RESULTS

IMPACT OF CHLORINE RESIDUAL

Findings

Stevens et al.[3] showed that the formation of chloroform, and presumably the other halogen-substituted DBPs, ceased when the FAC residual became zero, as shown in Figure 2. Thus, some FAC must be present throughout a precursor test for the determination to be valid. Remember, if the FAC residual is zero at the end of the incubation period, this does not mean that halogen-substituted DBPs will not have formed, it just means that the formation of halogen-substituted DBPs was *limited*, as shown in Figure 2.

The data shown in Figure 3 demonstrate that the FAC residual at the end of incubation does have an effect on [$TTHM_7$], but the effect is not linear over a wide range of FAC residuals. Calculation of the change in [$TTHM_7$] per unit change in FAC residual, the slope of the data shown in Figure 3, shows that the change in [$TTHM_7$] is lower, but not zero, at FAC residuals of about 3 mg/l and greater, as shown in Figure 4. Thus, having an FAC residual of about 3 mg/l is a good target for precursor evaluation studies. Of course, if a shorter incubation period were used, the necessary target FAC residual would change. In the presence of bromide ion, the bromine-incorpora-

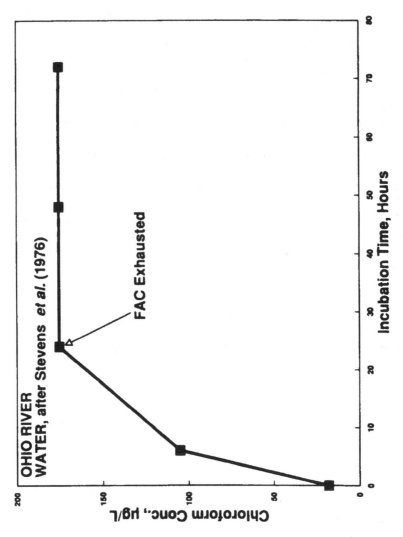

Figure 2 Influence of FAC residual on chloroform formation.[3]

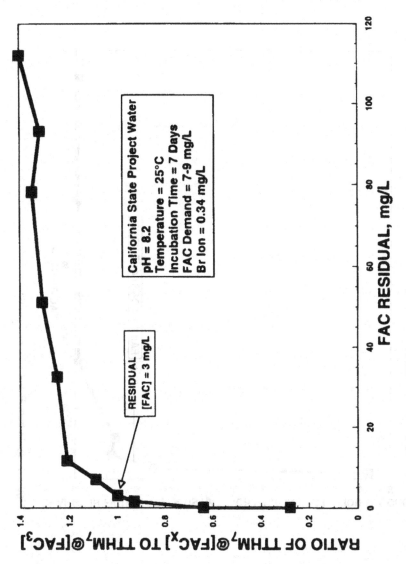

Figure 3 Impact of FAC residual on TTHM$_t$ concentration.

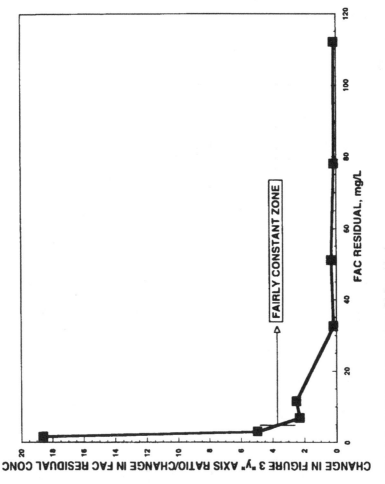

Figure 4   Δy/Δx from Figure 3 vs. FAC residual.

tion factor n Equation 1 is not constant as the FAC residual varies, but rises sharply at low FAC residuals, as shown in Figure 5.

Mechanisms

Why are the [$TTHM_t$] and n affected by FAC residuals above and below 3 mg/l? As the FAC residual becomes smaller the HOBr formed stays the same because it is formed first and is related to the bromide ion concentration, but the amount of HOCl goes down. Thus, the [$TTHMBr_t$], the numerator in Equation 1, is fairly constant, while the [$TTHMCl_t$] declines because of HOCl limitation. The lower [$TTHMCl_t$] causes the [$TTHM_t$], the denominator in Equation 1, to decrease, resulting in an increase in n, as shown in Figure 5.

At high FAC residuals, excess HOCl, the mob, is formed. This results in additional chlorine substitution by mass action and an increase in [$TTHMCl_t$]. Only a certain amount of HOBr can be formed, however, because of a fixed amount of bromide ion being present. Thus, the [$TTHMBr_t$] is fairly constant, while the [$TTHM_t$] increases, as shown in Figure 3. The net result of incubation in the presence of large excesses of FAC is to increase [$TTHM_t$] and decrease n, as shown in Figure 5.

IMPACT OF INITIAL [Br⁻]/ AVERAGE [Cl⁺] MOLAR RATIO

The initial [Br^-]/average [Cl^+] molar ratio in a THM formation test (THMFP) can be changed by having different values for either the numerator or the denominator, or both. All three situations will be discussed separately.

Constant Initial [Br⁻], Varying FAC Residual

Findings — The data from two different waters, shown in Figures 6 and 7, demonstrate that, when the average FAC dose is increased (to the left in the figures), decreasing the initial [Br^-]/average [Cl^+] molar ratio, n becomes smaller because the $TTHMBr_t$ concentration (the numerator in Equation 1) declined only slightly from about 2 μmol/l to about 1.6 μmol/l for the data shown in Figure 6 or rose only slightly from about 2.5 to about 3.0 μmol/l (i.e., 20%) for the data shown in Figure 7. On the other hand, the $TTHM_t$ concentration (the denominator in Equation 1) rose from about 1.8 μmol/l to about 2.7 μmol/l for the data shown in Figure 6 or rose from about 1.8 μmol/l about 2.3 μmol/L (i.e., 31%) for the data shown in Figure 7. These relatively small changes in the $TTHMBr_t$ concentrations combined with the relatively larger changes in the $TTHM_t$ and $TTHMCl_t$ concentrations resulted in a decline in n (moving left in the figures), less relative bromine substitution. Note, the "x" axis in Figure 6 is presented as a log scale just to spread out the data at

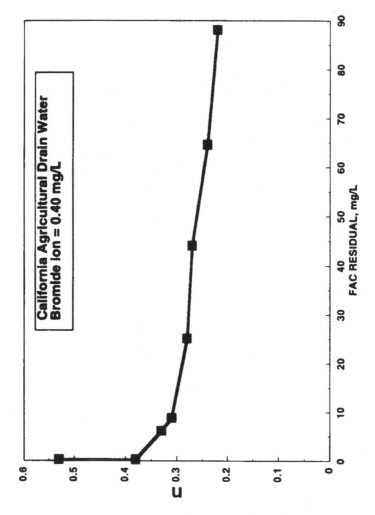

Figure 5   Variation in "n" with an increase in FAC concentration.

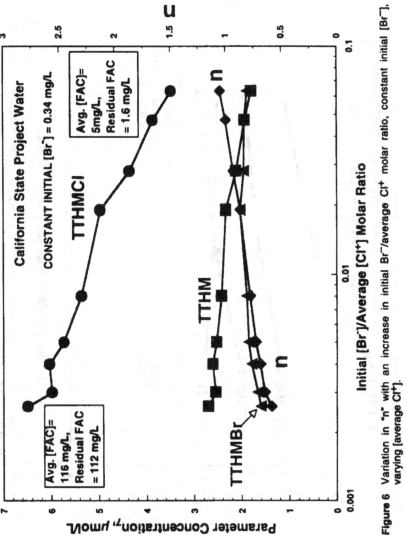

Figure 6 Variation in "n" with an increase in initial Br⁻/average Cl⁺ molar ratio, constant initial [Br⁻], varying [average Cl⁺].

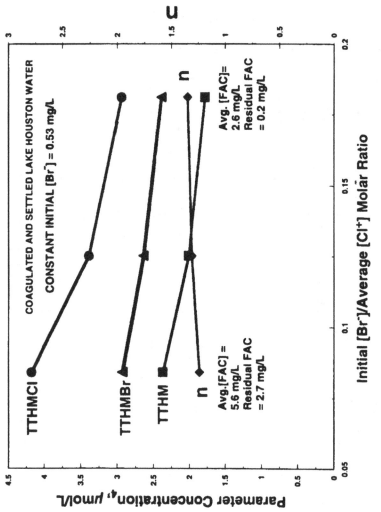

Figure 7 Variation in "n" with an increase in initial Br⁻/average Cl⁺ molar ratio, constant initial [Br⁻], varying [average Cl⁺].

the higher FAC doses; Figure 7 has a conventional arithmetic scale for the "x" axis.

Mechanisms — For the conditions shown in Figure 6, as the average FAC dose increases, the numbers in the mob (HOCl) increase and, thus, they can outcompete the stronger bullies (HOBr). Thus, the $TTHMBr_7$ concentration declined slightly as the FAC residual after incubation increased from 1.6 to 112 mg/l (right to left). This is because some active sites that could have been bromine substituted are now chlorine substituted instead because of mass action when the high doses of FAC were used. This did not occur for the conditions shown in Figure 7 as the FAC dose did not become high enough. In both cases, however, the increase in the numbers in the mob (HOCl) caused some more difficult to halogenate precursor sites to be chlorine substituted, increasing the $TTHMCl_t$ concentration greatly, with a corresponding increase in the $TTHM_t$ concentration. Because the $TTHMBr_t$ concentration changed slightly compared to the change in $TTHM_t$ concentration, n declined from right to left for the data shown in Figures 6 and 7.

Constant FAC Residual, Varying Initial [Br⁻]

Because in "standard" THMFP testing a constant FAC residual is recommended, the initial $[Br^-]$/average $[Cl^+]$ molar ratio in various samples would change because of a change in the numerator of the ratio, in contrast to the data shown in Figures 6 and 7. Data to illustrate this condition are presented and discussed below.

Findings — As the concentration of bromide ion increased, as shown in Figure 8, the $TTHMBr_4$ concentration increased greatly, while the $TTHMCl_4$ concentration only declined slightly. This combination of changes caused the $TTHM_4$ concentration to increase somewhat, but not as much as the $TTHMBr_4$. Thus, according to Equation 1, n, indicative of the degree of bromine incorporation, increased.

Mechanisms — Similarly to the situation shown in Figures 6 and 7, when the initial $[Br^-]$/average $[Cl^+]$ molar ratio increased, n became larger, as shown in Figure 8 (left to right), but in this case for a different reason. Here, as the numerator of the initial $[Br^-]$/average $[Cl^+]$ molar ratio increased, the denominator being nearly constant, the $TTHMBr_4$ concentration increased more rapidly than the $TTHM_4$ concentration; thus, n increased. The increase in the HOBr concentration that was produced at higher bromide ion concentrations resulted in more precursor sites becoming active to the increased number of bullies (HOBr) and thus, somewhat more $TTHM_4$ was formed. With HOBr being a better halogenating agent than HOCl, this resulted in an increase in $[TTHMBr_4]$ and a slight decrease in $[TTHMCl_4]$, fewer sites left over for the

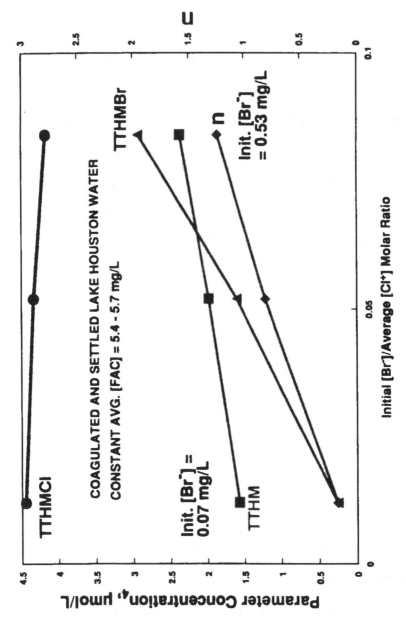

Figure 8   Variation in n with an increase in initial Br⁻/Average Cl⁻ molar ratio, constant [average Cl⁻], varying initial [Br⁻].

Table 1   Influence of Bromide Ion Concentration and Average FAC Dose on Bromine Substitution Aldrich Humic Acid in Deionized Water

Parameter	"Low" sample	"High" sample
Br⁻ (mg/l)	0.67	4.0
DOC (mg/l)	3.0	3.0
pH	6.3–7.2	6.7–7.1
FAC dose (mg/l)	21.2	126
FAC residual (mg/l)	11.2	95
Average FAC (mg/l)	16.2	110.5
Initial Br⁻/average Cl⁺ molar ratio	0.037	0.032
THM$_4$ species		
$\quad$ CHCl$_3$ (μmol/l)	0.26	0.17
$\quad$ CHBrCl$_2$ (μmol/l)	0.44	0.38
$\quad$ CHBr$_2$Cl (μmol/l)	0.52	0.49
$\quad$ CHBr$_3$ (μmol/l)	0.17	0.45
TTHM (μmol/l)	1.38	1.49
TTHMBr (μmol/l)	1.98	2.71
Bromine Incorporation factor, n	1.43	1.82

mob (HOCl). This has the net effect of enriching the TTHM with bromine and causing n to increase.

Constant Initial [Br⁻]/Average [Cl⁺] Molar Ratio, Varying Initial [Br⁻], and FAC Residual

Two samples in which Aldrich humic acid (AHA) was used as model precursor material were compared to determine if the *magnitude* of the numerator and denominator in the initial [Br⁻]/average [Cl⁺] molar ratio had any influence on bromine substitution and, thus, on n. One 5-mg/l AHA solution (DOC = 3.0 mg/l) was spiked with 0.67 mg/l of bromide ion ("low"); another was spiked with 4 mg/l of bromide ion ("high"). Both of these samples were then dosed with different amounts of FAC to produce an initial [Br⁻]/average [Cl⁺] molar ratio of nearly 0.03, which were then incubated for 4 d.

Findings — The data in Table 1 show that, although the "high" sample formed more bromoform, similar bromine substitution occurred, as measured by n, in spite of a nearly sixfold change in the absolute values of the numerator and denominator of the initial [Br⁻]/average [Cl⁺] molar ratio.

Mechanisms — If the molar concentrations of the two halogenating species, HOCl and HOBr, change together, the relative attack on the active sites of the precursor does not change, in contrast to what occurs if only the denominator (as shown in Figures 6 and 7) or the numerator (as shown in Figure 8) are varied. Thus, about the same degree of bromine incorporation, as represented by n, will occur.

IMPACT OF INCUBATION CONDITIONS

Time

Findings — To determine the kinetics of bromine substitution, THM formation data for 2 d from H. O. Banks water (outflow from the California Delta) were evaluated to determine the relative rate of TTHMBr and TTHMCl formation. The data shown in Figure 9 demonstrate that bromine incorporation is faster than chlorine substitution. This is indicated by the $[TTHMBr_{1h}]/[TTHMBr_{48h}]$ ratio (as a percentage) being much higher than the $[TTHMCl_{1h}]/[TTHMCl_{48h}]$ ratio (as a percentage) at any given time less than 48 h.

Mechanisms — HOBr is not only a stronger halogenating agent than HOCl, but the kinetics of the bromine-substitution reaction is faster than the chlorine-substitution reaction. Note that, as the TTHM concentration increases because of the continued formation of TTHMCl as time passes, n becomes smaller.

Chlorine Residual

Findings — When evaluating the formation of DBPs by the SDS method, the incubation period is usually shorter than when performing a precursor test, and the chlorination conditions are usually milder. A lower FAC residual present at the end of incubation will influence the bromine-substitution reaction. To illustrate, data for both TTHM and HAA5 formation in coagulated and settled Lake Austin water (C&SLAW)[12] were evaluated. As shown in Figure 10, they demonstrate that the $[TTHMBr_t]$ decreased only slightly as the FAC residual after incubation decreased from 4 mg/l (FP) to 0.7 mg/l (SDS), while the $[TTHMCl_t]$ decreased markedly, resulting in a higher value for n. Similarly, the $[HAABr2_t]$ changed little as the FAC residual after incubation was lowered, but the $[HAACl3_t]$ declined greatly, resulting in a higher value for n'(2/5), as shown in Figure 11.

Mechanisms — As has been seen previously, a lower FAC residual present at the end of incubation (FP vs. SDS) results in the denominator of the initial $[Br^-]$/average $[Cl^+]$ molar ratio to be lower. This influenced n and n'(2/5) in these data in the following manner. As the average FAC dose decreased, the numbers in the mob (HOCl) decreased and they could not compete with the stronger bullies (HOBr), so the bromine-substitution reaction was not affected very much. In contrast, the decrease in the mob (HOCl) caused a restriction in the precursor sites that could be chlorine substituted, decreasing the chlorine substitution greatly, with a corresponding decrease in the $TTHM_t$ and $HAA5_t$ concentrations. Thus, both n and n'(2/5) were higher in the FP samples as compared to the SDS samples, as shown in Figures 10 and 11, from left to right.

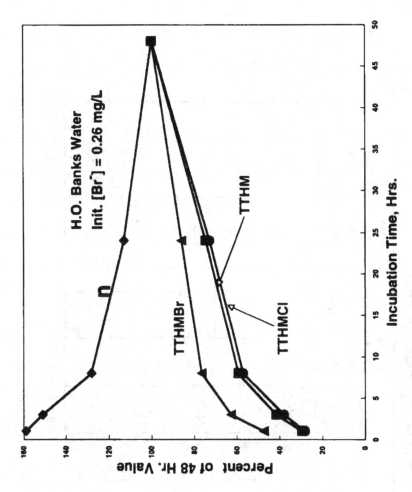

Figure 9   Kinetics of THM formation.

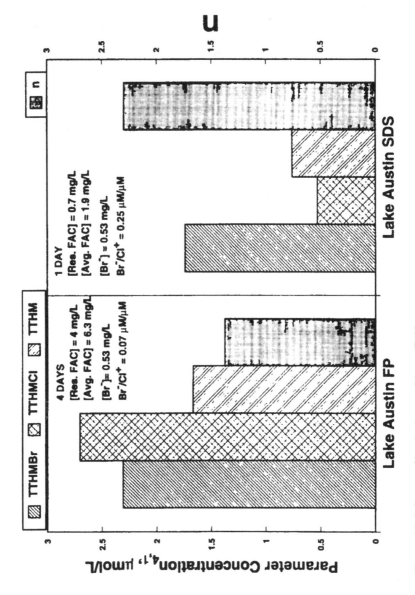

Figure 10 Influence of incubation conditions—THMs.

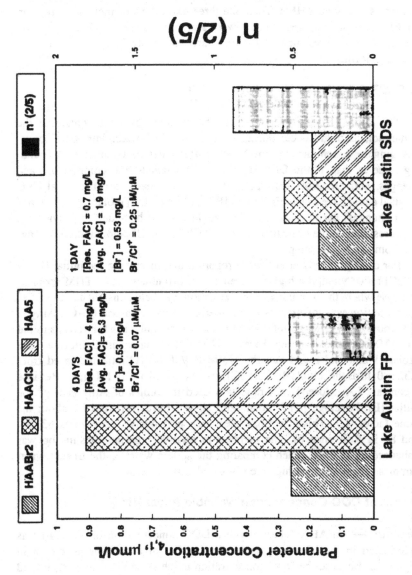

Figure 11   Influence of incubation conditions—HAAs.

Incubation time also influences the differences between FP and SDS results. As was seen from the data shown in **Figure 9,** n becomes smaller as time passes because of the faster formation of bromine-substituted THMs as compared to chlorine-substituted THMs. Thus, the three extra days of incubation time in the FP test would also result in a lower value of n, as seen in **Figure 10.** Data similar to those presented in **Figure 9** are not available for HAABr2 and HAACl3 formation.

IMPACT OF INITIAL [Br⁻]/[DOC] RATIO

The initial [Br⁻]/[DOC] ratio can be changed by either changing the numerator, as might occur during an episode of saltwater intrusion[24] or by changing the denominator, as might occur after water treatment.[11] A confounding issue is that, if the DOC concentration changes, the chlorine demand changes, and precursor tests performed with a nearly constant final FAC residual could have a changing initial [Br⁻]/average [Cl⁺] molar ratio (denominator changes) because less chlorine would need to be added to provide the target residual, at the same time the initial [Br⁻]/[DOC] ratio would be changing from sample to sample.

For example, Amy et al.[25] have reported that increasing the initial [Br⁻]/[DOC] ratio "yield[s] a higher percentage of brominated (sic) THM species" in a case where the bromide ion concentration was held constant and the DOC concentration was lowered using powdered activated carbon (PAC). Amy et al.[25] compared [THM-Br]/[THM-X] ratio on a weight basis for various initial [Br⁻]/[DOC] ratios ranging from 0.037 to 0.062 mg/mg in California State Project Water and found that the [THM-Br]/[THM-X] ratio increased from 0.32 to 0.48 as the initial [Br⁻]/[DOC] ratio increased nearly 1.7 times. Because a constant FAC dose/[DOC] ratio was used in studies of Amy et al.,[25] the initial [Br⁻]/average [Cl⁺] molar ratio should have increased by about the same factor, 1.7. This effect has been seen previously (see **Figures 63, 84,** and **85** in Reference 6). The increase in relative bromine substitution was enhanced in all cases when comparing the source water to the treated water samples. These relationships are discussed further below.

Constant DOC Concentration, Variable Initial [Br⁻]

Findings — An AHA solution with a DOC concentration of 3.0 mg/l was chlorinated in the laboratory under three different conditions as shown in **Figure 12.** Because the DOC concentration is the same in samples A, B, and C, none of them are precursor limited with respect to each other. Comparing samples A and B, having similar FAC residuals, shows that with about a sixfold increase in bromide ion concentration the initial [Br⁻]/[DOC] ratio and the initial [Br⁻]/average [Cl⁺] molar ratio also increased about sixfold and n rose markedly. Which ratio change, Br⁻/DOC or initial Br⁻/average

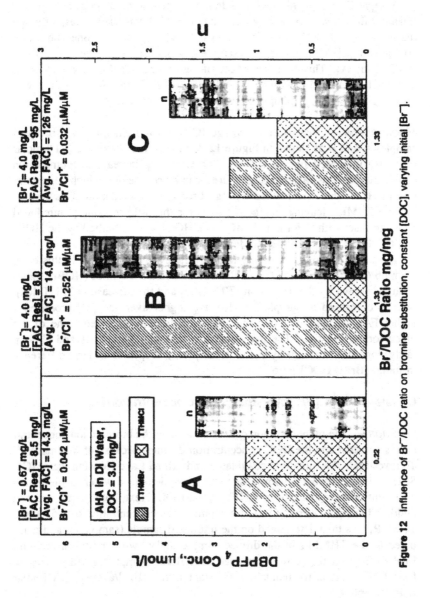

Figure 12 Influence of Br⁻/DOC ratio on bromine substitution, constant [DOC], varying initial [Br⁻].

Cl$^+$, caused the increase in n cannot be determined from the sample A and B data as shown in Figure 12.

Sample C, with as high a bromide ion concentration as in sample B was treated differently. Here, a very large average [Cl$^+$] was used. Thus, although the initial [Br$^-$]/[DOC] ratio in samples B and C were the same, the initial [Br$^-$]/average [Cl$^+$] molar ratio in sample C was nearly the same as in sample A (76% of A). This had the effect of permitting the formation of more TTHMCl, compared to sample B, and thus, more TTHM. As a result, n was only slightly higher in sample C, as compared to sample A.

Mechanisms — Because the average [Cl$^+$] did not change much between samples A and B as shown in Figure 12, the initial [Br$^-$]/average [Cl$^+$] molar ratio also increased sixfold with the corresponding increase in bromide ion concentration. This caused a large increase in n because of the high concentration of HOBr that was formed through bromide ion oxidation. This created much TTHMBr, leaving few active sites for the little remaining unreduced HOCl to react with to form TTHMCl. The HOCl was outbullied by the HOBr and the TTHMCl concentration declined markedly.

In sample C, however, a very large average [Cl$^+$] was used. In this case, the mob (HOCl) in sample C actually outcompeted the bullies (HOBr) somewhat, as shown by the decline in [TTHMBr] and the increase in [TTHMCl], both as compared to sample B. Also, the higher average [Cl$^+$] produced a higher FAC demand and this may have oxidized and altered some of the precursors active sites. The net result was that n in samples A and C were fairly close together (slightly higher in sample C), in spite of the large difference in initial [Br$^-$]/[DOC] ratio.

Constant Initial [Br$^-$], Variable DOC Concentration

Findings — The more common case is one where the bromide ion concentration is constant and the DOC concentration declines because of treatment.[6,25] To investigate this situation, coagulated and settled Lake Houston water (C& SLHW) was diluted 1:1 with nearly DOC-free deionized water to simulate a treatment process that removed 50% of the DOC. Next, both the diluted and undiluted samples were spiked to a bromide ion concentration of 0.53 mg/l. THMFP$_4$ was then determined on the undiluted sample (sample A) and compared to the THMFP$_4$ in the diluted sample under two conditions, constant final FAC residual (the recommended procedure—sample B) and increasing final FAC residual to maintain a constant initial [Br$^-$]/average [Cl$^+$]molar ratio (sample C).

The data shown in Figure 13 demonstrate that, when the final FAC residual is kept at about 4 mg/l (samples A and B), the [TTHMBr] is nearly constant, but the formation of TTHMCl is lower, in part because of less precursor material and in part because a lower FAC dose was required to provide the

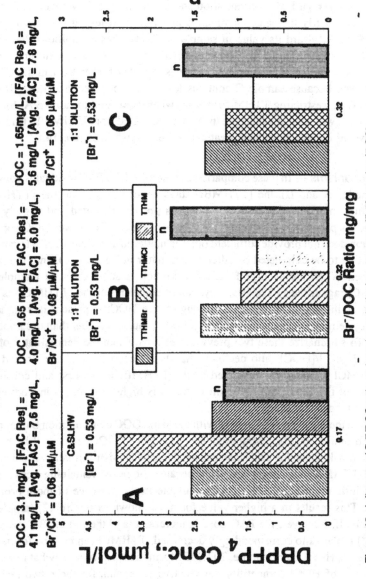

Figure 13 Influence of Br⁻/DOC ratio on bromine substitution, constant [Br⁻], varying [DOC].

FAC residual (lower FAC demand), thus a lower average [Cl$^+$]. These two factors, precursor limitation and lower chlorine demand (resulting in a higher initial [Br$^-$]/average [Cl$^+$] molar ratio), combined to cause a 51% higher value of n in sample B. When the final FAC residual is increased, however, such that the average FAC concentration and the initial [Br$^-$]/average [Cl$^+$] molar ratio are nearly the same (samples A and C), the resulting n value declined 8.5% back toward its value in sample A. Under these conditions, the effect of the change in initial [Br$^-$]/average [Cl$^+$] molar ratio has been removed.

The value of n in sample C is still somewhat higher than in sample A, however, because sample C contains less precursor (precursor limited) than sample A, restricting TTHM formation when compared to sample A. Similar behavior was seen at a field site using data collected from Utility # 11 (granular activated carbon adsorption treatment),[11] as shown in Figure 14.

Mechanisms — In these comparisons, when the final FAC residual is constant (samples A and B), the [TTHMBr] stays about constant but the [TTHMCl] declines markedly because the system is precursor limited and slightly less HOCl is present (lower average FAC concentration). The lower DOC concentration does not provide sufficient organic matter to allow the TTHM formation reaction to be completed (smaller denominator in Equation 1), and, thus, n is higher. When the average FAC concentration is about constant (samples A and C), again the [TTHMBr] stays about constant but the [TTHMCl] is still low, as in sample B, because, although more HOCl is present in sample C, the system is precursor limited (less TTHM); thus n is higher than in sample A.

To summarize these two previous sections, when the denominator of the initial [Br$^-$]/[DOC] ratio declines, the water becomes precursor limited and TTHMCl formation is restricted as the TTHMBr forms first and consumes many of the active precursor sites; thus n is higher, as shown in Figures 13 and 14, samples A and C.

On the other hand, in waters with the same DOC concentration but increasing bromide ion concentrations, the initial [Br$^-$]/[DOC] ratio will be higher and so will n, but for a different reason. Here, TTHMBr formation is enhanced and TTHMCl formation restricted because of fewer remaining active sites and little unreduced HOCl being available in the presence of more bromide ion. This results in a higher value for n, as shown in Figure 12, samples A and B. If, however, the FAC dose is increased, and the initial [Br$^-$]/average [Cl$^+$] molar ratio correspondingly decreased, TTHMCl can now be formed by the extra HOCl. Furthermore, TTHMBr formation will be somewhat restricted because of HOCl competition for reactive sites, and, for these two reasons, n will be smaller (compare samples B and C as shown in Figure 12). Thus, both the initial [Br$^-$]/[DOC] ratio and the initial [Br$^-$]/average [Cl$^+$] molar ratio influence halogen incorporation into THMs, but through different mechanisms.

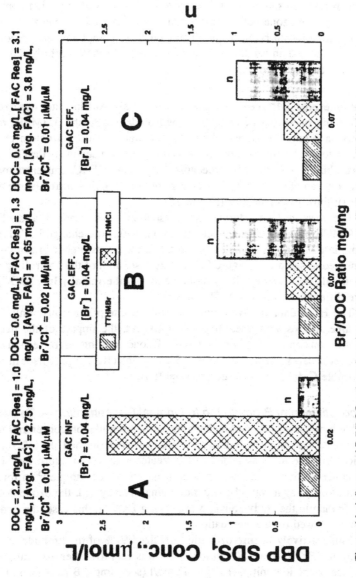

Figure 14 Influence of Br⁻/DOC ratio on bromine substitution, constant [Br⁻], varying [DOC].[11]

Vary both DOC Concentration and Initial [Br⁻]

Certain circumstances occur in which both the DOC concentration and the bromide ion concentration change between two samples that are being compared.[26] An example would be ozone treatment of bromide ion-containing water.[12,27] The ozone both destroys some DOC as well as oxidizing some of the bromide ion to bromate ion. Thus, properly comparing the precursor in an untreated and an ozonated sample must be performed carefully. This situation will be discussed below.

Unadulterated Samples — In a study where Lake Austin water was ozonated with 3 mg of O_3/mg of DOC and then biodegraded, the THMFP$_4$ results, shown in Figure 15, were difficult to interpret.[12] In the formation potential test, the loss in bromide ion during ozonation caused a decline in the [TTHMBr] (fewer bullies—HOBr), but, because the average FAC concentration was similar in samples A and B, little change in [TTHMCl] occurred (same mob—HOCl); thus, n was smaller (Figure 15).

The loss in DOC during biodegradation, 35%, caused a decline in [TTHMCl] because of precursor limitation, with little change in [TTHMBr] (same initial [Br⁻] in samples B and C); thus, n was higher in sample C as compared to sample B (Figure 15). In this case, both of the controlling ratios, the initial [Br⁻]/average [Cl⁺] molar ratio and the initial [Br⁻]/[DOC] ratio were different in samples A, B, and C.

The HAA data are even more difficult to interpret because only five of the nine species were measured. Nevertheless, the impact of the declining bromide ion concentration on the HAABr2 concentration can be seen in sample B from the data presented in Figure 16, as can the effect of precursor limitation in sample C, both as compared to sample A.

Reconstitution of Bromide Ion to Original Concentration — To avoid the problem of the influence of a changing initial bromide ion concentration after treatment, ozonation in this case, a study was performed on C&SLAW in which bromide ion was added to the treated samples to return the bromide ion concentration to that which was in the untreated sample.[13] In this case, in an effort to aggravate the situation being investigated, the ozone dose was higher than in the study above, 5 mg O_3/mg DOC in this case, the ozonation being followed by biodegradation.

This relatively intense ozonation oxidized 97% of the bromide ion; thus, the initial [Br⁻]/average [Cl⁺] molar ratio was very low after ozonation, 0.002, because of the low initial [Br⁻], 0.02 mg/l (see sample B as shown in Figure 17). This caused the TTHMBr formation to be minimized. The excess HOCl (bigger mob) created a relatively higher [TTHMCl], giving the appearance of the creation of TTHMCl precursor during the ozonation step, sample B. The

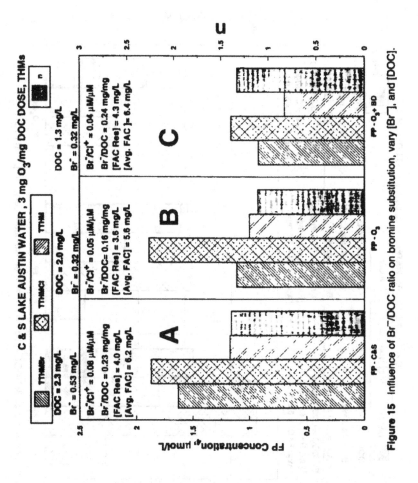

Figure 15 Influence of Br⁻/DOC ratio on bromine substitution, vary [Br⁻], and [DOC].

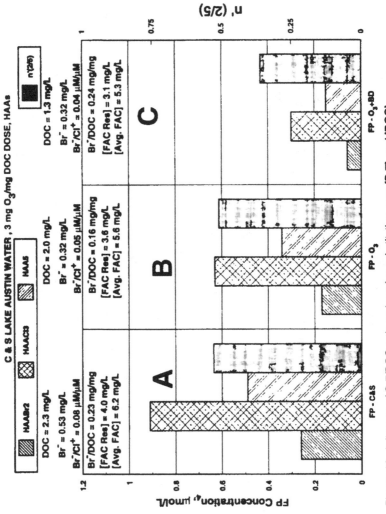

Figure 16   Influence of Br⁻/DOC ratio on bromine substitution, vary [Br⁻], and [DOC].

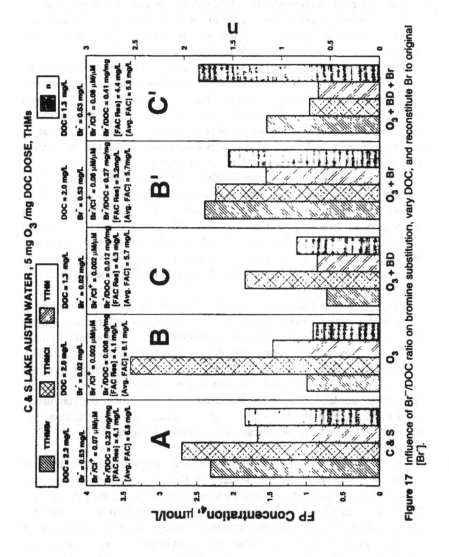

Figure 17 Influence of Br⁻/DOC ratio on bromine substitution, vary DOC, and reconstitute Br to original [Br⁻].

"extra" TTHMCl did not compensate for the great loss in TTHMBr, however, so [TTHM] was less, as was n, sample B compared to sample A (Figure 17).

After the biodegradation of the DOC (35%), the [TTHMCl] was lowered 46%, (precursor limitation), compared to the 27% lower for the [TTHMBr] (bullies stronger than mob); thus, n was somewhat higher after this treatment step, sample C compared to sample B, Figure 17.

This pattern of data, samples A, B, and C, was compared to the bromine substitution that occurred when bromide ion was added back to the original concentration before the FP test was performed, samples B' and C', as shown in Figure 17. In this case, [TTHMBr] returned to the same concentration as found in the C&SLAW (compare samples A and B' in Figure 17), indicating that the decline in [TTHMBr] in sample B was because of the loss in bromide ion, rather than actual oxidation of TTHMBr precursor.

The consumption of HOCl for oxidation of the additional bromide ion added to sample B' caused TTHMCl formation to be less than occurred without bromide ion addition (compare samples B and B' in Figure 17), eliminating the apparent formation of TTHMCl precursor. The concentrations of all of the parameters, TTHMBr, TTHMCl, and TTHM, were lower after biodegradation, sample C', Figure 17, but about in the same proportion; thus, n was not much higher in sample C' as compared to sample B', Figure 17, and actual precursor removal was demonstrated.

Because, as noted above, ozonation oxidizes bromide ion, the decline in the bromide ion concentration influenced the substitution of bromine into the HAAs as well. Without bromide ion re-addition, the very low initial [Br⁻]/ average [Cl⁺] molar ratio nearly completely prevented bromine substitution, sample B as shown in Figure 18. Sufficient precursor was available to produce more HAACl3 in the chlorinated ozonated water, sample B, than was in the C&SLAW, sample A, Figure 18. Based on these data alone, the conclusion could be drawn that HAACl3 precursor was created during ozonation, and that this precursor was biodegradable (compare samples A, B, and C in Figure 18).

When bromide ion was added to the treated samples, however, different results were obtained. Now, the initial [Br⁻]/average [Cl⁺] molar ratio was nearly the same in all three samples being compared, samples A, B', and C' as shown in Figure 18. Comparing the C&SLAW (sample A) with the corresponding ozonated sample (sample B') shows that HAACl3 precursor was *not* formed during ozonation and that a slight quantity of precursor was actually destroyed (compare samples A and B' in Figure 18). Thus, the pattern described in the previous paragraph was an artifact caused by the bromine loss. After biodegradation, a more balanced halogen-substitution pattern was seen once the bromide ion loss was corrected and, as with the THMs, actual loss in precursor was demonstrated (compare samples A and C' in Figure 18). Interpretation of these data is complicated by the inability to measure HAA9 because analytic standards for four HAAs were unavailable. Based on the

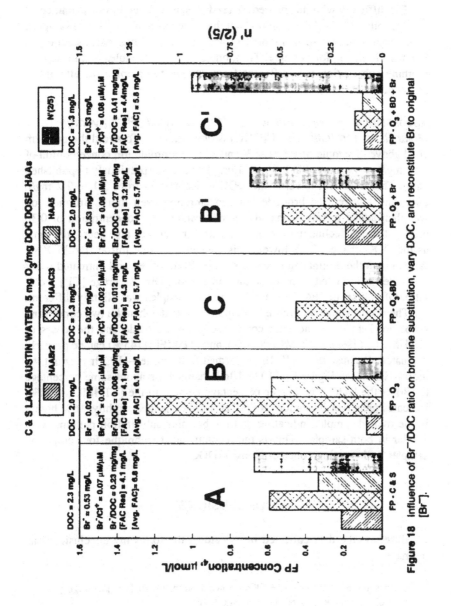

Figure 18 Influence of Br⁻/DOC ratio on bromine substitution, vary DOC, and reconstitute Br to original [Br⁻].

[TTHMBr] in sample C' in Figure 17, the [HAABr6] in sample C' in Figure 18 would be expected to be higher, if it could have been calculated.

The difficulty with this procedure used was, as pointed out by Summers,[28] that the initial $[Br^-]/[DOC]$ ratio is not the same in the reconstituted samples, 0.008 mg/mg in sample B, Figures 17 and 18, vs. 0.27 mg/mg in sample B', Figures 17 and 18, for example. Thus, one of the key variables that influences bromine substitution was different from sample to sample. An alternative procedure is outlined in the next section.

Reconstitution of Bromide Ion to Original Initial $[Br^-]/Average$ $[Cl^+]$ Molar Ratio and Initial $[Br^-]/[DOC]$ Molar Ratio — To avoid the problem cited above, a sample of C&SLHW spiked to a bromide ion concentration of 0.53 mg/l was ozonated at a 5 mg O_3/mg DOC dose (previously unpublished data). This treatment lowered the DOC concentration from 3.1 to 2.2 mg/l, 29%, and lowered the bromide ion concentration even more. Bromide ion was then added to this ozonated water to establish a concentration of 0.38 mg/l so that the decline in bromide ion concentration was also nearly the same as the DOC change, 29% lower. This created the same initial $[Br^-]/[DOC]$ ratio in both the original and treated sample. Both treated and untreated water were then chlorinated to produce the same initial $[Br^-]/average$ $[Cl^+]$ molar ratio. Of course, the FAC residual in the two samples was slightly different.

The results shown in Figure 19 demonstrate that, when the two controlling ratios are kept the same, true comparisons can be made and the losses of [TTHMBr4] (28%), [TTHMCl4] (54%), and [THMFP4] (45%) can be properly evaluated. Because the TTHMBr is formed first (bullies—HOBr beat out the mob—HOCl, see Figure 9), [TTHMBr4] changed the about same percentage (34%) as the DOC. The higher percentage decline in the [TTHMCl4] (54%) indicates precursor limitation. Note that n only changed slightly, 16% higher in the treated sample, indicating that the bromine-substitution patterns were similar in both samples. This is the recommended technique for comparing samples with varying concentrations of DOC.

CONCLUSIONS

Based on the data collected in this study, the following conclusions can be made:

1. In samples with the same DOC concentration, initial $[Br^-]/average$ $[Cl^+]$ molar ratio controls bromine substitution.
2. The degree of bromine substitution increases as the initial $[Br^-]/average$ $[Cl^+]$ molar ratio increases, either when the numerator of the ratio increases or the denominator of the ratio decreases. The mechanism resulting in an increase in bromine substitution is not the same in both cases, however.

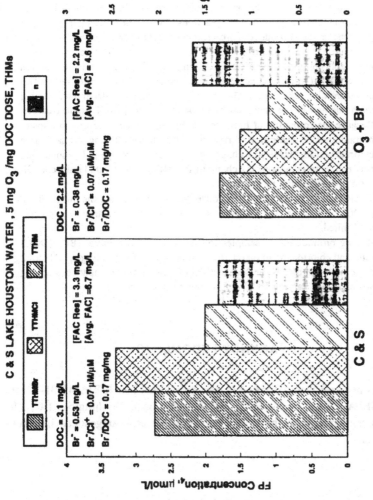

Figure 19 Influence of Br⁻/DOC ratio on bromine substitution, vary DOC, reconstitute Br to equal Br⁻/DOC, and initial [Br⁻]/average FAC ratios.

3. The magnitude of the bromide ion concentration or average FAC dose among samples does not tend to influence bromine substitution overall, as long as the initial [Br⁻]/average[Cl⁺] molar ratio is the same.

4. HOBr attacks more sites in precursors than HOCl, reacts with them faster than HOCl, and HOCl is reduced during the oxidation of bromide ion to HOBr. The reasons the bromine incorporation changes under different circumstances is a result of these reactions; thus, higher initial [Br⁻]/average [Cl⁺] molar ratios produce more bromine incorporation.

5. In addition to the effect of the initial [Br⁻]/average[Cl⁺] molar ratio, when DOC concentrations are different among samples, the initial bromide ion concentration to DOC concentration (initial [Br⁻]/[DOC] ratio) ratio influences bromine substitution. The cause of the change will depend on whether the numerator or the denominator in the initial [Br⁻]/[DOC] ratio changes.

6. Because the formation of bromate ion during ozonation results in a lower bromide ion concentration, formation potential tests having the same free chlorine residual result in a changing initial [Br⁻]/average [Cl⁺] molar ratio between samples, as well as a changing the initial [Br⁻]/[DOC] ratio, complicating interpretation of the results.

7. Treatment that lowers the DOC concentration, thus raising the initial [Br⁻]/ [DOC] ratio, causes the TTHM and HAA5 formation reaction to be precursor limited.

8. Adding bromide ion to samples to maintain the same initial [Br⁻]/[DOC] ratio and chlorinating to maintain the same initial [Br⁻]/average [Cl⁺] molar ratio in all samples is the best method of making true comparisons among samples with varying DOC and bromide ion concentrations.

ACKNOWLEDGMENTS

This research was supported in part by a Grant No. R-816449-01-1 from the USEPA Office of Research and Development, Washington, DC. The authors are grateful to the reviewers whose helpful suggestions improved the paper.

REFERENCES

1. Stevens, A. A. and Symons, J. M., Measurement of trihalomethane and precursor concentration changes, *J. Am. Water Works Assoc.*, 69(10), 546, 1977.

2. American Public Health Association, *Standard Methods for the Examination of Water and Wastewater*, 17th ed., Washington, DC, 1989.

3. Stevens, A. A., Slocum, C. J., Seegar, D. R., and Robeck, G. G., Chlorination of organics in drinking water, *J. Am. Water Works Assoc.*, 68(11), 615, 1976.

4. Miller, R. and Hartman, D. J., Feasibility Study of Granular Activated Carbon Adsorption and On-Site Regeneration, EPA-600/2-82-087A, US. Environmental Protection Agency, Cincinnati, 1982.

5. American Public Health Association, *Standard Methods for the Examination of Water and Wastewater,* 19th ed., Washington, DC, 1995.

6. Symons, J. M., Stevens, A. A., Clark, R. M., Geldreich, E. E., Love, O. T., Jr., and DeMarco, J., *Treatment Techniques for Controlling Trihalomethanes in Drinking Water,* EPA-600/2-81/156; Also available from the American Water Works Association, Denver, 1981.

7. Symons, J. M., Krasner, S. W., Simms, L. A., and Sclimenti, M., Measurement of THM and precursor concentrations revisited: the effect of bromide ion, *J. Am. Water Works Assoc.,* 85(1), 51, 1993.

8. Shukairy, H. M. and Summers, R. S., Organic Precursor Removal and DBP Speciation, presented at the 206th Annual Meeting of the American Chemical Society, Chicago, August 23–25, 1993; also Chapter 14 of this text.

9. Pourmoghaddas, H., Stevens, A. A., Kinman, R. N., Dressman, R. C., Moore, L. A., and Ireland, J. C., Effect of bromide ion formation of HAAs during chlorination, *J. Am. Water Works Assoc.,* 85(1), 82, 1993.

10. Cipparone, L. A., Diehl, A. C., and Speitel, G. E., Jr., The Effect of Biodegradable Organic Carbon Removal on the Bacterial Regrowth Potential and Disinfection By-Product Formation of Ozonated Water, Proc. 1993 Annual Conference, Water Research, American Water Works Association, June 6–10, 1993, 529.

11. Metropolitan Water District of Southern California and Montgomery, J. M., Consulting Engineers, Inc., Disinfection By-Products in United States Drinking Waters, final report for the United States Environmental Protection Agency and Association of Metropolitan Water Agencies, Metropolitan Water District Water Quality Division, LaVerne, CA, November 1989.

12. Speitel, G. E., Jr., Symons, J. M., Diehl, A. C., Sorensen, H. W., and Cipparone, L. A., Effect of ozone dosage and subsequent biodegradation on removal of DBP precursors, *J. Am. Water Works Assoc.,* 85(5), 86, 1993.

13. Symons, J. M., Speitel, G. E., Jr., Diehl, A. C., and Sorensen, H. W., Jr., Precursor control in waters containing bromide, *J. Am. Water Works Assoc.,* 86(6), 48, 1994.

14. Gould, J. P., Fitchorn, L. E., and Urheim, E., Formation of brominated trihalomethanes: extent and kinetics, in *Water Chlorination, Environmental Impact and Health Effects,* Vol. 4, Jolley, R. L., Brungs, W. A., Cotruvo, J. A., Cumming, R. B., Mattice, J. S., and Jacobs, V. A., Eds., Ann Arbor Science, Ann Arbor, MI, 1983, chap. 12.

15. Stevens, A. A., U.S. Environmental Protection Agency, Cincinnati, personal communication, 1984.

16. Henderson, J. E., Peyton, G. R., and Glaze, W. H., A convenient liquid-liquid extraction method for the determination of halomethanes in water at the parts-per-billion level, in *Identification and Analysis of Organic Pollutants in Water,* Keith, L. H., Ed., Ann Arbor Science, Ann Arbor, MI, 1976, 105.

17. Koch, B., Crofts, E. W., Schimpff, W. K., and Davis, M. K., Analysis of Halogenated Disinfection By-Products, Proc. American Water Works Association Water Quality Technology Conference, St. Louis, November 13–17, 1988, 429.

18. Krasner, S. W., Sclimenti, M. J., and Hwang, C. J., Experiences with Implementing a Laboratory Program to Sample and Analyze for Disinfection By-Products in a National Study, presented at the American Water Works Association Water Quality Technology Conference, St. Louis, November 13–17, 1988.

19. American Public Health Association, *Standard Methods for the Examination of Water and Wastewater,* 18th ed., Washington DC, 1992.

20. Pfaff, J. D., Brockhoff, C. A., and O'Dell, J. W., The Determination of Inorganic Anions in Water by Ion Chromatography, Method 300.0 A&B, Environmental Monitoring and Systems Laboratory, Cincinnati, December 1989.

21. Kuo, C. Y., Krasner, S. W., Stalker, G. A., and Weinberg, H. S., Analysis of Inorganic By-Products in Water by Ion Chromatography, presented at the American Water Works Association Water Quality Technology Conference, San Diego, CA, November 11–15, 1990, 503.

22. Dionex Corp., Document No. 034656, Revision 1, March 16, 1992.

23. Pourmoghaddas, H., Stevens, A. A., Kinman, R. N., Dressman, R. C., Moore, L. A., and Ireland, J. C., Effect of bromide ion on formation of HAAs during chlorination, *J. Am. Water Works Assoc.,* 85(1), 82, 1993.

24. Krasner, S. W., McGuire, M. J., Jacangelo, J. G., Patania, N. L., Reagan, K. M., and Aieta, E. M., The occurrence of disinfection by-products in US drinking water, *J. Am. Water Works Assoc.,* 81(8), 41, 1989.

25. Amy, G. L., Tan, L., and Davis, M. K., The effects of ozonation and activated carbon adsorption on trihalomethane speciation, *Water Res.,* 25(2), 191, 1991.

26. Summers, R. S., Benz, M. A., Shukairy, H. M., and Cummings, L., Effect of separation processes on the formation of brominated THMs, *J. Am. Water Works Assoc.,* 85(1), 88, 1993.

27. Siddiqui, M. S. and Amy, G. A., Factors affecting DBP formation during ozone-bromide reactions, *J. Am. Water Works Assoc.,* 85(1), 63, 1993.

28. Summers, R. S., personal communication, Cincinnati, 1993.

29. Haag, W. R. and Hoigné, J., Ozonation of bromide-containing waters: kinetics of formation of hypobromous acid and bromate, *Environ. Sci. Technol.,* 17(5), 261, 1983.

30. Koch, B., Krasner, S. W., Sclimenti, M. J., and Schimpff, W. K., Prediction of the formation of DBPs by the simulated distribution system, *J. Am. Water Works Assoc.,* 83(10), 62, 1991.

31. Morgen, E. M., Scarpino, P., and Summers, R. S., Measurement of Biodegradable Dissolved Organic Carbon in Drinking Water, *Proceedings of 1990 Annual Conference,* American Water Works Association, Cincinnati, 1990, 573.

32. Reckhow, D. A. and Singer, P. C., The removal of organic halide precursors by preozonation and alum coagulation, *J. Am. Water Works Assoc.,* 76(4), 151, 1984.

CHAPTER 5

Removal of Chloroform from Drinking Water Using High-Energy Electron Beam Irradiation

Fei T. Mak, William J. Cooper, Charles N. Kurucz, Michael G. Nickelsen, and Thomas D. Waite

ABSTRACT

The use of the high-energy electron beam process has been shown to be an effective way to control chloroform in drinking water treatment. The major factors that affected the removal efficiency were solute concentration and carbonate ion alkalinity. Solute concentration was studied at concentrations that far exceed those expected in treated drinking water; however, the results at high concentration show that this process may be applicable for other purposes such as hazardous waste site remediation. Based on the bimolecular rate constants of $CHCl_3$ and the three reactive species formed in irradiated water, it was shown that the e_{aq}^- was the reactive intermediate responsible for the destruction of the $CHCl_3$ in aqueous solutions. At high carbonate ion concentrations, where the $OH\cdot$ reacts rapidly with the CO_3^{2-}, the effective "steady-state" concentration of the e_{aq}^- was increased and more efficient removal of the $CHCl_3$ was observed. The addition of 3% by weight of kaolin clay did not affect the removal efficiency of the $CHCl_3$ in any of the experiments. It is suggested that this process may be applicable to many organic reaction by-products in drinking water treatment.

INTRODUCTION

Chloroform (trichloromethane) is classified as a group B2 carcinogen,[1] i.e., a probable human carcinogen based on sufficient evidence from animal studies and inadequate evidence from human studies. Chloroform has been

1-56670-136-8/96/$0.00+$.50

shown to affect the nervous system and to cause tumors in the liver and kidneys of test animals.[1] The general population is exposed to chloroform by inhalation, consumption of food, and primarily by ingestion of drinking water that has been disinfected with chlorine.

Chlorination has been used as a drinking water disinfectant since the early 1900s. However, chlorine reacts with naturally occurring humic substances in the water and forms trihalomethanes, a group of compounds that include chloroform, chlorodibromomethane, bromodichloromethane, and bromoform. These chemicals are regulated in drinking water at a maximum contamination level (MCL) of 100 μg l^{-1} as a yearly average for all trihalomethanes combined. The international regulations set by the World Health Organization allow an MCL of 30 μg l^{-1} for just chloroform.

Given the assumption that chlorination is one of the best methods for disinfection and that it will continue to be used, the concentration of chloroform in drinking water needs to be controlled by either removing the chemical precursors, i.e., the humic substances, before chlorination or by removing the chloroform after chlorination. Because there are few economically acceptable methods available for the removal of humic substances from water, the removal of chloroform after chlorination has been of considerable interest. Methods that have been studied for this purpose are aeration, activated carbon adsorption, and ozonation.[2-5]

Aeration, or air stripping, in which a substance is transferred from liquid to gas phase, has been used effectively to reduce the concentration of taste- and odor-producing compounds and certain organic chemicals.[2] However, there are concerns regarding the cost of keeping air emissions within the regulated MCLs.

Activated carbon adsorption has also been studied extensively for water treatment. Again, this process does not destroy the hazardous chemicals, but transfers them by adsorption onto powdered or granular activated carbon.[3] There is some concern that, when the MCL is lowered, more frequent changes of carbon may become necessary in this process, thus driving the cost prohibitively high.

Ozone's usefulness in disinfecting polluted water was first recognized in 1886. The direct reaction of ozone with chloroform is very slow ($<0.1\ M^{-1}s^{-1}$). When the process is enhanced using UV light and hydrogen peroxide to enhance hydroxyl radical (OH$\cdot$) formation, its applicability is increased.

A possible alternative to the more traditional treatment methods is the use of heterogeneous catalysis using TiO_2 and photolysis.[6,7] These studies have shown that $CHCl_3$ was destroyed. The destruction of the $CHCl_3$ proceeded to completion with the formation of CO_2 and chloride ion. To our knowledge this process has not been studied at pilot scale for its application in drinking water treatment.

A recent study suggested that the use of high-energy electrons was effective in destroying all four of the trihalomethanes.[8] The objective of the present study was to develop a better understanding of the innovative high-energy electron beam irradiation treatment process as an effective way to remove chloroform from various waters. The experimental design included studies of various water qualities using pH adjustment to examine the effect of carbonate and bicarbonate concentrations and addition of kaolin clay to study the effect of the presence of suspended solids on the removal of chloroform.

METHODS AND MATERIALS

ELECTRON BEAM RESEARCH FACILITY (EBRF)

All experiments were carried out at the EBRF, located at the Miami Dade Central District Wastewater Treatment Plant on Virginia Key in Miami. The electron beam consists of a 1.5-MeV accelerator with an insulated-core transformer (ICT) power supply. The beam current is variable between 0 and 50 mA, which allows for absorbed doses of 0 to 800 krad. The electron beam is scanned at 200 Hz to cover an area approximately 114 cm (40 in.) wide and 5 cm (2 in.) high. Influent streams are exposed to this beam at 460 l (120 gal) per minute. Five resistance temperature devices (RTDs) are mounted in the influent and effluent stream immediately before and after the beam, and a computer reads and records the temperature difference. A change of 1°C is equivalent to 1 cal g^{-1} of absorbed energy, which corresponds to a dose of approximately 418 krad (4.18 kGy). The facility has been described in more detail elsewhere.[9,10]

EXPERIMENTAL PROCEDURE

All chemicals were used without further purification. The chloroform was of 99% purity (Burdick & Jackson Laboratories, Inc.). 1,1,2-Trichloroethane was used as an internal standard for quantification (Aldrich Chemicals, 98% purity). Pentane was trihalomethane analysis grade (Baxter).

Solutions of chloroform for large-scale batch experiments were prepared in potable water in either an 11,500-l (3000 gal) or a 23,000-l (6000 gal) tank truck, using a pulsed-diaphragm metering pump and in-line mixer injection system. To quantitatively determine the destruction efficiency of chloroform in a treated groundwater, several variables were studied. Initial solute concentration was studied at three levels: nominally 0.84, 8.4, and 84 μM (100, 1000, and 10,000 μg l^{-1}). The water quality was varied by adding solids and/or changing the pH. The alkalinity of the water was determined by titrating a 50-ml sample against 0.105 M HCl to a pH of 4.5. The treated water as received typically has a carbonate alkalinity of 44 to 48 mg l^{-1} at a pH of about 9, with a ratio of carbonate/bicarbonate ion of 0.2 (Table 1). To adjust the carbonate/bicarbonate ion ratio, hydrochloric acid was added to lower the

Table 1   Water Quality of Potable Water
            Used in Experiments

Parameter	Concentration
pH	8.46–9.55
DO[a] (mg l^{-1} O_2)	3.5–5.6
DOC[b] (mg l^{-1} C)	3.9–6.0
Alkalinity (mg l^{-1} $CaCO_3$)	44.1–47.8
NO_3^- (mg l^{-1} as N)	0.29–0.36
SO_4^{2-} (mg l^{-1})	20.1–21.5
Br^- (mg l^{-1})	0.28–0.30
Cl^- (mg l^{-1})	40.9–42.2
NH_2Cl (mg l^{-1} as Cl_2)	2.0–3.0

[a] DO = dissolved oxygen.
[b] DOC = dissolved organic carbon.

pH to 7, where the bicarbonate concentration was >99%, and then to pH 5 to remove the carbonate alkalinity. To determine the effect of inert solids on chloroform removal, kaolin clay was added at 3% w/w to solutions at pH 5 and 7. The addition of the clay resulted in a lowering of the pH from 9 to 7; therefore, the effect of inert solids was not determined at pH 9. Samples were collected in 47-ml vials, with a 1-min delay between influent and effluent sampling, for each dose. The bottles were then immediately chilled on ice. Analysis of these samples by gas chromatography used a liquid-liquid extraction procedure.[11] To keep the concentration of chloroform within the linear working range of the chromatographic standard curve, the 1000 μg l^{-1} samples were diluted one-to-ten by adding 4.2 ml of the sample to 38 ml of purged distilled water. The 10,000-μg l^{-1} samples were diluted one-to-hundred by adding 0.42 ml of the sample to 42 ml of purged distilled water. The 100-μg l^{-1} samples were not diluted; 5 ml of the sample was removed and discarded. After dilution, all samples were treated as follows: 3.0 ml of pentane was added to the sample, and the vial sealed. This leaves 2.0-ml headspace to allow the solvent and sample to mix thoroughly. The 1,1,2-trichloroethane solution was then added as an internal standard. The vials were agitated at 300 rpm for 3 min on a model G2 Gyrotory Shaker (New Brunswick Scientific). The pentane layer was allowed to separate from the water layer, and then transferred to a 2.0-ml autosampler vial; 2 μl of the extract was the injected into a Hewlett-Packard 5890A gas chromatograph, equipped with a 30-m DB-624 fused silica column (J & W Scientific) and an electron capture detector. The signal was analyzed using ELAB data acquisition software (OMS Tech).

RESULTS AND DISCUSSION

Irradiation of water by high-energy electrons results in the formation of two reducing species: the aqueous electron, e_{aq}^-, and the hydrogen atom, $H\cdot$; and one oxidizing species: the hydroxyl radical, $OH\cdot$, according to Equation 1:

$$H_2O \xrightarrow{\wedge\wedge} [2.6] \, e_{aq}^- + [0.6] \, H\bullet + [2.7] \, OH\bullet + [0.45] \, H_2$$
$$+ [0.7] \, H_2O_2 + [2.7] \, H_3O^+ \qquad (1)$$

The numbers in brackets are G values, the number of radicals, molecules, or ions formed or destroyed for every 100 eV of energy absorbed by a system. For Equation 1, these G values are 10^{-7} after the high-energy electron has passed through the solution. Chloroform reacts rapidly with the aqueous electron (1 to $3 \times 10^{10} \, M^{-1}s^{-1}$), and to a lesser extent with the hydrogen atom ($1.1 \times 10^7 \, M^{-1}s^{-1}$) and hydroxyl radical ($1 \times 10^7 \, M^{-1}s^{-1}$).[12-14] Therefore, based on the G values in Equation 1 and the bimolecular reaction rate constants, the destruction of chloroform appears to result mainly from the reaction with the aqueous electron. Reaction intermediates may react more efficiently with the hydrogen atom and hydroxyl radicals and, therefore, all of these reactive species may be important in the destruction of chloroform in natural systems.

The experimental design consisted of a total of 24 experiments, three influent $CHCl_3$ concentrations at three different pHs with each experiment being replicated twice. The addition of the kaolin clay decreased the pH of the water to 7; therefore, only pH 5 and 7 were studied in the presence of clay and no replicates were conducted.

Each experiment consisted of four doses including a zero control dose. At each dose four samples were collected before and after irradiation and triplicate or quadruplicate $CHCl_3$ analyses were conducted using the chromatographic procedure outlined above. The overall experimental design was completely randomized with respect to all of the variables. Because a maximum of six experiments could be run in any one week, the study was completed over a 4-week period.

CHLOROFORM REMOVAL

Figure 1 illustrates the removal efficiency of chloroform at pH 9, one of the three influent pH levels studied, at the three influent solute concentrations. These results were obtained in the absence of the inert kaolin clay and they demonstrate that $CHCl_3$ was effectively removed from solution at all three influent concentrations. The only difference was that, as the solute concentration increased, the dose, i.e., the energy, required to attain 99% removal, also increased.

The second variable studied was the effect of influent pH on removal efficiency. In the aqueous solutions studied the alkalinity was approximately 44 to 48 mg l^{-1} as $CaCO_3$. Therefore, changes in pH affected the speciation of the carbonate/bicarbonate ion as well as the absolute concentration at the lower pH. Figure 2 shows the removal of chloroform at a nominal influent concentration of 83.7 μM, at three pH levels, 9, 7 and 5. Figures 1 and 2 demonstrate that the removal efficiency was greater at pH 9 when compared

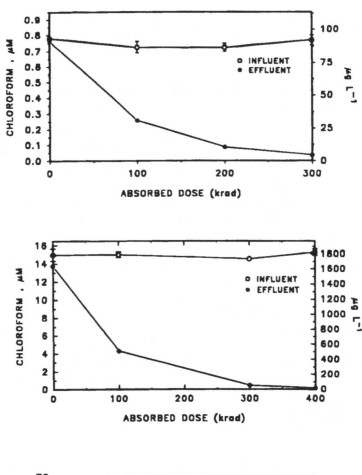

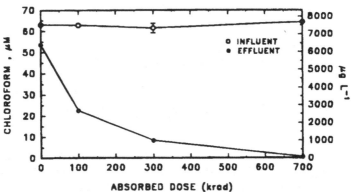

Figure 1   Chloroform removal in the absence of clay at pH 9 for the three solute influent
concentration. Each point is the mean of four analyses. The error bars are
the standard deviation from the mean; where no error bars are shown the
value was smaller than the data point.

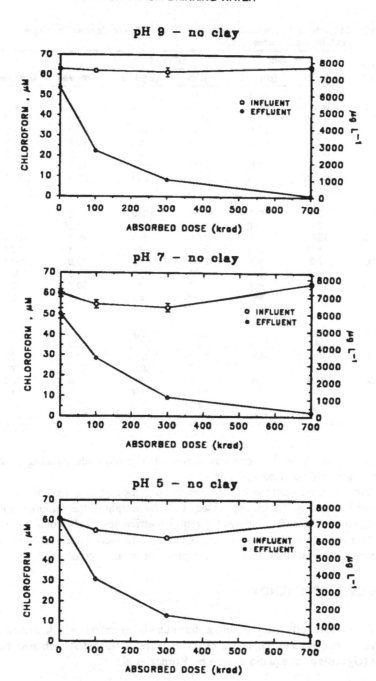

Figure 2 Chloroform removal in the absence of clay at 7000 to 8000 μg l⁻¹ for the three pH levels. Each point is the mean of four analyses. The error bars are the standard deviation from the mean; where no error bars are shown the value was smaller than the data point.

Table 2   Chloroform Removal Efficiency at Three Influent Concentrations at pH 5 in the Absence of Clay

Dose (krad)	Influent (μM)	Influent ($\mu g\ l^{-1}$)	Effluent (μM)	Effluent ($\mu g\ l^{-1}$)	% removed	G value
0	0.819	97.8	0.805	96.1	1.72	—
	0.747	89.2	0.718	85.7	3.94	—
100	0.789	94.2	0.596	71.1	24.5	0.002
	0.741	88.5	0.384	45.8	48.2	0.003
200	0.813	97.1	0.464	55.4	43.2	0.002
	0.736	87.9	0.244	29.1	66.8	0.002
300	0.787	94.0	0.151	18.0	80.8	0.002
	0.759	90.6	0.154	18.4	79.7	0.002
0	13.7	1640	11.1	1330	18.8	—
	8.76	1040	7.92	945	9.66	—
100	13.0	1550	7.04	840	45.9	0.058
	8.28	988	5.38	642	35.1	0.028
300	12.3	1470	2.11	252	82.9	0.033
	8.60	1030	1.08	129	87.5	0.024
400	12.5	1490	1.28	153	89.8	0.027
	8.31	992	0.200	23.9	97.6	0.020
0	60.8	7260	60.8	7260	—	—
	44.0	5250	39.4	4710	10.4	—
100	54.9	6550	30.8	3680	43.9	0.233
	50.5	6030	32.8	3920	35.1	0.171
300	51.4	6140	13.2	1580	74.3	0.123
	48.9	5840	12.0	1430	75.5	0.119
700	59.7	7130	3.91	467	93.5	0.077
	46.2	5510	4.18	499	91.0	0.058

to the other two pHs. This result is consistent with our understanding of the process and will be discussed below.

Another major variable studied was the presence of an inert solid in the form of kaolin clay. Kaolin was added at 3% by weight to the solutions prior to irradiation. Figure 3 compares the removal efficiency of chloroform at pH 7 with and without the clay present. Qualitatively there appeared to be no difference in removal efficiency in the presence or absence of solids.

REMOVAL EFFICIENCY

The removal efficiency for chloroform can be quantitatively described in terms of two constants, G_D and k. G values were calculated at each absorbed dose (G_D) using the equation given by Kurucz et al.:[9,10]

$$G_D = \frac{[\Delta R]\ N_A}{D \times 6.24 \times 10^{17}} \qquad (2)$$

Table 3   Chloroform Removal Efficiency at Three Influent Concentrations at pH 7 in the Absence of Clay

Dose (krad)	Influent (µM)	Influent (µg l⁻¹)	Effluent (µM)	Effluent (µg l⁻¹)	% removed	G value
0	0.962	115	0.860	103	10.6	—
	0.827	98.7	0.730	87.1	11.8	—
100	0.926	111	0.369	44.1	60.2	0.005
	0.837	99.9	0.279	33.3	66.6	0.005
200	0.920	110	0.153	18.3	83.3	0.004
300	0.965	98.0	0.072	8.60	92.6	0.003
	0.821	115	0.140	16.7	83.0	0.002
400	0.801	95.6	BMDL		>99.9	0.002
0	10.2	1220	8.70	1040	14.5	—
	8.45	1010	7.65	913	9.33	—
100	9.26	1110	4.02	480	56.6	0.051
	8.08	965	3.33	398	58.8	0.046
300	8.40	1003	0.836	99.8	90.0	0.024
	7.95	949	0.812	96.9	89.8	0.023
400	8.18	978	0.451	53.8	94.5	0.019
	7.95	949	0.733	87.5	90.8	0.017
0	60.1	7180	50.4	6020	16.0	—
	51.4	6140	46.5	5550	9.66	—
100	55.0	6570	28.5	3410	48.2	0.256
	50.0	5970	25.1	2990	49.4	0.237
300	53.4	6380	9.39	1120	82.6	0.142
	47.1	5620	12.3	1470	73.9	0.112
700	64.6	7710	2.14	255	96.7	0.086
	46.5	5550	1.63	195	96.5	0.062

where D is the dose in kilorads; $[\Delta R]$ is the difference in organic solute concentration, in moles per liter, at dose D and the zero dose sample obtained after the water had passed through the system (i.e., after irradiation sampling point); 6.24×10^{17} is the conversion factor from krads to 100 eV l⁻¹; and, N_A is Avogadro's number, 6.02×10^{23}.

Tables 2 to 4 summarize the data for the removal of chloroform under the different experimental conditions, with replicate experiments without clay. Tables 5 and 6 summarize the data for chloroform removal in the presence of clay.

The G value, an estimate of chloroform removal efficiency, increased as the influent pH increased. A possible explanation for this observation results from the presence of the carbonate alkalinity in the aqueous solutions.

The bimolecular rate constants for the reaction of the OH· and CO_3^{2-} and HCO_3^- are 3.9×10^8 M^{-1} s^{-1} and 8.5×10^6 M^{-1} s^{-1}, respectively. In the solutions used for these experiments at pH 9, the total alkalinity (Table 1) is approximately 20% CO_3^{2-}. At this concentration and pH the CO_3^{2-} acts as an efficient sink for OH·. We believe that the net result of this is to increase the

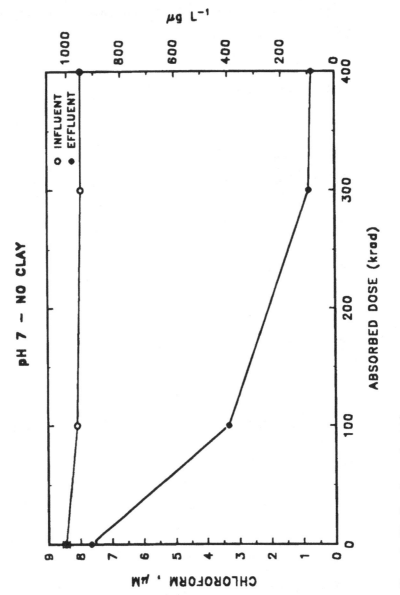

Figure 3 (Continued on next page.)

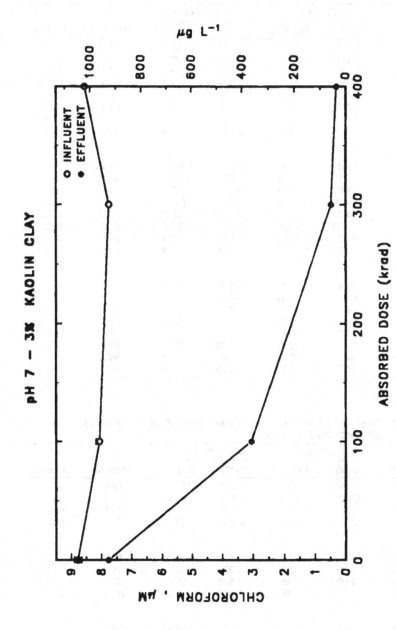

Figure 3 The effect of the presence of inert solids at pH 7 on removal efficiency of chloroform. Each point is the mean of four analyses. The error bars are the standard deviation from the mean; where no error bars are shown the value was smaller than the data point.

Table 4   Chloroform Removal Efficiency at Three Influent Concentrations at pH 9 in the Absence of Clay

Dose (krad)	Influent		Effluent		% removed	G value
	(μM)	(μg l^{-1})	(μM)	(μg l^{-1})		
0	0.779	93.0	0.766	91.4	1.621	—
	0.840	100	0.841	100	—	—
100	0.727	86.8	0.259	30.9	64.4	0.005
	0.733	87.5	0.329	39.3	55.1	0.004
200	0.725	86.6	0.087	10.3	88.0	0.003
	0.770	91.9	0.120	14.3	84.4	0.003
300	0.770	91.9	0.038	4.54	95.1	0.002
	0.792	94.6	0.049	5.85	93.8	0.002
0	15.0	790	13.7	1640	8.44	—
	8.12	969	7.01	837	13.7	—
100	15.0	1790	4.29	512	71.4	0.103
	9.20	1100	3.61	431	60.7	0.054
300	14.5	1730	0.440	52.5	97.0	0.045
	9.24	1100	0.500	59.6	94.6	0.028
400	15.3	1830	0.159	18.9	99.0	0.036
	9.27	1110	0.087	10.4	99.1	0.022
0	59.4	7090	51.9	6200	12.5	—
	63.1	7530	53.6	6400	15.1	—
100	59.2	7070	17.9	2140	69.8	0.398
	62.9	7510	22.5	2690	64.2	0.389
300	57.0	6810	1.86	222	96.7	0.177
	61.5	7340	8.38	1000	86.4	0.171
700	54.5	6510	0.361	43.1	99.3	0.075
	64.3	7680	0.630	75.2	99.0	0.088

Table 5   Chloroform Removal Efficiency at Three Influent Concentrations at pH 5 with 3% Kaolin Clay

Dose (krad)	Influent		Effluent		% removed	G value
	(μM)	(μg l^{-1})	(μM)	(μg l^{-1})		
0	0.988	118	0.864	103	12.6	—
100	0.698	83.3	0.412	49.2	41.0	0.003
200	0.798	95.3	0.319	38.1	60.1	0.002
300	0.766	91.5	0.184	21.9	76.0	0.002
0	6.40	764	5.51	658	13.9	—
100	6.09	727	3.37	402	44.6	0.026
300	4.90	585	0.834	99.6	83.0	0.013
400	5.74	685	0.613	73.2	89.3	0.012
0	119	14,200	97.3	11,600	18.2	—
100	121	14,500	66.9	7,990	44.7	0.520
300	120	14,300	23.6	2,810	80.3	0.309
700	110	13,100	4.60	549	95.8	0.145

Table 6 Chloroform Removal Efficiency at Three Influent Concentrations at pH 7 with 3% Kaolin Clay

Dose (krad)	Influent (μM)	Influent ($\mu g\ l^{-1}$)	Effluent (μM)	Effluent ($\mu g\ l^{-1}$)	% removed	G value
0	0.796	95.0	0.693	82.7	13.0	—
100	0.776	92.6	0.262	31.3	66.3	0.005
200	0.712	85.0	0.137	16.4	80.7	0.003
300	0.676	80.7	0.063	7.52	90.6	0.002
0	8.78	1,050	7.78	929	11.4	—
100	8.06	962	3.05	364	62.1	0.048
300	7.76	926	0.488	58.3	93.7	0.023
400	8.56	1,020	0.305	36.4	96.4	0.020
0	117	13,400	113	13,500	3.25	—
100	72.9	8,700	37.9	4,530	48.0	0.338
300	115	13,800	12.2	1,460	89.4	0.331
700	95.3	11,400	1.03	123	98.9	0.130

effective concentration of the e_{aq}^-, by minimizing the recombination reaction between the e_{aq}^- and $OH\cdot$. The increased concentration of e_{aq}^- resulted in an increased removal efficiency of the chloroform at pH 9.

The second quantitative estimate used for evaluating removal efficiency of the $CHCl_3$ was the dose constant, k. The dose constant is the slope of the line obtained by plotting ln [R] vs. dose, where [R] is the concentration of $CHCl_3$ in the after irradiated solution. Figures 4 to 6 show a few typical plots.

The doses required to remove 50, 90, and 99% ($D_{0.50}$, $D_{0.90}$, and $D_{0.99}$) of the chloroform for all 24 experiments were calculated. These data were used as the basis of the statistical analysis and also to estimate the dose required to remove chloroform under all of the experimental conditions (Table 7):

$$D_{0.50} = \frac{\ln 2}{k}$$

$$D_{0.90} = \frac{\ln 10}{k} \tag{3}$$

$$D_{0.99} = \frac{\ln 100}{k}$$

As expected from the data obtained for the G values, the dose required to remove the $CHCl_3$ decreased with increasing pH and increased with increasing solute concentration.

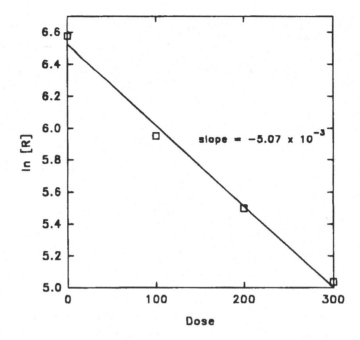

Figure 4   Chloroform at a target concentration of 0.84 μM (100 μg l^{-1}) at pH 5: ln [R] vs. D.

STATISTICAL DATA ANALYSES

A three-factor analysis of variance was performed on the $D_{0.99}$ values using the general linear model (GLM) procedure of Minitab. The GLM procedure was used to handle the unbalanced factorial design which resulted from not being able to run the pH 9 and 3% clay experimental condition. As can be seen by the results of the analysis of variance shown in Table 8, no replicate differences or main effects due to the presence of clay were significant at α = 0.05 and the interaction terms were marginally significant at α = 0.08. The significant interactions indicate that the effects of pH and the presence of clay depended on the concentration of chloroform. Note that other interactions could not be tested with the unbalanced data available.

The adjusted means and standard deviations of the $D_{0.99}$ values of the individual experimental factors (main effects) are given in Table 9. The required dose generally increases with chloroform concentration although there is not a significant difference between 100 and 1000 μg l^{-1}. It is also clear that the required dose (to achieve 99% removal) decreased with increasing pH. The difference between the 0 and 3% clay means was not significant. The means and standard deviations of the $D_{0.99}$ values for combinations of concentration and pH and in the presence of clay are given in Table 10. The

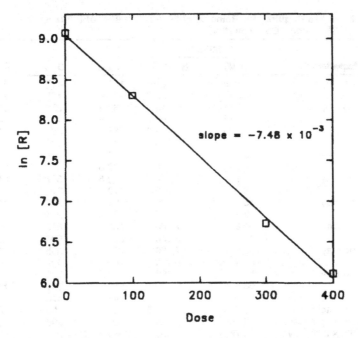

Figure 5 Chloroform at a target concentration of 8.4 μ*M* (1000 μg l⁻¹) at pH 7: ln [R] vs. D.

effect of pH appears to be stronger for high concentrations of chloroform, that is, the reduction in required dose as pH increased was larger for high concentrations. The presence of clay appeared to have no effect on removal.

IMPLICATIONS FOR DRINKING WATER TREATMENT

We have shown that chloroform was effectively destroyed using high-energy electron beam irradiation. In our previous study we showed that all of the trihalomethanes were effectively destroyed.[8] No halogenated organic reaction by-products were produced and a halide ion mass balance was obtained.[8] Therefore, we believe that it is possible to speculate on the ability of the process to remove other disinfection by-products using this method as well. Some of the compounds of interest are the haloacetonitriles, haloacetic acids, chlorophenols, and other miscellaneous compounds, e.g., chloral hydrate.[16] Table 11 lists the available reaction rate constants of the other disinfection by-products with the reactive transient species from Equation 1.[17-24] Based on the comparison of the of rate constants (Table 11) with those of chloroform and the aqueous electron, hydroxyl radical, and hydrogen atom, the high-energy electron beam process would be expected to effectively destroy chloral hydrate, trichloroacetonitrile, monochloroacetic acid, 2,4,6-trichloro-

Table 7   Doses Required for Removal of 50, 90, and 99% of Influent
           Chloroform Concentration

Experiment	Influent (μM)	($\mu g\ l^{-1}$)	$D_{0.50}$	$D_{0.90}$	$D_{0.99}$
pH 5, no clay	0.819	97.8	131	437	873
	0.747	89.2	137	455	910
pH 7, no clay	0.962	115	83.2	276	553
	0.827	98.7	36.8	122	245
pH 9, no clay	0.779	93.0	68.5	228	455
	0.840	100	72.8	242	484
pH 5, no clay	13.7	1,640	125	416	832
	8.76	1,050	77.3	257	514
pH 7, no clay	10.2	1,220	92.5	307	615
	8.45	1,010	113	377	754
pH 9, no clay	15.0	1,790	61.9	206	411
	8.12	969	64.4	214	428
pH 5, no clay	60.8	7,260	184	610	1,220
	44.0	5,250	210	696	1,392
pH 7, no clay	60.1	7,180	155	515	1,030
	51.4	6,140	148	490	980
pH 9, no clay	59.4	7,090	99.2	330	659
	63.1	7,530	112	371	743
pH 5, 3% kaolin clay	0.988	118	142	470	940
pH 7, 3% kaolin clay	0.796	95.0	88.6	294	589
pH 5, 3% kaolin clay	6.40	764	120	398	796
pH 7, 3% kaolin clay	8.78	1,050	83.4	277	554
pH 5, 3% kaolin clay	119	14,200	157	522	1,045
pH 7, 3% kaolin clay	117	14,000	107	356	711

phenol, and, to a lesser extent, the trichloroacetate ion and monobromoacetic acid. Additional studies are needed to quantitatively assess the applicability of this process to disinfection by-product control; however, it would appear that this process would have wide applicability in water treatment.

The cost of treatment using high-energy electron beam irradiation has been previously estimated.[8]

SUMMARY AND CONCLUSIONS

High-energy electron beam irradiation successfully decomposed chloroform in treated groundwater. The removal of chloroform was assumed to be mostly due to reaction with the aqueous electron, based on the reaction rate constants.[12-14] The doses required to remove 50, 90, or 99% of the chloroform from water increased as the solute concentration increased. Removal efficiency was better at low doses and at high pH levels, especially for the high concentrations. The presence of inert solids in the solution does not appear to affect removal. In drinking water treatment the presence of solids in treated water is rare; however, the experiments with clay indicate that this process would have applicability to waters containing solids, i.e., surface waters containing solids and for another application, hazardous waste site remediation.

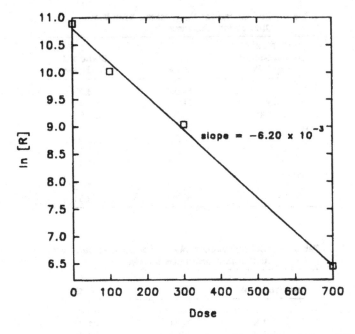

Figure 6 Chloroform at a target concentration of 84 μM (10,000 $\mu g\ l^{-1}$) at pH 9: ln [R] vs. D.

Table 8 Analysis of Variance for $D_{0.99}$ vs. Concentration, pH, and Clay Content for Experiments with Chloroform

Source	df	Seq SS	Adj SS	Adj MS	F	p
Concentration	2	656,282	225,014	112,507	8.00	0.007
pH	2	693,757	696,376	348,188	24.75	0.000
Clay	1	11,664	13,827	13,827	0.98	0.343
Replicate	1	2,178	2,178	2,178	0.15	0.701
Concentration * pH	4	123,808	161,342	40,335	2.787	0.075
Concentration * Clay	2	109,990	109,990	54,995	3.91	0.052
Error	11	154,754	154,754	14,069	—	—
Total	23	1,752,433	—	—	—	—

Note: df = degrees of freedom, Seq SS = sequential sum of squares, Adj SS = adjusted sum of squares, Adj MS = adjusted mean square, F = F-distribution, and p = p-value.

Based on the reaction rate constants with the reactive transient species, high-energy electron beam irradiation appears to be promising for the removal of other disinfection by-products, besides chloroform, from drinking water. Future work will include using a kinetic modeling program for predicting removal efficiencies, and for obtaining a more-detailed description of reaction rates and yields for the destruction of chloroform using high-energy electron beam irradiation.

Table 9   Cell Means for $D_{0.99}$ Values from the Chloroform Experiments

Nominal concentration ($\mu g \, l^{-1}$)	Mean	Standard deviation
100	640.8	53.23
1,000	585.0	53.23
10,000	859.8	53.23
pH		
5	932.4	43.08
7	655.6	43.08
9	497.5	58.48
Clay		
0%	727.7	27.96
3%	662.7	59.31

Table 10   Cell Interaction Means for $D_{0.99}$ Values from the Chloroform Experiments

Conc/100 * pH	Mean	Standard deviation
		71.96
1 * 5	922.0	71.96
1 * 7	476.7	99.34
1 * 9	523.6	71.96
10 * 5	707.9	71.96
10 * 7	634.9	99.34
10 * 9	412.1	71.96
100 * 5	1167.3	71.96
100 * 7	855.2	99.34
100 * 9	556.7	
Conc/100 * Clay		
1 * 0%	586.7	48.42
1 * 3%	694.9	94.81
10 * 0%	592.3	48.42
10 * 3%	577.6	94.81
100 * 0%	1004.0	48.42
100 * 3%	715.5	94.81

Table 11   Reaction Rates of Disinfection By-Products with Reactive Transient Species ($M^{-1}s^{-1}$)

Compound	e_{aq}^-	OH·	H·
Chloroform	3.0×10^{10}	5.0×10^6	1.1×10^7
Trichloroacetonitrile	3.2×10^{10}	3.9×10^7	—
Chloral hydrate	1.2×10^{10}	3.1×10^9	—
Monochloroacetic acid	6.9×10^9	4.3×10^7	6.5×10^3
Monobromoacetic acid	—	—	2.4×10^8
Trichloroacetate ion	8.5×10^9	—	—
2,4,6-Trichlorophenol	—	1.2×10^{10}	1.1×10^9

Note: — Reaction rates not available.

ACKNOWLEDGMENTS

The editorial assistance of L. Anita Holloway was greatly appreciated. The cooperation of the Miami-Dade Water and Sewer Authority was essential in the completion of the work. This research was supported by the National Science Foundation under Grants CES-8714640 and BCS-9108033, the Environmental Protection Agency under Cooperative Agreement CR-816815-01-0, and Grant R-816932-01-0. The work described in this paper has not been reviewed by the U.S. Environmental Protection Agency and therefore the contents do not necessarily reflect the views of the Agency and no official endorsement should be inferred.

REFERENCES

1. Agency for Toxic Substances and Disease Registry, Toxicological Profile for Chloroform, Atlanta, 1989.
2. Adams, J. Q. and Clark, R. M., Evaluating the cost of packed tower aeration and GAC for controlling selected organics, *J. Am. Water Works Assoc.*, 83(1), 49, 1991.
3. Njam, I. N., Snoeyink, V. L., Lykins, B. W., Jr., and Adams, J. Q., Using powdered activated carbon: a critical review, *J. Am. Water Works Assoc.*, 83(1), 65, 1991.
4. Langlais, B., Reckhow, D. A., and Brink, D. R., *Ozone in Water Treatment: Application and Engineering*, Lewis Publishers, Chelsea, MI, 1991, 27.
5. Hoigne, H. and Bader H., Rate constants of reactions of ozone with organic and inorganic compounds in water. I, *Water Res.*, 17, 173, 1983.
6. Pruden, A. L. and Ollis, D. F., Degradation of chloroform by photoassisted heterogeneous catalysis in dilute aqueous suspensions of titanium dioxide, *Environ. Sci. Technol.*, 17, 626, 1983.
7. Korman, C., Bahnemann, D. W., and Hoffman, M. R., Photolysis of chloroform and other organic molecules in aqueous TiO_2 suspensions, *Environ. Sci. Technol.*, 25, 494, 1991.
8. Cooper, W. J., Cadavid, E. M., Nickelsen, M. G., Lin, K., Kurucz, C. N., and Waite, T. D., Removing THMs from drinking water using high-energy electron-beam irradiation, *J. Am. Water Works Assoc.*, 85(9), 106, 1993.
9. Kurucz, C. N., Waite, T. D., Cooper, W. J., and Nickelsen, M. G., High energy electron beam irradiation of water, wastewater and sludge, in *Advances in Nuclear Science and Technology*, Vol. 22, Lewins, J. and Becker, M., Eds., Plenum Press, New York, 1991, 1.
10. Kurucz, C. N., Waite, T. D., Cooper, W. J., and Nickelsen, M. G., Full-scale electron beam treatment of hazardous wastes—effectiveness and costs, in *Proc. 45th Industrial Waste Conference, Purdue University*, Lewis Publishers, Chelsea, MI, 1991, 539.
11. Mehran, M. F., Slifker, R. A., and Cooper, W. J., A simplified liquid-liquid extraction method for analysis of trihalomethanes in drinking water, *J. Chrom. Sci.*, 22, 241, 1984.

12. Hart, E. J., Gordon, S., and Thomas, J. K., Rate constants of hydrated electron reactions with organic compounds, *J. Phys. Chem.*, 68, 1271, 1964.

13. Neta, P., Fessenden, R. W., and Sculer, R. H., An electron spin resonance study of the rate constants for reaction of hydrogen atoms with organic compounds in aqueous solution, *J. Phys. Chem.*, 75, 1654, 1971.

14. Anbar, M., Meyerstein, D., and Neta, P., Reactivity of aliphatic compounds towards hydroxyl radicals, *J. Chem. Soc. B.*, 742, 1966.

15. Lin, K., Cooper, W. J., Nickelsen, M. G., Kurucz, C. N., and Waite, T. D., The decomposition of aqueous solutions of phenol using high energy electron beam irradiation—a large-scale study, Appl. Rad. Isotopes (in press).

16. Krasner, S. W., McGuire, M. J., Jacangelo, J. G., Patania, N. L., Reagan, K. M., and Aieta, E. M., The occurrence of disinfection by-products in US drinking water, *J. Am. Water Works Assoc.*, 81, 41, 1989.

17. Lal, M., Schoeneich, C., Moenig, J., and Asmus, K.-D., Rate constants for the reactions of halogenated organic radicals, *Int. J. Radiat. Biol.*, 54, 773, 1988.

18. Anbar, M. and Hart, E. J., The reaction of haloaliphatic compounds with hydrated electrons, *J. Phys. Chem.*, 69, 271, 1965.

19. Erikson, T., Henglein, A., and Stockhausen, K., Pulse radiolytic oxidation of chloral hydrate in oxygenated and deoxygenated aqueous solutions, *J. Chem. Soc., Faraday Trans. 1*, 69, 337, 1973.

20. Hayon, E. and Allen, A. O., Evidence for two kinds of "H atoms" in the radiation chemistry of water, *J. Phys. Chem.*, 65, 2181, 1961.

21. Adams, G. E., Boag, J. W., Currant, J., and Michael, B. D., Absolute rate constants for the reaction of the hydroxyl radicals with organic compounds, in *Pulse Radiolysis*, Ebert, M., Keene, J. P., Swallow, A. J., and Baxendale, J. H., Eds., Academic Press, New York, 1965, 131.

22. Jortner, J. and Rabani, J., The decomposition of chloroacetic acid in aqueous solutions by atomic hydrogen. I. Comparison with radiation chemical data, *J. Phys. Chem.*, 66, 2078, 1962.

23. Neta, P., Fessender, R. W., and Schuler, R. H., An electron spin resonance study of the rate constants for reaction of hydrogen atoms with organic compounds in aqueous solutions, *J. Phys. Chem.*, 75, 1654, 1971.

24. Draper, R. B., Fox, M. A., Pelizzetti, E., and Serpone, N., Pulse radiolysis of 2,4,6-trichlorophenol: formation, kinetics, and properties of hydroxytrichloro-cyclohexadienyl, trichlorophenoxyl, and dihydroxytrichlorocyclohexadienyl radicals, *J. Phys. Chem.*, 93, 1938, 1989.

CHAPTER 6

Disinfection By-Product Precursor Removal from Natural Waters Using Gamma Radiation to Simulate an Innovative Water Treatment Process

William J. Cooper, Khatereh L. Sawal, Yael S. Hoogland, RoseAnn Slifker, Michael G. Nickelsen, Charles N. Kurucz, and Thomas D. Waite

ABSTRACT

Bench-scale (batch) studies have been conducted, using a ^{60}Co-γ irradiation system, to simulate the high-energy electron beam flow-through process, for the removal of natural organic matter in groundwater. The results indicate that ^{60}Co-γ irradiation, or the high-energy electron beam process, would effectively remove natural organic matter, also referred to as disinfection by-product precursors, from natural waters as a first step in drinking water treatment. This conclusion is based on a significant reduction in dissolved organic carbon concentration in the treated samples when compared to the untreated control. The ultraviolet/visible (UV/vis) spectra from 200 to 500 nm showed a significant reduction in absorption in the region below 400 nm with increased treatment, indicating that the nature of the organic matter in the samples had also changed considerably. The addition of a relatively high concentration of chlorine to samples after treatment resulted in a marked reduction in total trihalomethane (THM) concentration when compared to those in the chlorinated, no-dose control. The THMs were determined after a 144-h reaction time. The reduction in THM formation, in the irradiated samples, indicated that free chlorine could be used in the normal disinfection concentration range, 1 mg l^{-1}, without fear of exceeding disinfection by-product maximum contamination levels (MCLs) mandated under the Safe Drinking Water Act.

INTRODUCTION

Chlorine, the most commonly used primary disinfectant in drinking water treatment, is also used to provide residual disinfection capability in the distribution system. When waters containing naturally occurring humic substances (precursors) are chlorinated, reaction (disinfection) by-products (DBPs) that may compromise the chemical quality of the drinking water are formed. The most common of these DBPs are the four THMs. THMs are regulated at an MCL of 100 μg l^{-1} because they are suspect carcinogens.[1] Because of the ease of analysis they can also be used as a convenient measure of other DBPs. An increase in THM formation usually parallels an increase in other DBPs.

The increasingly stringent regulations being promulgated for DBPs make it imperative that new technologies be investigated and evaluated that address as many of the new requirements as possible. It is likely that one process alone will not solve all of the problems and afford an economical alternative that would not require a major redesign of water treatment plants. Two general strategies exist for the treatment of THMs and other DBPs. The first is the removal of precursor material prior to chlorination, and the second option is to destroy the DPBs once they are formed. This paper reports initial bench-scale batch studies on the removal (destruction) of precursors, measured by dissolved organic carbon (DOC) and ultraviolet absorption from 200 to 500 nm, as one treatment option. Although we have shown that the electron beam process is effective for THM control,[2,3] we feel the removal of precursors is the preferred alternative for consideration in future water treatment plant design or for upgrading existing plants. We base this assumption on the flexibility of the process with applications that include, disinfection,[4] regulated organic chemical control in contaminated source water,[5-14] precursor control,[15] assistance in microflocculation,[16] and taste and odor control. Therefore, the most logical placement of this process would be as the first unit process in the treatment system.

This study was designed to obtain preliminary data to examine the applicability of the process to precursor removal. Groundwater obtained from the Biscayne aquifer was used for the study.

EXPERIMENTAL

GROUNDWATER

Samples of groundwater, from the Biscayne Aquifer, were collected from well A at Florida International University, Miami. The Biscayne Aquifer is a highly productive aquifer that has a pH of 7 and is saturated in carbonate alkalinity, approximately 190 mg l^{-1}. Typically, the chloride ion concentration is 70 mg l^{-1}.

^{60}Co-γ IRRADIATION

All irradiations were conducted at a 5000-Ci ^{60}Co-γ irradiation facility, located at the University of Miami Medical School.

The ^{60}Co-γ source was located in the center of a 200 × 200 cm platform. Enscribed on the platform are concentric circles varying from 10 to 100 cm in radius. These concentric circles served as a reference when samples were irradiated. The relationship of distance and time to give a predetermined applied dose has been established using standard radiation monitoring equipment (this is routinely done by the facility managers), and confirmation of the instrumental method was performed by using chemical dosimetry, the Fricke Dosimeter.[17]

The amount of energy from a ^{60}Co-γ source (or a high-energy electron beam) absorbed by an irradiated material per unit mass is called dose. A common unit of dose is the rad, defined as the energy absorption of 100 ergs/g of material.

A linear regression of a ln/ln plot of distance vs. applied dose was generated to determine the dose rate at any distance from the ^{60}Co source:

$$\ln[\text{absorbed dose}] = -1.958 * \ln[\text{distance}] + 13.356$$

Where [absorbed dose] is in kilorads per minute, and [distance] is in centimeters. Samples were placed at the appropriate distance for a preset time to yield the desired dose, 0, 500, 1000, and 2000 krad.

The groundwater (and dosimetry) samples were irradiated in EPA 47-ml Teflon® lined screw-cap vials with no headspace and chilled prior to and after irradiation.

THM FORMATION POTENTIAL

Samples were chlorinated before and after irradiation, at a total chlorine concentration of 10, 15, or 20 mg l^{-1}. (Chlorine concentrations were determined using the iodometric titration method.[18]) After chlorination the samples were allowed to react for a 6-d period (144 h) in the dark, at room temperature, ~23°C. The samples were then quenched using 30 mg $Na_2S_2O_3$ in each 47-ml reaction vial and stored at 4°C until analysis.

ANALYTICAL PROCEDURES

The DOC was determined using a Shimadzu Model TOC-5000, total organic carbon analyzer. UV/vis spectra were obtained using a Hewlett-Packard Model 8452A Diode Array Spectrophotometer, in 1-cm quartz cells.

The THMs were determined by gas chromatography using liquid-liquid extraction.[1,2,19] A 5-ml automatic pipette was used to remove 5.0 ml of the sample from each 47-ml vial and 3.0 ml of pentane (THM analysis grade) was immediately added and the vial resealed. The 2.0-ml headspace allowed the solvent and sample to mix thoroughly, which is required for effective extraction. The internal standard (1,1,2-trichloroethane) was then injected into the aqueous layer of the sample vial. A model G-2 Gyrotary Shaker (New Brunswick Scientific) was adapted to accommodate eight horizontally mounted extraction vials. The samples were agitated at 300 rpm for 3 min. The pentane layer was allowed to separate from the water (approximately 1 min) then was transferred to 2.0-ml amber crimp top autosampler vials; 2 μl of the extracted pentane mixture was injected into a Hewlett-Packard Model 5890 gas chromatograph equipped with a 30 × 0.539 mm DB-624 bonded phase Megabore® column (J&W Scientific) and electron capture detector. Helium carrier gas was supplied at 6 ml min^{-1}, and nitrogen make-up gas at 40 ml min^{-1}. The injector was maintained at 200°C, the detector at 250°C, and the column oven programmed from 30°C (1-min hold) to 100°C (6°C min^{-1}, 2 min-hold) to 170°C (20°C min^{-1}, 5-min hold). The resulting chromatogram was processed via an HP3392A integrator or ELAB data acquisition software (OMS Tech).

RESULTS AND DISCUSSION

Both high-energy electrons and γ-rays result in a similar decomposition of water, which leads to the formation of three highly reactive species; the reducing, aqueous electron (e_{aq}^-), and hydrogen atom (H·), and the oxidizing, hydroxyl radical (OH·). The chemistry of the irradiation of pure water by γ-rays, at 10^{-7} seconds after the passage of high-energy electrons, is described by the following equation:[20]

$$H_2O \rightsquigarrow e_{aq}^- (2.6) + H·(0.55) + OH·(2.7) + H_2(0.45)$$
$$+ H_2O_2(0.71) + H_3O^+(2.7) \qquad (1)$$

The relative concentration of the species is given by the G value, in parentheses in Equation 1, which is the number of species formed by the absorption of 100 eV of energy. The primary difference between the two processes is that the ^{60}Co-γ source delivers relatively low dose rates. The dose rates used in this study ranged from between 5.36 × 10^{16} and 3.33 × 10^{18} eV l^{-1} s^{-1}, at 100 and 10 cm from the source, respectively. The electron beam system delivers 3.12 × 10^{22} to 1.00 × 10^{24} eV l^{-1} s^{-1}, at an absorbed dose of 25 and 800 krad, respectively. The experiments conducted for this study at the ^{60}Co-γ source were batch experiments, whereas those conducted for other studies by our group at the Electron Beam Research Facility were flow-through studies (120 gpm). Although there may be some subtle differences between the batch and flow-through processes, we feel that studies using ^{60}Co can be used to establish the feasibility of the process, prior to large scale studies.

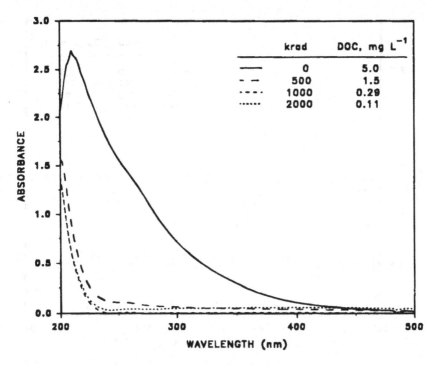

Figure 1   UV/vis spectra of treated and untreated groundwater; DOC = 5.0 mg l^{-1}.

DISSOLVED ORGANIC CARBON REMOVAL

Two groundwater samples were collected at two different times of the year. Figures 1 and 2 show the UV/vis spectra of the water samples irradiated at several doses. The DOC of each water before and after irradiation is also tabulated in the figures. The initial DOC was 5.0 and 6.1 mg l^{-1}, for the samples obtained during the summer and winter, respectively.

The use of high-energy electrons or γ-rays to reduce or destroy the DOC from natural waters results from the reactions of the reactive transient species formed in irradiated aqueous solutions. If this is the only process that results in the removal of the DOC, then a general equation describing this process can be formulated as follows:

$$\frac{d[DOC]}{dt} = k_e[DOC][e^-] + k_H[DOC][H\cdot] + k_{OH}[DOC][OH\cdot] \qquad (2)$$

Where [DOC] represents the concentration of the dissolved organic carbon and the respective reactive species concentration, moles per liter, and the second-order rate constant of the reaction, k, molar concentration per second. The relative distribution of the transient reactive species is given in Equation

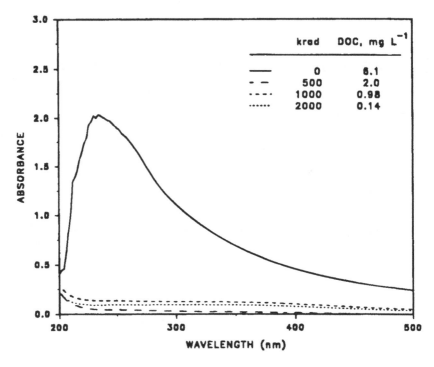

Figure 2 Ultraviolet/visible spectra of treated and untreated groundwater; DOC = 6.1 mg l^{-1}.

1 by the G value. The estimated concentration of each transient reactive species can be calculated for pure water at any dose using the following equation:

$$1 \text{ Mrad} = 1.04 \times 10^{-3} \text{ mol } l^{-1} = 1.04 \text{ m}M \qquad (3)$$

for a chemical with a $G = 1$. Therefore, by irradiating samples at different doses the removal efficiency can be determined. Ultimately this information would be used to determine the economic feasibility of the process for a specific application.

A quantitative method for evaluating removal efficiency has been used to provide the dose required to obtain any treatment objective. By plotting the ln[DOC] vs. dose a straight line was obtained (Figure 3). The slope of the line, which is the dose constant, k, can then be used as a means of comparison between different chemicals or in this case different samples of groundwater, i.e., DOC. The dose constants for the two samples were -0.00193 and -0.00186, for the water with 5.0 and 6.1 mg l^{-1} DOC, respectively. These constants were close and indicated that the removal efficiency from the two samples was similar. Further studies should be conducted to extend this to

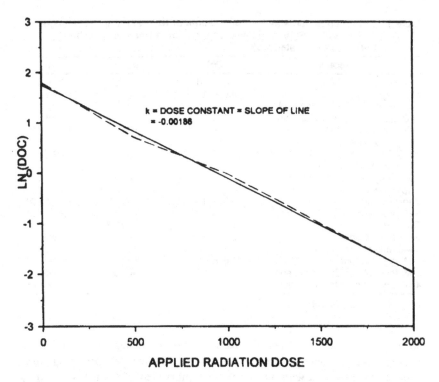

Figure 3 Determination of dose constant for the water sample with DOC = 6.1 mg l⁻¹.
(- - - - connects data points, —— was linear regression.)

groundwater with higher DOC and to determine the effect of factors such as alkalinity and pH on removal efficiency.

To evaluate the relative importance of the three reactive species on the removal of DOC requires bimolecular rate constants. In studies to better understand the reactions of O_3 with DOC, the OH· bimolecular rate constant has been estimated.[21] The most recent studies of Peyton indicate that the value is approximately 2.3 to 2.4 × 10⁸ $M^{-1}s^{-1}$ as C. To our knowledge no bimolecular rate constants have been published for the reaction of the hydrated electron or the reaction of the H· with DOC.

It is possible to speculate that both the e_{aq}^- and the H· are also involved in reactions that lead to the destruction of the DOC based on rate constants of compounds that are likely to be present as part of the DOC, i.e., humic substances. The UV/visible spectra of the samples changed with increasing irradiation with a marked reduction in the absorbance throughout the wavelength range 200 to 400 nm. These changes are, in a general sense, associated with the $n \rightarrow \pi^*$ and $\pi \rightarrow \pi^*$ electronic transitions found in compounds containing carbonyl and aromatic factions. Therefore, it is reasonable to assume that the moieties responsible for this absorption are mainly those associated

Table 1   Compounds that May Be Components of Natural Humic Substances and Reaction Rate Constants with the Three Reactive Species Formed in Irradiated Water[22]

Compound	OH·	e_{aq}^-	H·
Phenol	1.0×10^{10}	4.3×10^{10}	1.7×10^9
Phenoxy ion	9.6×10^9	4.0×10^8	NF
o-Dihydroxybenzene (catechol)	1.1×10^{10}	NF	NF
m-Dihydroxybenzene (resorcinol)	1.2×10^{10}	NF	NF
p-Dihydroxybenzene (hydroquinone)	1.0×10^{10}	NF	1.3×10^9
1,3,5-Trihydroxybenzene (phloroglucinol)[23]	$>1.0 \times 10^{10}$	NF	NF
1,2,3-Trihydroxycyclo-hexadiene (pyrogallol)	NF	NF	2.5×10^9
Benzoic acid	4.3×10^9	1.2×10^{10}	9.2×10^8
Benzaldehyde	4.4×10^9	NF	1.4×10^9
Quinone	1.2×10^9	2.3×10^{10}	8.3×10^9
2-Methoxyphenol	2.0×10^{10}	NF	NF
3-Methoxyphenol	3.2×10^{10}	NF	NF
4-Methyoxyphenol	2.6×10^{10}	NF	NF
3-Methyl-1,2-dihydroxybenzene	1.6×10^{10}	2.0×10^7	NF
4-Methyl-1,2-dihydroxybenzene	NF	NF	1.5×10^9

Note: NF = Not found.

with "phenolic-like" compounds. Reaction rate constants for several likely analogs of those phenolic-like compounds found in naturally occurring humic substances consistent with the above are summarized in Table 1. From these data it can be seen that the rate constants for e_{aq}^- and those of the H· are comparable to those of the OH·. Therefore, by inference, it is reasonable to assume that both can contribute to the mineralization of the DOC. Further, data summarized in Table 1, e.g., ionization of phenol and similar organic function groups, suggests that, if the e_{aq}^- is of importance, pH may affect the removal of DOC from natural waters. Additional studies should be undertaken to determine if this indeed occurs.

A third method of evaluating the effectiveness of the process would be to determine the effect of increasing treatment dose on the formation of THMs in chlorinated samples. Samples that had been irradiated and a zero dose control were chlorinated with increasing concentrations of chlorine. Low concentrations of chlorine <5 mg l^{-1} resulted in very little THM formation, presumably in part from the reduction of the HOCl by the H_2O_2 that was formed during irradiation. Therefore, three higher concentrations were employed. Figure 4 summarizes the data for the formation of the total THMs from the four samples. Table 2 summarizes the concentrations of the individual THMs. It is evident that increasing irradiation resulted in a decreased THM concentration. These results suggest that, at the chlorine concentrations required to provide a disinfectant residual in the distribution system, minimal formation of disinfection by-products would result. We can speculate, based on the THM concentrations, that these would be well within the proposed limits.

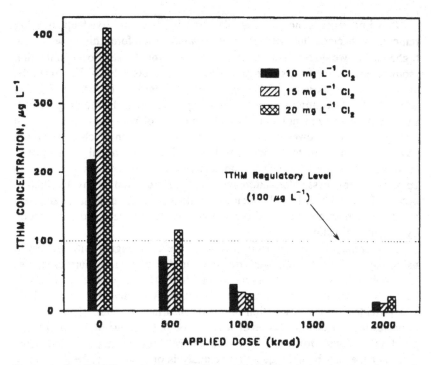

Figure 4 Total THM formation (reaction time = 144 h) of treated and untreated groundwater (DOC = 5.0) at several doses.

Table 2 The Concentration of Individual THMs in Chlorinated Groundwater at Different Chlorine Concentrations and Different Dose

Cl_2 (mg l^{-1})	Dose (krad)	THM formation (μg l^{-1})			
		$CHCl_3$	$CHBrCl_2$	$CHBr_2Cl$	$CHBr_3$
10	0	176[a] ± 2.8[b]	35.2 ± 0.17	6.30 ± 0.16	0.39 ± 0.01
	500	20.0 ± 0.79	37.8 ± 1.7	16.8 ± 0.90	2.72 ± 0.18
	1000	10.7 ± 0.43	17.9 ± 0.10	7.17 ± 0.07	1.38 ± 0.03
	2000	3.54 ± 0.01	5.86 ± 0.02	3.04 ± 0.02	0.82 ± 0.01
15	0	315 ± 25	57.2 ± 0.32	8.72 ± 0.09	0.35 ± 0.01
	500	29.4 ± 0.65	27.9 ± 0.08	8.42 ± 0.11	0.80 ± 0.02
	1000	11.7 ± 0.05	11.1 ± 0.05	3.32 ± 0.01	0.43 ± 0.01
	2000	4.61 ± 0.00	4.85 ± 0.01	1.54 ± 0.01	0.25 ± 0.01
20	0	347 ± 5.7	54.2 ± 0.19	7.66 ± 0.14	0.32 ± 0.01
	500	52.9 ± 0.29	50.4 ± 0.16	11.4 ± 0.05	1.02 ± 0.02
	1000	10.7 ± 0.35	10.2 ± 0.21	2.95 ± 0.01	0.43 ± 0.01
	2000	8.43 ± 0.08	9.91 ± 0.13	1.89 ± 0.06	0.41 ± 0.02

[a] Mean value of four readings.
[b] Standard deviation.

It is important to note that at 500 krad we have observed up to 5 log removal of bacteria and viruses in wastewater. Therefore, this dose, or the higher doses, would result in a treated water that would meet most disinfection requirements. One of the stable radiologic by-products is H_2O_2. For example, at 500 krad a maximum of 17 mg l^{-1} H_2O_2 would be formed in distilled water. At this concentration, the H_2O_2 would provide some disinfecting capability through the treatment plant. The concentration of H_2O_2 was not measured during this study; however, in practice the final addition of chlorine would have to be sufficient to "titrate" the H_2O_2 prior to the distribution system. However, because of the disinfection capabilities of the process, there would be no need for excess chlorination, i.e., above that required for the distribution system. Thus, a 0.5- to 1-mg l^{-1} residual would provide the necessary safeguard for the distribution system and presumably result in little or no disinfection by-product formation.

In summary, the results from this study are encouraging with respect to using high-energy electron beam treatment in drinking water. Further studies are required using a flow-through system under differing conditions likely to be encountered at different treatment facilities. A preliminary cost analysis was provided for the destruction of THMs in drinking water.[2] That analysis indicates that the cost of this process would be in the range of $1.50 to $2.00 per 1000 gallons. However, these data are very preliminary and additional work is necessary before a quantitative analysis of costs could be developed.

ACKNOWLEDGMENTS

The editorial assistance of L. Anita Holloway was greatly appreciated. The cooperation of the Miami-Dade Water and Sewer Authority was essential in the completion of the work. This research was supported by the National Science Foundation under Grants CES-8714640 and BCS-9108033, the Environmental Protection Agency under Cooperative Agreement CR-816815-01-0, and Grant R-816932-01-0. The work described in this paper has not been reviewed by the U.S. Environmental Protection Agency and therefore the contents do not necessarily reflect the views of the Agency and no official endorsement should be inferred.

REFERENCES

1. Agency for Toxic Substances and Disease Registry, Toxicological Profile for Chloroform, Atlanta, 1989.
2. Cooper, W. J., Cadavid, E. M., Nickelsen, M. G., Lin, K., Kurucz, C. N., and Waite, T. D., Removing THMs from drinking water using high-energy electron-beam irradiation, *J. Am. Water Works Assoc.*, 85(9), 106, 1993.

3. Mak, F. T., Cooper, W. J., Kurucz, C. N., Nickelsen, M. G., and Waite, T. D., Chapter 5 of this text.

4. Farooq, S., Kurucz, C. N., Waite, T. D., and Cooper, W. J., Disinfection of wastewaters: high energy electron vs gamma irradiation, *Water Res.*, 27, 1177, 1993.

5. Kurucz, C. N., Waite, T. D., Cooper, W. J., and Nickelsen, M. G., High energy electron beam irradiation of water, wastewater and sludge, in *Advances in Nuclear Science and Technology*, Vol. 22, Lewins, J. and Becker, M., Eds., Plenum Press, New York, 1991, 1.

6. Kurucz, C. N., Waite, T. D., Cooper, W. J., and Nickelsen, M. G., Full-scale electron beam treatment of hazardous wastes—effectiveness and costs, *Proc. 45th Industrial Waste Conference, Purdue University*, Lewis Publishers, Chelsea, MI, 1991, 539.

7. Nickelsen, M. G., Cooper, W. J., Waite, T. D., and Kurucz, C. N., Removal of benzene and selected alkyl substituted benzenes from aqueous solution utilizing continuous high-energy electron irradiation, *Environ. Sci. Technol.*, 26, 144, 1992.

8. Cooper, W. J., Nickelsen, M. G., Meacham, D. E., Waite, T. D., and Kurucz, C. N., High energy electron beam irradiation: an innovative process for the treatment of aqueous-based organic hazardous wastes, *J. Environ. Sci. Health*, A27(1), 219, 1992.

9. Cooper, W. J., Nickelsen, M. G., Meacham, D. E., Waite, T. D., and Kurucz, C. N., High energy electron beam irradiation: an advanced oxidation process for the treatment of aqueous based organic hazardous wastes, *Water Pollut. Res. J. Canada*, 27, 69, 1992.

10. Cooper, W. J., Meacham, D. E., Nickelsen, M. G., Lin, K., Ford, D. B., Kurucz, C. N., and Waite, T. D., The removal of tri- (TCE) and tetrachloroethylene (PCE) from aqueous solution using high energy electrons, *J. Air Waste Mgmt. Assoc.*, 43, 1358, 1993.

11. Kurucz, C. N., Waite, T. D., and Cooper, W. J., The Miami electron beam research facility: a large scale wastewater treatment application, *Radiat. Phys. Chem.*, 45, 299, 1995.

12. Wang, T., Waite, T. D., Kurucz, C. N., and Cooper, W. J., Oxidant reduction and biodegradability improvement of paper mill effluent by irradiation, *Water Res.*, 28, 237, 1994.

13. Nickelsen, M. G., Cooper, W. J., Lin, K., Kurucz, C. N., and Waite, T. D., High energy electron beam generation of oxidants for the treatment of benzene and toluene in the presence of radical scavengers, *Water Res.*, 28, 1227, 1994.

14. Lin, K., Cooper, W. J., Nickelsen, M. G., Kurucz, C. N., and Waite, T. D., The decomposition of aqueous solutions of phenol using high energy electron beam irradiation—a large scale study, *Appl. Radiat. Isot.* (in press).

15. Arai, H., Arai, M., and Sakumoto, A., Exhaustive degradation of humic acid in water by simultaneous application of radiation and ozone, *Water Res.*, 20, 885, 1986.

16. Wang, T., Waite, T. D., Kurucz, C. N., and Cooper, W. J., Sludge dewatering and conditioning by gamma and high energy electron beam radiation, *Water Environ. Fed.*, submitted.

17. Spinks, J. W. T. and Woods, R. J., *An Introduction to Radiation Chemistry*, John Wiley & Sons, New York, 1990.

18. Standard Methods for the Examination of Water and Wastewater, 16th ed., APHA, AWWA AND WEF, Washington, D.C., 1985.

19. Mehran, M. F., Slifker, R. A., and Cooper, W. J., A simplified liquid-liquid extraction method for analysis of trihalomethanes in drinking water, *J. Chrom. Sci.*, 22, 241, 1984.

20. Buxton, G. V., Greenstock, C. L., Helman, W. P., and Ross, A. B., Critical review of rate constants for reactions of hydrated electrons, hydrogen atoms and hydroxyl radicals ($\cdot OH/\cdot O^-$) in aqueous solution, *J. Phys. Chem. Ref. Data*, 17, 513, 1988.

21. Peyton, G. R., Illinois State Water Survey, personal communication, March 1994.

22. Ross, A. B., Bielski, B. H. J., Buxton, G. V., Cabelli, D. E., Greenstock, C. L., Helman, W. P., Huie, R. E., Grodkowski, H., Neta, P., and Mailard, W. G., NIST Standard Reference Database 40, NDRL/NIST Solution Kinetics Database, Version 2.0, Gaithersburg, MD, 1994.

23. Wang, D., György, I., Hildebrand, K., and von Sonntag, C., Free radical induced oxidation of phloroglucinol. A pulse radiolysis and EPR study, *J. Chem. Soc. Perkin Trans. 2*, 145, 1994.

PART III

Ozonation and Brominated Disinfection By-Products

CHAPTER 7

An Overview of Ozonation Disinfection By-Products

Howard S. Weinberg and William H. Glaze

ABSTRACT

Ozone use in drinking water treatment is expanding as consumers demand a higher level of water quality, and this has sparked several new studies to investigate ozone disinfection by-products (DBPs). The major DBPs resulting from ozone treatment of surface or groundwaters have been identified as low molecular weight aliphatic aldehydes, in particular formaldehyde and acetaldehyde, the dialdehyde glyoxal, and the keto-aldehyde methyl glyoxal. Other partial oxidation by-products with carbonyl functionalities include glyoxylic, keto-malonic, and pyruvic acids. Hydrogen peroxide and organic peroxides have also been found in plants using ozone but appear to be removed, as are the aldehydes, by filters that possess an active biomass. Bromide, in raw waters, engages ozone in a complex cycle in which both organic bromide and inorganic bromate are end products. This paper will review a recent large-scale study of the occurrence and formation of ozone DBPs and show how studies of these by-products can be used to control their formation in finished water.

INTRODUCTION

Water treatment plants throughout the world are turning more and more to ozonation as a means to improve their drinking water quality. In the U.S., many utilities employ ozone to help them meet the requirements of the safe drinking water act and the new disinfectants and disinfection by-products rule. As a powerful oxidant, ozone is bound to engage the raw water natural organics and some of the inorganic components (in particular the bromide ion) in a series of reactions that ultimately produce a variety of by-products that are likely to be more polar (organics) and in higher oxidation states (organics and inorganics) than their precursors. The truth is that relatively little is known of the identity of by-products formed by the interaction of ozone with natural water components, and even less has been studied of their occurrence in treatment plants employing ozone. This is a result, in part, of the limited analytical capability for extracting and determining the polar organic fraction from the aqueous matrix, some components of which are expected to be very polar and/or unstable and therefore not suited to standard preconcentration techniques. Clearly, a chemical methodology is needed to target ozonation by-products of health concern and for application in field studies to determine their occurrence in actual drinking waters.

To address these issues, our group combined with others in the water treatment industry to form a consortium project ultimately supported by the AWWA Research Foundation, with majority contributions from six utilities. The project consisted of two interconnected components: a laboratory phase in which analytical methods for targeted ozonation by-products were developed and/or optimized and a survey phase in which ozone by-products were analyzed in water samples taken from selected treatment plants that utilized ozonation.

LABORATORY STUDIES

In the laboratory-scale research of the consortium project, studies were carried out to determine what factors affected the levels of specific classes of ozone by-products that were identified in advance as important ones for study. These studies used primarily a local water source (University Lake, Orange County, NC) and in some cases waters from supplies in other parts of North America. The specific classes of by-products that were targeted are as follows: aldehydes, mixed functional aldo and keto acids, organic and inorganic brominated by-products, hydrogen peroxide, organic peroxides, and epoxides. In addition, this paper reports the results of carboxylic acids targeted by studies outside the framework of the consortium project. A flow chart summarizing the routes by which these by-products might be formed, in the absence of bromide, is presented in Figure 1. Most of the laboratory studies were carried out by ozonating the water in a 5-l glass reactor but in a few cases a 70-l stainless steel reactor was used. Ozone, generated from an air supply, was added to the water at selected doses to achieve various ozone:TOC ratios. In some studies, the effect caused by the variation of other water quality parameters such as pH and alkalinity was studied.

CARBONYL COMPOUNDS

Background

It has been known for many years that ozonation of natural water produces polar organic compounds such as aldehydes and carboxylic acids,[1-4] presum-

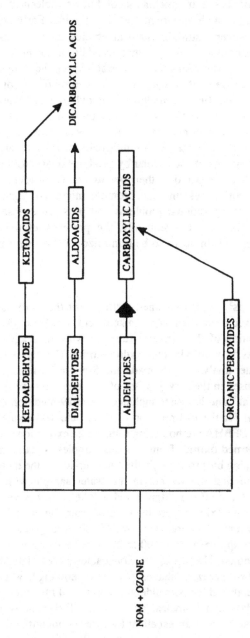

Figure 1 Pathways for organic by-products formed during the ozonation of drinking water.

ably from the oxidative fragmentation of higher molecular weight natural organic compounds such as humic and fulvic acids. Early studies revealed the presence of several straight chain aldehydes such as butanal, pentanal, and heptanal. Later studies showed that, in addition, lower molecular weight aldehydes such as formaldehyde and acetaldehyde, the dialdehyde glyoxal, and the keto-aldehyde methyl glyoxal are formed by ozonation.[5] The consortium project also identified the diketone dimethyl glyoxal.[6] The origin of these ozonation by-products is not known, but one presumes they are formed by the addition of ozone to either unsaturated side chains or aromatic functionalities of natural organic matter, which has a multiplicity of organic structural units.

The major concerns of these polar by-products relates to their high degree of biodegradability compared to their precursors. High levels of these compounds in distribution systems can promote microbial regrowth which, in turn, could lead to operational problems and possibly expose consumers to gastrointestinal diseases. Consequently, the presence of these by-products resulting from ozonation needs to be minimized in treatment plant effluents.

Analysis

Several groups[5-9] have examined methods for the analysis of aldehydes, ketones, and mixed functional aldo and keto acids and have settled on aqueous phase derivatization of the carbonyl function with the reagent O-(2,3,4,5,6-pentafluorobenzyl)-hydroxylamine hydrochloride (PFBHA) originally used for drinking water analysis by Yamada and Somiya.[7] This is the method that was eventually used in the survey phase of the consortium project as modified by Sclimenti et al.[9] and has been included in the 19th edition of the *Standard Methods* manual for the examination of water and wastewater. This project also adapted the PFBHA method using solid-phase extraction (SPE) techniques to isolate the formed oximes from aqueous samples containing low levels of carbonyl-containing by-products. In the final version of the methods, carbonyl compounds were first derivatized in the water sample with PFBHA, then passed through the SPE cartridge at pH 4 where they were adsorbed. The PFBHA-derivatized oximes were then eluted from the cartridge with a minimum volume of methyl t-butyl ether (MtBE) so as to give an extraction ratio (water:MtBE) of approximately 100 to 250, thus enhancing the sensitivity of the analytical method. The mixed functional acids were isolated by derivatizing the ether extract with diazomethane.[6] A scheme showing how a single aqueous sample can be analyzed for both aldehyde and mixed functional acids is shown in Figure 2. Mass spectrometric analysis of PFBHA-derivatized carbonyl functionalities has proven an excellent tool for the identification of unknown by-products due to the distinct fragment of PFBHA oximes at a charge to mass ratio (m/z) of 181.[5] Derivatized, mixed functional, or oxo acids can be identified by the occurrence of an additional ion fragment at m/z = 59 due to the methylated carboxylate group. Using this method of characterization,

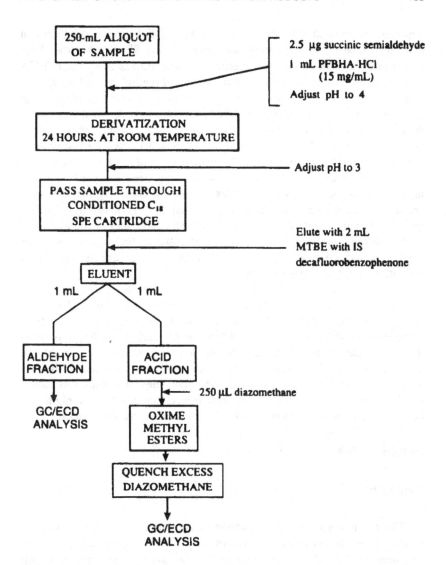

Figure 2 Scheme for simultaneous analysis of neutral and acid carbonyl-containing DBPs.

20 by-products containing carbonyl functionalities have been identified in ozonated drinking water.[10]

Laboratory studies of the consortium project examined the recovery of these compounds from different matrix waters and sought to establish sound quantification criteria for the aldehydes and mixed functional carbonyl compounds. The method promises to be of use in field studies where the extraction cartridge could be shipped to a remote laboratory for extraction and analysis;

however, the method was used for only a few of the surveys conducted in this study. One of the most important results of the SPE work came about accidentally, as a result of some preliminary field studies using the PFBHA method. It was found that a commercial sample of methyl glyoxal being used as a standard in the field work (and by other laboratories) was not pure, as presumed. This led to erroneously high levels of methyl glyoxal in some of the early reports from this project, as well as other studies of aldehyde formation after ozonation. As a result of this discovery, it became clear that primary standards of the oximes are needed for optimization of the PFBHA method.[8]

Low molecular weight carboxylic acids have been analyzed by direct injection of quenched aqueous samples into an ion chromatographic system set up for the analysis of anions using a sodium hydroxide eluent.[12] Current method detection limits are 10 μg/l for formic, acetic, and oxalic acids and method development continues using preconcentration techniques to increase sensitivity and using capillary zone electrophoresis for quicker and better resolution of the acid by-products.

The source for these ozonation by-products has been investigated by another team of researchers who compared the ozonation of natural water to that of ozonated fractions of natural organic matter from the same water.[11] Their results indicate fulvic acid and hydrophilic fractions of NOM as being the major aldehyde precursors and that fulvic acid is partially responsible for some of the mixed functional acid by-products. Oxo acids appear to be promoted at higher pH compared to acids and aldehydes suggesting OH radical involvement in the former case yet these mixed functional acids might also be expected to be transitional states between dialdehydes and diacids.

PEROXIDES AND EPOXIDES

Background

The chemistry of ozone interaction in natural waters is responsible for short-lived peroxygenated intermediates as well as hydrogen peroxide. Hydrogen peroxide has often been considered as an intermediate product of the ozonation of organic compounds and, although its presence could catalyze further decomposition of ozone, a terminal residual may, under certain circumstances, exist in plant effluents. Hydrogen peroxide is a compound with known health effects and the European community has suggested tolerance limits for its presence in distributed water at less than 100 μg/l.[13]

Organic epoxides are well known alkylating agents that have been identified in many cases as genotoxic or carcinogenic materials. While they are not expected to be major products of ozonation reactions, they may certainly be formed as reaction intermediates and their isolation and identification would be of use in obtaining a better understanding of aqueous ozonation chemistry.

The potentially trace levels of epoxides produced requires a very systematic isolation, concentration, and specific analytical approach.

Analysis

The peroxides were analyzed by a modified method in which a peroxidase enzyme reacts with the peroxide in the presence of a hydrogen donor (p-hydroxyphenylacetic acid) causing the latter to form a fluorescent dimer.[14] This method also detects certain organic hydroperoxides which were quantified as a group and monitored in ozonated water, including through treatment plants. There are some indications in the literature[15,16] that individual organic peroxides can be detected using this method of dimerization which follows separation of the individual components in the eluent of a reversed-phase high-performance liquid chromatographic analysis.

Some initial work was also undertaken to establish an analytical procedure for stabilizing epoxides and making them amenable to chromatographic separation. The method involved the isolation and derivatization of the epoxides with anhydrous hydrogen chloride to form the chlorohydrins which were then further derivatized with heptafluorobutyrylimidazole to make them detectable at 1-μg/l levels in a gas chromatograph with electron capture detection. Further development of this method is, however, required before it can be used in field studies.

BROMINATED BY-PRODUCTS

Background

The laboratory study of the consortium project also included a detailed investigation of brominated by-products formed by the ozonation of waters containing bromide ion. This is of concern since some American water supplies have rather high natural bromide levels, and previous studies have shown that bromide may either be oxidized to bromate ion or to hypobromous acid. The latter may then react with natural organic matter (NOM) to form brominated organics such as bromoform, brominated acids, bromoacetonitriles, etc. Bromination of NOM is thought to occur primarily by HOBr, whereas oxidation to bromate involves primarily the conjugate base, hypobromite.[17] A scheme showing the different pathways by which brominated by-products are formed during ozonation is presented in Figure 3. Note that hydroxy radicals formed during ozone decomposition contribute to bromate formation via BrO radicals and bromite.[18] A systematic study was carried out of the by-products obtained when University Lake water, to which various amounts of bromide had been added, was oxidized with ozone (ozone:TOC ≈ 1).[19] In addition to the brominated organics listed above, the samples after ozonation were analyzed for dissolved, adsorbable (on granular activated carbon) organic bromide (DOBr)

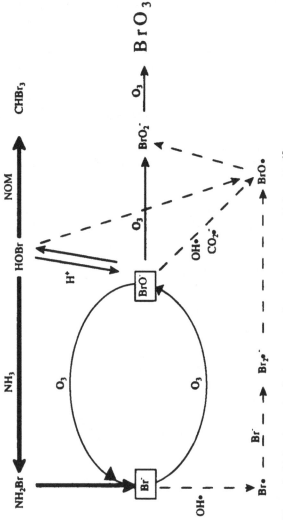

Figure 3 Molecular and radical pathways of ozone interaction with bromide.[18]

and bromate (by ion chromatography). Also, in the course of this study of ozonation with enhanced levels of bromide ion, there was found a group of previously unidentified, labile compounds identified as bromohydrins.[20] Subsequently, we have shown that these bromohydrins are due in part to bromination of impurities in the two solvents (pentane and M*t*BE) used for extraction. However, when the extraction process was improved and with the use of quality-assured control blanks, there were some indications that these components were still present as ozone by-products albeit at much lower amounts than originally thought.

In general, the accurate recovery and quantification of brominated, organic compounds is very dependent on their preservation in transit to the analytical laboratory. Ammonium sulfate was the most effective quenching agent for hypobromous acid, the active brominating species resulting from the interaction of inorganic bromide with ozone. Also, it was found that bromate does appear to form at significant levels with elevated bromide levels and its concentration is enhanced at high pH. The presence of ammonia, on the other hand, tends to tie-up the hypobromous acid and reduce the formation of these brominated by-products. These compounds are also formed by reaction of the natural water with aqueous bromine, irrespective of prior ozonation, and, since HOBr is formed in the process, by chlorination in the presence of high bromide levels. However, the nature of their precursors remains a mystery although it has been assumed for some years now that most DBPs are derived from the humic content of NOM. In the case of bromohydrin by-products, naturally occurring terpenes would be the more likely precursor.[20]

The issue of bromate formation during ozonation has drawn particular attention recently during the negotiation of regulations for DBPs in drinking water.[22] For the time being, a maximum contaminant level (MCL) of 10 μg/l bromate in finished water will be set due to limitations in practical quantification. However, the lifetime cancer risks associated with these levels has been determined by the U.S. EPA to be too high for long-term acceptance. With this in mind, an analytical approach was developed that would permit bromate quantification at 1 μg/l or lower in waters containing excess amounts of other anions.

Analysis

The organic by-products were analyzed in two fractions: neutrals which include trihalomethanes, haloacetonitriles, and haloketones were extracted from aqueous solutions at ambient pH using either pentane or hexane;[23] haloacetic acids were extracted at low pH (<2) with M*t*BE and derivatized with diazomethane to form the esters.[24] Both fractions were analyzed by gas chromatography using electron capture detection. For the analysis of bromate, aqueous samples were not preserved with any of the salts typically used in quenching OBr⁻. The samples were injected directly into an ion chromatographic system set up to resolve low microgram per liter levels of bromate from milligram per liter levels of chloride.[25] Studies undertaken subsequent to the cited project achieved lower detection limits. The methodology utilized selective anion removal by sample pretreatment followed by preconcentration of a 5-ml aqueous sample on a short concentrator column. The retained anions, including bromate, are then flushed onto the analytical column, and with the aid of suppressed conductivity bromate has been detected down to 0.5 μg/l in waters containing several milliequivalents of anions.[26] This method has allowed quantification of bromate in ozonated waters containing 0.05 mg/l bromide which were previously thought to be bromate-free.

SURVEY OF OZONE DBPs IN TREATMENT PLANTS

ALDEHYDES

Viewed as a whole, the results from the consortium project indicate several fundamentals relating to aldehyde formation and removal during the preparation of drinking water using ozonation;

1. Aldehyde formation is predominantly a *function of applied ozone dose and total organic carbon content* of the ozone contactor influent water, with the levels of concentrations formed decreasing in the following order: formaldehyde > glyoxal > acetaldehyde > methyl glyoxal > C_3–C_{10} monoaliphatic aldehydes.
2. *The nature of the NOM* does appear to play a role in determining aldehyde formation potential since different waters sources exposed to the same ozone:TOC ratio at the same pH produce different levels of by-products. This is illustrated in Table 1 which indicates aldehyde formation (using the same analytical method) resulting from ozonation at different plants throughout this survey[27] as well as those obtained from an independently run pilot plant study.[28] Pilot plant E2, for example, produced 27 μg/l formaldehyde at an ozone:TOC of 0.7 mg/mg whereas at a ratio of 0.6 at full-scale plant G the level is less than half this amount (12 μg/l). Pilot plant C1, on the other hand, at an ozone:TOC ratio of 1.5 produces only 12.9 μg/l formaldehyde. The variability exhibited could be explained by a closer scrutiny of each plant's operational efficiency (e.g., ozone transfer efficiency) as well as the kinetic potential of ozone interaction with the constituents of the raw water which could be a function of pH. However, it is quite conceivable that the NOM components of each utility's source water may vary in structural character thus affecting its potential to be degraded by ozone into the aldehyde by-products. Recent studies[29] with laboratory controlled semi-batch ozonations of raw waters from different water sources throughout the U.S. have further highlighted this NOM nature/by-product dependency. In particular, two waters with similar TOC values (15.6 and 13.8 mg/l C, respectively) subjected to the same ozonation conditions produced 568 and

Table 1 Levels of Major Aldehyde By-Products in the Ozone Contactor Effluents of Drinking Water Treatment Plants (Both Full and Pilot Scale)

Plant ID	O₃:TOC (mg/mg)	pH of ozonation	Formaldehyde (μg/l)	Acetaldehyde (μg/l)	Glyoxal (μg/l)	Methyl glyoxal (μg/l)
PP A[28]	0.5	NA	6	2.8	1	2.8
PP A[28]	1.4	NA	6.2	2.2	4.2	1.2
PP A[28]	2.2	NA	8.4	3.2	5	6.4
PP B[27]	1.1	8.8	65	25	9.2	24
PP C1[27]	1.5	6.7	12.9	NA	7.6	10
PP C2[27]	1.08	6.8	3.9	bdl	4.6	1.9
PP D[27]	1.3	7.9	13	3.6	9	5.2
PP E[27]	0.7	7.48	27	4.9	15	NA
FS A[27]	1.24	7.4	26	5.9	9.5	7.4
FS B[27]	1.3	NA	16	6.8	NA	NA
FS C[27]	0.2	9.9	4.1	15	bdl	1
FS D[27]	0.5	7.4	13	3.7	8.9	9
FS E1[27]	0.56	6.6	19.2	NA	8.7	8.0
FS E2[12]	0.5	6.4	19.8	14	5.53	10.7
FS F[27]	0.68	7.6	3.3	1.8	6.6	2.5
FS G[27]	0.6	7.9	12	10	9.8	7

Note: PP indicates a pilot-plant study using a countercurrent flow-through ozone contactor; FS indicates a full-scale drinking water facility; NA indicates data not
available; bdl indicates below detection limit.

80 μg/l formaldehyde. While the pH of the latter at 8.0 was slightly higher
than the former (7.4), it appears that the nature of the organic material is
the predominantly determining factor. When the raw waters from these
plants were reconstituted in the laboratory after concentration on XAD resins
and again ozonated under the same conditions, the latter plant water produced
higher levels of formaldehyde (on a micrograms of aldehyde per milligrams
of TOC basis) than previously. The former water produced similar concentrations under both conditions. Since the extracted water contained a higher
proportion of hydrophobic NOM than the raw water, it is apparent that the
TOC content of both waters comprises a different distribution of NOM.

3. Beyond this NOM character dependency, there appears to be a range of
concentrations within which these by-products form. *Aldehyde yields* when
normalized to both TOC and applied ozone dose ranged from 2.4 to 20
μg/l. This suggests that both TOC and/or ozone dose may be limiting factors
in the production of aldehydes. Other studies using reconstituted organic
material from raw waters[29] illustrated limiting values of ozone:TOC ratios
for maximum aldehyde formation. Ozonations at higher ratios than these
values even showed a net loss in aldehyde formation. It is quite likely that,
at these higher oxidant doses, further oxidation of aldehydes is taking place
producing mixed functional acidic aldehydes or carboxylic acids (see Figure
1). Indeed another pilot plant study showed that, in increasing the ozone:TOC
ratio from 0.2 to 1 on a water with dissolved organic carbon of 3.2 mg/l
C, the rate of formation of glyoxylic, pyruvic, and keto-malonic acids, was
faster and produced larger concentrations than any of the four major
aldehydes.[30]

4. As *pH* affects the mechanism of ozone reaction in water, it might be expected
   that aldehyde formation might reflect this. It is generally accepted that most
   aldehyde formation occurs when molecular ozone attacks an unsaturated
   site between two carbon atoms via the Criegee mechanism.[31] However, as
   the pH at which ozonation occurs increases, ozone decomposition into OH
   radicals is accelerated and, although their oxidation potential is higher than
   molecular ozone, they have not shown a tendency toward aldehyde formation. On the contrary, the radicals might be expected to promote further
   oxidation of aldehydes to their carboxylic acids. In this way, one might
   expect to see an overall reduced formation of aldehydes at high pH. This
   indeed was the findings of Zhou et al.,[32] who showed that at ozone:TOC
   ratios above 1 ozonation of fulvic acid reduced each of three targeted
   aldehydic compounds when the pH of the reaction was increased from 5
   to 9. This was also confirmed in the work of Schechter and Singer,[29] who
   ozonated extracts of NOM concentrates in the range of pH from 5.5 to 8.5
   and in the work of Andrews and Huck,[11] who ozonated two surface waters
   at pH 6 and 8.

In terms of potential health effects relating to the persistency of aldehydes
in distribution systems for drinking water, it is becoming increasingly clear
that the presence of aldehydes in distributed waters can be detrimental to the
quality of consumers' waters. This is due to the direct relationship of aldehyde
concentration to assimilable organic carbon levels which in the absence of a
disinfectant residual can encourage microbial regrowth in distribution systems.
Indeed, once formed by ozonation, aldehydes will persist in the water and
may continue to form if the water is further exposed to other oxidants.[32] Table
2 illustrates this effect in treatment plants where the distribution systems were
sampled. In each case, there are some levels of aldehydes present in consumers'
drinking waters. In some waters, these appear to dissipate as a function of
retention time (plant E2) possibly, but not exclusively, due to volatilization.
In others, particularly those showing lasting disinfectant residuals (plants D
and G), the levels persist to the consumer's tap. In view of the potential health
effects related to aldehyde persistency, it is preferable to incorporate treatment
processes prior to terminal disinfection that can reduce by-product levels
produced during ozonation. Different filtration modes have been investigated
for this purpose and the consensus is that filtration rate and degree of biological
activity on the filter media are the controlling factors in determining effectiveness of aldehyde removal. Some of these effects are shown in Table 3. It
appears that, when operated in a biological mode, aldehyde removal is impres-
sive (greater than 75% but mostly above 90%) with the glyoxals being the
components removed the most. For the types of filters in biological mode
shown in this table, it seems that a slower filtration rate achieves highest
removal of aldehydes. This has been confirmed by Zhou et al.[32] and it is
assumed that in this mode a combination of adsorption of aldehyde to the
filter surface followed by biodegradation of the adsorbed aldehyde takes

AN OVERVIEW OF OZONATION DISINFECTION BY-PRODUCTS   177

Table 2 Persistence of Aldehydes in Distribution Systems

Plant	Postdisinfectant	Residence time (h)	Disinfectant residual (mg/l)	Formaldehyde (µg/l) PE	DS	Acetaldehyde (µg/l) PE	DS	Glyoxal (µg/l) PE	DS	Methyl glyoxal (µg/l) PE	DS
FSA	Chlorine	5	0.6	bdl	bdl	2.7	1.3	2.1	bdl	bdl	bdl
FSD	Chloramine	5.5	2.22	13	12	7.1	5.8	7.8	6.4	8	7
		7	2.37		13		6.6		7.1		7
		8	2.26		14		6.8		7.9		7
FSE1	Chlorine dioxide	10	bdl	na	5.1	na	na	bdl	1.0	2.0	2.0
FSE2	Chlorine dioxide	2	bdl	4.32	4.77	14	14.5	1.0	0.71	3.5	3.29
		8	bdl		1.28		8.4		0.49		1.32
		12	bdl		0.58		6.6		0.48		0.5
FSG	Chlorine	3	0.35	12	13	-11	13	11	12	9	9
		10	0.2		13		11		12		10

Note: bdl—below detection limit; na—not available; PE—sampling point at plant effluent following postdisinfection; DS—sampling point in the distribution system.

Table 3   Effect of Filtration Mode on Aldehyde Removal

Plant	Filter type	Filtration rate (gpm/ft²)	Evidence of bioactivity	% Total alde-hyde removal
PPB	GAC	12	Observable colony growth	95%
PPC1	GAC	5	3 months operation, no disinfectant applied ahead of the filters	75%
	Anthracite	5	None	10% increase
PPE	Anthracite/sand	3.1	All filters backwashed with dechlorinated filter effluent, carbon in full-scale plant operation for 18 months, 91% AOC removal by GAC	13%
	Anthracite/sand then GAC	1.0		97%
	GAC/sand	1.0		93%
FSA	GAC/sand	2.0	On-line for 1 month, no application of chlorine ahead of filters	81%
FSD	Anthracite/sand	6.3	None, chlorine applied ahead of filters	46% increase
FSE1	GAC	3.9	No recent backwashing, >50% removal of AOC	92%

place more effectively at the higher contact time provided for by the slower filtration rate.

A word of caution is in order here. Although biological filtration has been shown to be effective in reducing aldehyde concentrations produced by ozonation, there remain aldehyde precursors unaffected by ozone and subsequent treatment processes, some of which could react with postdisinfectants to produce additional aldehydes or chlorinated aldehydes which may persist in the distribution systems. It is, therefore, apparent that the future of ozone use for improved drinking water quality will be based to a large extent on the ability to optimize ozonation processes to remove all disinfectant by-product precursors while providing a residual disinfectant in distributed drinking water.

MIXED FUNCTIONAL AND SATURATED CARBOXYLIC ACIDS

Carbonyl moieties are expected to be prevalent among the ozonation by-products of natural waters. As oxidation proceeds, aldehydes may be expected to form carboxylic acids, ketoaldehydes will form keto acids (e.g., methyl glyoxal to pyruvic acid), and dialdehydes will produce aldoacids (e.g., glyoxal to glyoxylic acid). Some evidence for these pathways is shown by the reduction in aldehyde levels at high ozone doses compared to low doses on the same source water discussed under aldehydes above. This suggests that at higher ozone doses levels of acidic by-products will likely accumulate in the treated water. In addition, the presence of some of these acids may contribute to accelerated decomposition of ozone. Glyoxylic acid, for example, has been identified as both an initiator and promotor of OH radical chain reactions at low pH[33] and has been found at some quite considerable concentrations in ozonated solutions of fulvic acid under these conditions.[34] There are also some published reports of levels of 1 μmol of each of glyoxylic and keto-malonic acid produced during a pilot-plant study utilizing ozone:TOC dose ratios of 1:1.[30] In these tests, mixed functional acids were produced at higher levels than those of aldehydes. In the limited application of the developed methodology for these acids in treatment plants, the levels in ozone contactor effluents were similar to those for aldehydes. Treatment plant effluents showed no trace of these compounds when biological filtration was employed after ozonation.[6]

Low molecular weight carboxylic acids were measured in the plant effluent and distribution system of a full-scale operating treatment plant using chlorine as postdisinfectant. At the plant effluent, formate was present at 128 μg/l, acetate at 108 μg/l, and oxalate at 197 μg/l. Three hours into the distribution system these levels had dropped to 33 μg/l, below detection limit, and 170 μg/l, respectively. These numbers may again give rise to the concern of microbial proliferation in distribution systems. Certain plate-count bacteria such as *Pseudomonas* and *Spirillium* present in the distribution system can utilize these carboxylic acids as nutrients.[35]

PEROXIDES

These oxidation by-products appear in most mechanisms involving ozone attack on organic precursors. Their presence in water so treated may, however, be short lived since the organic peroxides are mostly unstable and will decompose into hydrogen peroxide and a carbonyl-containing product. It is no surprise, therefore, that levels of hydrogen peroxide in ozone contactor effluents far exceeded those of organic peroxides in those plants surveyed. Figure 4a illustrates the fate of hydrogen peroxide through several plants while Figure 4b illustrates the trends for organic peroxides. It is apparent that peroxides produced during ozonation are efficiently removed through filtration but in most instances they are reformed during postdisinfection. In those plants where distribution systems were monitored, most of the peroxides appear to dissipate by the time the water reaches the consumer's tap. A review of all peroxide data in ozone contactor effluents available from various surveys indicates a range of hydrogen peroxide from 14 to 72 μg/l and organic peroxides from 2.9 to 22 μg/l with the highest values obtained at highest ozone doses. That most of these are removed during biological filtration but again form during subsequent disinfection suggests, as in the case of aldehydes, that the oxidation of NOM into smaller biodegradable components is incomplete during single-stage ozonation. Some of the nonbiodegradable carbon is, therefore, a precursor to polar organic by-products and this suggests the potential presence of a

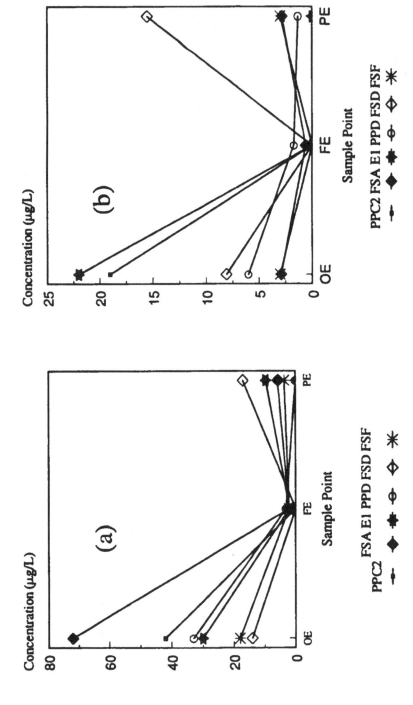

Figure 4 Trends in peroxides through treatment plants; (a) hydrogen peroxide, (b) organic peroxides.

Table 4   Brominated By-Products Resulting from Ozonation

Plant	Bromide concentration (μg/l)	Ozone dose (mg/l)	pH	By-products in ozone contactor (μg/l)			
				BrO_3^-	CNBr	$CHBr_3$	Bromo-picrin
PPB	800	9.2	9.4	20	bdl	bdl	bdl
PPD	220	3.5	8.0	27	7.8	0.4	bdl
	2000	3.5	7.9	141	14	12	1.0
	3700	3.5	7.95	120	26	73	1.2
	3700	2.0	8.0	135	15	NA	0.7
PPF	260	4	6	8	9.6	bdl	NA
	1000	4	7	8	7.0	bdl	NA
	1000	4	7.9	24	9.4	bdl	NA
	1500	4	7.9	32	14	2.6	NA
FSA	20	3.0	7.4	5.0	7.8	bdl	NA
FSF	50	2.0	7.6	8	1.9	bdl	NA
FSG	280	1.5	7.9	10	NA	0.32	bdl

Note: bdl—below detection limit; NA—not available.

whole array of partial oxidation by-products as yet unidentified that persist in the distribution system.

BROMINATED BY-PRODUCTS

In the ten-utility survey of this study, bromide levels in raw waters ranged from less than 0.01 to 0.28 mg/l. This study was fairly representative of U.S. waters as a whole since the average bromide concentration has been assessed at 0.1 mg/l.[36] Similar studies in Europe assess a range of 0.01 to 0.2 mg/l.[37] The major bromine-containing by-products resulting from ozonation of a variety of surface groundwaters in this study were bromate, bromoform, and cyanogen bromide and the pathways by which they are likely to be formed is shown in Figure 3. Additional brominated organic by-products produced via the nondissociated HOBr pathway included dibromoacetonitrile, dibromoacetone, bromohydrins, and brominated acetic acids. The levels of these by-products is affected by the operational parameters of ozonation. However, once formed they are quite persistent throughout the treatment plant and may even be increased in concentration by postdisinfection. The effect of bromide level, ozone, dose, and pH on brominated by-product formation in a selection of plants is illustrated in Table 4. Some extrapolations from this table and corroborated by other studies indicate the complexity of the ozone-bromide interaction and reveal an incomplete understanding of the mechanistic pathways. The tenfold increase in bromide levels between plants FSA and PPD yields a less than sixfold increase in bromate but no appreciable difference in organic bromides (similar ozone:TOC ratios), in spite of the higher operating pH of plant PPD which would promote OH radical mechanisms and thereby divert bromide to bromate through the radical mechanism illustrated in Figure 3. At lower ozone:TOC ratios, the NOM-bromide pathway might be expected to influence the course of reactions and bromate levels would decrease while organic bromides would increase. However, in this study, the competing mechanisms were only apparent at high bromide-spiked levels in a pilot-plant operation. One of the sets of operation parameters at pilot plant A emphasized this. At identical bromide levels (3.7 mg/l) and pH of operation (8.0), a lower 2 mg/l ozone dose yielded higher bromate levels (135 μg/l) than the higher 3.5 mg/l ozone dose (120 μg/l). Brominated organic by-products increased in proportion to the ozone dose. These results introduce the concept of a limiting factor in the formation of bromate and this could be influenced by ozone transfer efficiency, contact time, or ozone residual. The latter factor appears to be the most influential in determining bromate levels in finished waters.[38] The influence of OH radical chemistry on bromate formation was further emphasized through pilot plant experiments in which hydrogen peroxide was dosed at selected ozone levels. A 0.5-mg/l dose of hydrogen peroxide added subsequent to a 4-mg/l ozone dose on a 3-mg/l TOC, 1.2-mg/l bromide water increased bromate levels from 32 μg/l (without peroxide) to 45 μg/l.

In view of the current concern over bromate levels in drinking waters, several control scenarios have been investigated in pilot plant tests. One of these utilizes the addition of ammonia to remove active bromine from the ozone-bromide cycle as bromamine as shown in Figure 3. This effect was shown in the treatment scenario described in the previous paragraph. In the presence of ammonia for the ozone peroxide test, bromate was reduced from 45 to 36 μg/l. Unfortunately, bromamine is oxidized by ozone to nitrate and bromide; the latter is then recycled and again oxidized by ozone. When all the ammonia is utilized in this fashion, bromate will form without any control strategy so that the only effect of ammonia was to delay bromate formation. Other bromate control strategies include curtailing the mechanism by which they are formed (lowering pH, radical quenching) or removal once formed (chemical reduction or physical treatment). The effect of pH is seen for treatment plant PPF in Table 4. At a 1-ppm bromide level, pH 7 gave a bromate formation of 8 μg/l while at pH 7.9 this was raised to 24 μg/l due to increased presence of OH radicals.

An experiment to quench radicals from the reaction utilized varying levels of alkalinity adjusted by addition of sodium carbonate. Results were somewhat mixed: 100 mg/l as calcium carbonate alkalinity at pH 8 produced 26 μg/l bromate compared to only 12 μg/l at 25 mg/l alkalinity. When raised to 200 mg/l, bromate was decreased to 19 μg/l. It is apparent from these results that carbonate cannot be used to quench radical mechanisms for bromate formation. Indeed, there is evidence from rate constants[39] that carbonate radicals formed from interaction of carbonate with OH radicals can themselves propagate bromate formation. Experiments with other OH radical quenching agents such as tert-butanol may provide definitive answers to the question of mechanistic pathways. Chemical reduction of bromate might involve the use of sulfur(IV) compounds or iron(II) salts. Although there are no published results for the latter, a 6:1 molar ratio of sulfite to bromate has been shown to achieve up to 50% bromate removal at pH 8.[40] Physical treatments however are even less clearly documented. There are some reports showing the effective use of activated carbon for the removal of bromate in organic pure solutions.[40] However, when applied to real waters, it appears that the competition for active sites on the carbon by NOM is a limiting factor in bromate removal which is minimal.[37] Studies are currently proceeding by various research groups to remove bromide from the ozone contactor influent by nanofiltration techniques. This approach, however, seems counterproductive to the rationale or ozone treatment since both energy-demanding processes attempt to remove DBP precursors and it would probably not be economical to utilize both processes in the same treatment plants. With regards to the overall perspective of brominated by-products resulting from ozonation, much research is still needed to fill a large gap in our accumulated knowledge in this field. This is underlined in the area of brominated by-products by the discovery that as little as 8% of the total dissolved organic bromide content of ozone contactor effluent may have been identified up to now.[19] There is no doubt that an overview of ozonation DBP identification such as this serves to highlight our limited understanding in the field. The next generation of rules and laws governing the quality of our drinking water will, therefore, best be served by continued studies of optimized oxidation and disinfection processes and by dedicated studies of their effects on the chemical quality of that water.

ACKNOWLEDGMENTS

The authors are grateful for the financial support of the AWWA Research Foundation (AWWARF) and the following utilities: East Bay Municipal Utility District, Hackensack Water Company, Los Angeles Department of Water and Power, Metropolitan Water District of Southern California, Portland Bureau of Water Works, and Ville de Laval, Quebec, Canada. This chapter includes material published in 1993 by the AWWA Research Foundation and American Water Works Association in the report entitled, "Identification and Occurrence of Ozonation By-Products in Drinking Water." The comments and views detailed in this paper may not necessarily reflect the views of the AWWA Research Foundation, its officers, directors, affiliates, or agents.

REFERENCES

1. Lawrence, J., Tosine, H., Onusaka, F. I., and Comba, M. E., The ozonation of natural waters: product identification, *Ozone Sci. Eng.*, 2, 55, 1980.
2. Anderson, L. J., Johnson, J. D., and Christman, R. F., The reaction of ozone with isolated aquatic fulvic acid, *Org. Geochem.*, 8(1), 65, 1985.

3. Killops, S. D., Volatile ozonization products of aqueous humic material, *Water Res.*, 20(2), 153, 1986.

4. Schalekamp, M., Pre- and intermediate oxidation of drinking water with ozone, chlorine and chlorine dioxide, *Ozone Sci. Eng.*, 8(2), 151, 1986.

5. Glaze, W. H., Koga, M., and Cancilla, D., Ozonation by-products. II. Improvement of an aqueous-phase derivatization method for the detection of formaldehyde and other carbonyl compounds formed by the ozonation of drinking water, *Environ. Sci. Technol.*, 23(7), 838, 1989.

6. Scheller, N. A., Modification of the PFBHA method: from LLE to SPE and examination of matrix effects upon derivatization, M.S. thesis, University of North Carolina, Chapel Hill, NC, 1993.

7. Yamada, H. and Somiya, I., The determination of carbonyl products in ozonated water by the PFBOA method, *Ozone Sci. Eng.*, 11(2), 127, 1989.

8. Cancilla, D. A. et al., Characterization of the *O*-(2,3,4,5,6-pentafluorobenzyl)-hydroxylamine hydrochloride (PFBOA) derivatives of some aliphatic mono- and dialdehydes and quantitative water analysis of these aldehydes, *J. AOAC Int.*, 75, 842, 1992.

9. Sclimenti, M. J., Krasner, S. W., Glaze, W. H., and Weinberg, H. S., Ozone disinfection by-products: optimization of the PFBHA derivatization method for the analysis of aldehydes, in *Proc. American Water Works Association Water Quality Technology Conference*, San Diego, CA, 1990.

10. Le Lacheur, R. M. et al., Identification of carbonyl compounds in environmental samples, *Environ. Sci. Technol.*, 27(13), 2743, 1993.

11. Andrews, S. A. and Huck, P. M., Using fractionated natural organic matter to quantitate organic by-products of ozonation, *Ozone Sci. Eng.*, 16(1), 1, 1994.

12. Weinberg, H. S., Oxidation and disinfection by-products progress, report to the Ecole Polytechnique, Montréal, 1993.

13. Deutscher Verein des Gas- und Wasserfachs. 1988. Eschborn.

14. Lazrus, A. L., Kok, G. L., Gitlin, S. N., Lind, J. A., and McLaren, S. E., Automated fluorometric method for hydrogen peroxide in atmospheric precipitation, *Anal. Chem.*, 57, 917, 1985.

15. Hellpointer, E. and Gab, S., Detection of methyl, hydroxymethyl, and hydroxyethyl hydroperoxides in air and precipitation, *Nature*, 337, 631, 1989.

16. Kurth, H.-H. et al., A high performance liquid chromatography system with an immobilized enzyme reactor for detection of hydrophilic organic peroxides, *Anal. Chem.*, 63, 2586, 1991.

17. Haag, W. R. and Hoigné, J., Ozonation of bromide-containing water: kinetics of formation of hypobromous acid and bromate, *Environ. Sci. Technol.*, 17, 261, 1983.

18. von Gunten, W. and Hoigné, J., Bromate formation during ozonation of bromide containing water, *Proc. 11th Ozone World Congress*, San Francisco, 1993; also Chapter 8 of this text.

19. Glaze, W. H., Weinberg, H. S., and Cavanagh, J. E., Evaluating the formation of brominated DBPs during ozonation, *J. AWWA*, 85 (1), 96, 1993.

20. Cavanagh, J. E. et al., Ozonation by-products: identification of bromohydrins from the ozonation of natural waters with enhanced bromide levels, *Environ. Sci. Technol.*, 26, 1658, 1992.

21. Richardson, S. D. et al., Multispectral identification of brominated organic by-

products formed by ozonation of natural waters, presented at the Annual Symposium of the American Society of Mass Spectrometry, Washington, DC, 1992.

22. Regli, S., Draft D/DBP rule language, USEPA Office of Groundwater and Drinking Water, Washington, DC, 1993.

23. Koch, B., Crofts, E. W., Schimpff, W. K., and Davis, M. K., Analysis of halogenated disinfection by-products by capillary chromatography, in *Proc. AWWA Water Quality Technology Conference,* St. Louis, 1988.

24. Chinn, R. and Krasner, S. W., A simplified technique for the measurement of halogenated organic acids in drinking water by electron capture gas chromatography, in *Proc. of the 28th Pacific Conference on Chemistry and Spectroscopy,* Pasadena, CA, 1989.

25. Kuo, C.-Y., Krasner, S. W., Stalker, G. A., and Weinberg, H. S., Analysis of inorganic disinfection by-products in ozonated drinking water by ion chromatography, in *Proc. American Water Works Association Water Quality Technology Conference,* San Diego, 1991.

26. Weinberg, H., Preconcentration techniques for bromate analysis in ozonated waters, *J. Chromatog.,* A671(1 and 2), 141, 1994.

27. Weinberg, H. S., Glaze, W. H., Krasner, S. W., and Sclimenti, M. J., Formation and removal of aldehydes in plants that use ozonation, *J. AWWA,* 85(5), 72, 1993.

28. Miltner, R. J., Shukairy, H. M., and Summers, R. S., Disinfection by-product formation and control by ozonation and biotreatment, *J. AWWA,* 84(11), 53, 1992.

29. Schechter, D. S. and Singer, P. C., Comparative formation of ozonation by-products in real and model waters, Proc. American Water Works Association Annual Conference, San Antonio, TX, 1993.

30. Xie, Y. and Reckhow, D. A., Formation of keto-acids in ozonated drinking water, *Ozone Sci. Eng.,* 14(3), 269, 1992.

31. Bailey, P. S., *Ozonation in Organic Chemistry,* Vol. 1, Academic Press, New York, 1978.

32. Zhou, X., Reckhow, D. A., and Tobiason, J. E., Formation and removal of aldehydes in drinking water treatment processes, in *Proc. AWWA Water Quality Technology Conference,* Toronto, 1993.

33. Staehelin, J. and Hoigné, J., Decomposition of ozone in the presence of solutes acting as promoters and inhibitors of radical chain reactions, *Environ. Sci. Technol.,* 19, 1206, 1985.

34. Xiong, F., Croué, J.-P., and Legube, B., Long-term ozone consumption by aquatic fulvic acids acting as precursors of radical chain reactions, *Environ. Sci. Technol.,* 26(5), 1059, 1992.

35. van der Kooij, D., The effect of treatment on assimilable organic carbon in drinking water, in *Proc. 2nd National Conference on Drinking Water,* Pergamon Press, Elmsford, NY, 1987.

36. Amy, G. et al., Survey of bromide ion concentrations in drinking waters sources and impact on DBP formation, AWWA Research Foundation, Denver, 1994.

37. Legube, B. et al., Bromide/bromate survey on different European water utilities, preprint of the International Workshop on Bromate and Water Treatment, Paris, November, 1993.

38. Krasner, S. W. et al., Formation and control of bromate during ozonation of waters containing bromide, *JAWWA*, 85(1), 73, 1993.
39. Buxton, G. V., Greenstock, C. L., Helman, W. P., and Ross, A. B., Critical review of rate constants for oxidation of hydrated electrons, hydrogen atoms and hydroxyl radicals ($\cdot OH/\cdot O^-$) in aqueous solutions, *J. Phys. Chem. Ref. Data*, 17(2), 513, 1988.
40. Prados-Ramirez, M. J., Ciba, N., and Bourbigot, M. M., Available techniques for reducing bromate in drinking water, Preprint of International Workshop on Bromate and Water Treatment, Paris, November, 1993.

CHAPTER 8

Ozonation of Bromide-Containing Waters: Bromate Formation through Ozone and Hydroxyl Radicals

Urs von Gunten and Jürg Hoigné

INTRODUCTION

REGULATORY CONSIDERATIONS

Bromate is a disinfection by-product of the ozonation of bromide-containing waters. Kurokawa et al.[1] have shown a carcinogenic effect of bromate in animal experiments which has created a need for regulating this compound in drinking water. The discussion of a regulatory standard for bromate in drinking water is governed by three main factors: (1) toxicity of bromate, (2) detection limit in a natural water matrix, and (3) disinfection safety criteria.

Disinfection by-products are of great concern mainly if they are suspected carcinogens. However, disinfection safety still has priority in drinking water treatment when the health effects of disinfection by-products are not entirely known.[2] The bromate concentration in drinking water associated with an excess lifetime cancer risk of 10^{-5} is 3 µg/l based on a linearized multistage model for a consumption of 2 l/d by a 70-kg adult.[3] Based on previous analytical feasabilities, the World Health Organization (WHO)[3] recommended a provisional guideline value of 25 µg/l drinking water. The EPA[4] introduced a maximum contaminant level for bromate at 10 µg/l which is in part based on the practical quantitation limit. However, all these recommendations might be subject to change. According to Legube et al.[5] a detection limit in drinking water of about 2 µg/l was found in several European laboratories. Hautmann[6] showed that it is possible to measure bromate as low as 0.25 µg/l in a river water matrix applying special preconcentration techniques. The decision of the WHO to set a provisional guideline value of 25 µg/l may postpone the actual bromate discussion temporarily, because it has been shown

1-56670-136-8/96/$0.00+$.50

Table 1   Bromide Concentrations in Selected Natural Waters

Water	Br⁻ (μg/l)	Ref.
Ocean	65,000	8
Lakes		
Lake Kinneret (Israel)	1,900	9
Lake Zürich (Switzerland)	3	10
Lake Konstanz (Switzerland)	3	10
Rivers		
Sweden (300 samples)	8–240	11
Rhine		
Basel	70	12
Karlsruhe	150	13
Netherlands	70–500	14
Five major U.S. rivers	4–80	15
Groundwaters		
Biscayne Aquifer (Florida)	160–250	16
Sweden (several sources)	16–80	11
Mineral waters (Switzerland)		
Aqui	1,000	17
Henniez	130–200	17

that only waters with high bromide levels (>200 μg/l) will be of concern for standard ozone treatments under these circumstances.[7] However, the tolerable bromate concentration will likely be subject for further discussions, and some basic studies are required to provide possibilities for minimization of bromate production in drinking water treatment.

BROMIDE CONCENTRATIONS IN NATURAL WATERS

As shown in Table 1,[8-17] bromide concentrations in natural waters vary from a few micrograms per liter up to several milligrams per liter. In coastal groundwaters used for drinking water production, elevated bromide concentrations are frequently attributable to infiltration of ocean water (e.g., Biscayne Aquifer in Table 1). The major sources of bromide in inland waters are related to local geological situations (e.g., mineral waters), natural fractionation (e.g., Lake Kinneret), and anthropogenic bromide emissions (e.g., potassium mining). The increasing bromide concentration as a function of the flow distance of the Rhine River is mainly due to soda production, potassium, and coal mining.

CHEMICAL REACTIONS

Molecular Ozone Mechanism

The reactions controlling the well-known and critically tested mechanism for direct interaction of molecular ozone with bromide are given in Table 2.[18-22] They predict bromate formation as a function of bromide concentration, pH, and ammonia concentration. A schematic representation of the molecular ozone mechanism is represented in Figure 1 by the solid lines. All the reaction

Table 2   Reactions and Rate Constants for the Molecular Ozone Mechanism

Number	Reaction	k or pK$_a$ (20°C)	Ref.
1	$O_3 + Br^- \rightarrow O_2 + OBr^-$	160 M^{-1} s^{-1}	18–21
2	$O_3 + OBr^- \rightarrow 2O_2 + Br^-$	330 M^{-1} s^{-1}	18–21
3a	$O_3 + OBr^- \rightarrow BrO_2^- + O_2$	100 M^{-1} s^{-1}	18–21
3b	$O_3 + HOBr \rightarrow BrO_2^- + O_2 + H^+$	≤0.013 M^{-1} s^{-1}	21
4	$BrO_2^- + O_3 \rightarrow BrO_3^-$	>10^5 M^{-1} s^{-1}	21
5	$HOBr \leftrightarrows H^+ + OBr^-$	9 (8.8)	21
6	$HOBr + NH_3 \rightarrow NH_2Br + H_2O$	8·10^7 M^{-1} s^{-1}	20
7	$O_3 + NH_2Br \rightarrow Y$	40·M^{-1} s^{-1}	22
8	$Y + 2O_3 \rightarrow 2H^+ + NO_3^- + Br^- + 3O_2$	$k_8 \gg k_7$	22
9	$NH_4^+ \leftrightarrows H^+ + NH_3$	9.3	

Note: Y are unknown products that react in later reactions.

rate constants are given for 20°C. Because the activation energies for most of the reactions are unknown, it is impossible to predict bromate formation at other temperatures. In pilot plant experiments performed on surface waters, a 20% increase of bromate was found for an increase of 10°C.[23]

In a previous study, we performed a kinetic simulation to predict bromate formation by molecular ozone as a function of its decisive parameters.[7] An example of a model calculation accounting for reactions 1 to 5 (in Table 2) is presented in Figure 2. Ozonation transforms bromide efficiently into hypobromous acid and hypobromite. For the rate constants listed above, 77% of the hypobromite formed then reacts through Reaction 2 back to bromide and 23% through Reaction 3a to bromate. The oxidation of protonated hypobromite

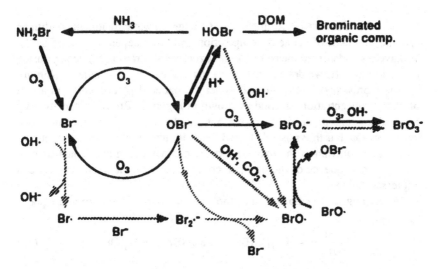

Figure 1   Reactions occurring upon ozonation of bromide-containing solutions. Reactions for molecular ozone (solid lines) and OH radicals (dashed lines) are shown.

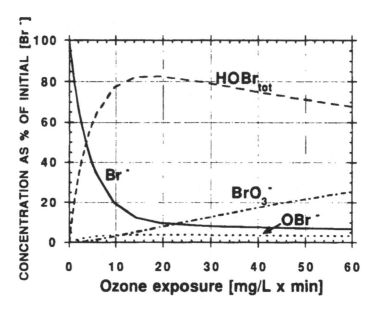

Figure 2 Model calculation for an extended ozonation of a bromide-containing water at pH 7.5 (accounting for Reactions 1 to 9, absence of OH radical reactions). Concentrations of different species are given in percent of initial [Br⁻] vs. the ozone exposure (c·t value).[7]

(hypobromous acid) by ozone is very slow (Reaction 3b) and therefore does not contribute to bromate formation.

Influence of Ozone Concentration — Changes in water treatment conditions can alter the rates of ozone decomposition and other depletions significantly. To have a standardized measure for the ozone concentration (c) acting during a reaction time (t) we defined the *ozone exposure* as the *c·t-value (mg/l·min)*. This is represented by the integral under the ozone depletion curve in a plot of ozone concentration vs. time as shown in Figure 3. Bromate formation by molecular ozone proceeds through reactions that are all first order in ozone concentration. It is thus possible to make standardized plots of bromate concentration as a function of the ozone exposure characterizing the ozonation process. The dependence of the bromate concentration on the ozone exposure is expressed below.

According to Reactions 1 to 3, the rate equation of hypobromite equals

$$\frac{d[OBr^-]}{dt} = [O_3]\{k_1[Br^-] - k_2[OBr^-] - k_{3_a}[OBr^-]\} \qquad (10)$$

According to Figure 2 we can assume steady state (ss) for [OBr⁻] for a large range of the ozonation process:

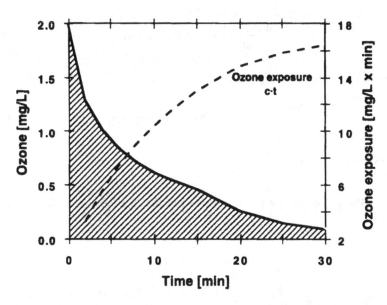

Figure 3   Typical ozone depletion curve in a batch-type reactor (sample from Lake Zürich spiked with 1 mg/l bromide; ozone dose 2 mg/l, pH 8). The area under the curve represents the ozone exposure (c·t value) which is represented by the dashed line.[24]

$$[Br^-] = [OBr^-]_{SS} \cdot \frac{k_2 + k_{3a}}{k_1} = [OBr^-] \cdot c \qquad (11)$$

The mass balance of bromine species gives

$$[Br^-]_0 = [Br^-] + [HOBr] + [OBr^-] + [BrO_3^-] \qquad (12)$$

or

$$[Br^-]_0 = [OBr^-] \cdot \left(1 + c + \frac{[H^+]}{K_{HOBr}}\right) + [BrO_3^-] = [OBr^-] \cdot c' + [BrO_3^-] \qquad (13)$$

thus

$$[OBr^-]_{SS} = \frac{[Br^-]_0 - [BrO_3^-]}{c'}, \text{ with } c' = \frac{k_2 + k_{3a}}{k_1} + \frac{[H^+]}{K_{HOBr}} + 1 \quad (14)$$

The formation of bromate can be expressed by the following differential equation:

$$\frac{d[BrO_3^-]}{dt} = k_3[OBr^-]_{ss}[O_3] \tag{15}$$

substituting $[OBr^-]$ by Equation 14:

$$\frac{d[BrO_3^-]}{dt} = \frac{k_3[O_3]}{c'}([Br^-]_0 - [BrO_3^-]) \tag{16}$$

or in integrated form:

$$[BrO_3^-] = [Br^-]_0\left(1 - e^{-\frac{k_3[O_3]t}{c'}}\right) \sim [Br^-]_0\frac{k_3[O_3]t}{c'} = c''[O_3]\cdot t \tag{17}$$

where

$$c'' = \frac{k_3}{k_1\cdot K_{HOBr}}\left(K_{HOBr}(k_2 + k_3) + k_1[H^+] + k_1\cdot K_{HOBr}\right)\cdot [Br^-]_0$$

The bromate concentration is thus a direct function of the ozone exposure. However, the approximative formula for the calculation of bromate has restrictions for its application at high pH (>7.5) because the steady-state approximation for $[OBr^-]$ (Equation 11) is no longer valid. Furthermore, it does not account for the induction period where $[OBr^-]_{ss}$ has to be built up.

The ozone exposure concept does not account for OH radical reactions. However, it still can be used as a tool to compare data relative to the molecular ozone mechanism. Plots of the bromate formation as a function of time for different ozonation conditions might not show the important differences because the information about the ozonation process get lost. If the ozone exposure is used as parameter, drastic changes in the mechanism suddenly become evident.

Influence of pH — Whereas Reaction 1 is independent of pH, the overall rates of Reactions 2 and 3 decrease with decreasing pH due to masking of hypobromite through protonation: the pK_a of hypobromous acid is approximately 9* (Equation 5). In the pH range of 7 to 8 only 1 to 10% of $[HOBr]_{tot}$ (in the form of OBr^-) takes part in reactions with molecular ozone.

Influence of Ammonia — If ammonia is present, hypobromite is masked by formation of monobromamine (Equation 6). In the pH range of drinking water

---

* Experiments at varying ionic strengths showed that the pKa of HOBr has to be set at 9 to give the best fit of our data.[24]

Table 3 Reactions and Rate Constants for the Hydroxyl Radical Mechanism

Number	Reaction	k or pK$_a$ (20°C)	Ref.
19	$Br^- + OH \rightleftharpoons BrOH^-$	$10^{10}\ M^{-1}\ s^{-1}$; $3.3 \cdot 10^7\ s^{-1}$	25
20	$BrOH^- \rightarrow Br + OH^-$	$4.2 \cdot 10^6\ s^{-1}$	25
21	$Br + Br^- \rightarrow Br_2^-$	$10^{10}\ M^{-1}\ s^{-1}$	25
22	$Br_2^- + Br_2^- \rightarrow Br_3^- + 2Br^-$	$2 \cdot 10^9\ M^{-1}\ s^{-1}$	26
23	$Br_2^- + OBr^- \rightarrow BrO + 2Br^-$	$8 \cdot 10^7\ M^{-1}\ s^{-1}$	27
24	$OH + OBr^- \rightarrow BrO + OH^-$	$4.5 \cdot 10^9\ M^{-1}\ s^{-1}$	27
25	$OH + HOBr \rightarrow BrO + H_2O$	$2 \cdot 10^9\ M^{-1}\ s^{-1}$	27
26	$CO_3^- + OBr^- \rightarrow BrO + CO_3^{2-}$	$4.3 \cdot 10^7\ M^{-1}\ s^{-1}$	28
27	$CO_3^{2-} + OH \rightarrow OH^- + CO_3^-$	$3.9 \cdot 10^8\ M^{-1}\ s^{-1}$	29
28	$HCO_3^- + OH \rightarrow OH^- + HCO_3$	$8.5 \cdot 10^6\ M^{-1}\ s^{-1}$	30
29	$HCO_3^- \rightleftharpoons H^+ + CO_3^{2-}$	10.3	
30	$HCO_3 \rightleftharpoons H^+ + CO_3^-$	9.6	31
31	$2BrO + H_2O \rightarrow BrO^- + BrO_2^- + 2H^+$	$4.9 \cdot 10^9\ M^{-1}\ s^{-1}$	27
32	$BrO_2^- + OH \rightarrow BrO_2 + OH^-$	$2 \cdot 10^9\ M^{-1}\ s^{-1}$	27
33	$BrO_2^- + CO_3^- \rightarrow BrO_2 + CO_3^{2-}$	$1.1 \cdot 10^8\ M^{-1}\ s^{-1}$	27
34	$2BrO_2 \rightleftharpoons Br_2O_4$	$1.4 \cdot 10^9\ M^{-1}\ s^{-1}$; $7 \cdot 10^7\ s^{-1}$	13
35	$Br_2O_4 + OH^- \rightarrow BrO_3^- + BrO_2^- + H^+$	$7 \cdot 10^8\ M^{-1}\ s^{-1}$	27

treatment this reaction is relatively fast. The amino group in monobromamine is slowly oxidized by ozone to nitrate. This oxidation also releases bromide according to Reactions 7 and 8. After most of the ammonia is depleted, bromate is formed according to Reactions 1 to 5. Thus ammonia causes a lag time for the formation of bromate. For molar ratios of $HOBr/NH_4^+ > 1$ $NHBr_2$ can be expected which reacts four times more slowly with ozone than NH_2Br.[22] Therefore, further reactions in the bromine-ammonia system involve $NHBr_2$ and NBr_3 such as[22]

$$2H_2O + NHBr_2 + NBr_3 \rightarrow N_2 + 3H^+ + 3Br^- + 2HOBr \qquad (18)$$

Although the importance of this reaction has not been verified by experiments yet in ozonated systems, we can assume that such reactions alter the rate of bromate formation and lead to a decrease of the NO_3^- yield with respect to the initial ammonium concentration.

Influence of other Parameters — The decomposition of ozone during water treatment leads to the formation of secondary oxidants such as OH radicals. The reactions of these strong oxidants also have to be considered because they might contribute significantly to the overall oxidation processes.

Critical compilations of relevant reaction and rate data for interactions of OH radicals with bromide and oxybromine species have been published previously.[25-31] The oxidation of bromide by hydroxyl (OH) radicals was investigated by pulse radiolysis experiments.[25] For a pH between 7 and 8 the results can be summarized by Reactions 19 to 21 in Table 3.[25-31] The first step of the interaction of bromide with hydroxyl radicals (Equation 19) is the reversible formation of an intermediate product $BrOH^-$ quantified by an equilibrium constant of about 300 M^{-1}. Reaction 19 is mainly important for high bromide concentrations as they occur at the beginning of the ozonation process (Figure 2). At this stage, Br_2^- is reproduced (Equations 19 to 21) as fast as it is used up. For our experimental conditions ($[Br^-] \leq 1.25 \cdot 10^{-5} M$) we can assume a steady state for $[BrOH^-]$. The rate law for the disappearance of OH radicals through Reactions 19 and 20 can then be formulated as

$$- \frac{d[OH]}{dt} = \frac{k_{+19}k_{20}}{k_{-19} + k_{20}} [OH][Br^-] = k_{eff}[OH][Br^-]$$

The apparent second-order rate constant here becomes $k_{eff} = 1.1 \cdot 10^9 \, M^{-1} \, s^{-1}$. Thus the effective rate of OH radical consumption is about a factor of 10 slower than expected from Reaction +19. The reversibility of Reaction 19 leads to a limitation of the production of $BrOH^-$ and thus Br_2^-, which either disproportionates (Equation 22) or further oxidizes OBr^- (Equation 23) or is reduced by organic matter. The latter mechanism was not systematically studied and thus it is not possible to quantify this reaction here.

For radical reactions to proceed toward higher oxidation states of bromine, it seems according to the known reactions[25-31] that hypobromite is needed as a further reactant. In solutions containing bromide and ozone, hypobromite accumulates as a primary intermediate. In addition to the oxidation by ozone, a direct oxidation of bromide to hypobromite with OH radicals seems to be possible. This could be shown in γ-irradiation experiments (^{60}Co, N_2O-saturated solutions). The key reaction for this pathway might be

$$Br_3^- + H_2O \rightarrow HOBr + H^+ + 2Br^- \tag{22a}$$

The importance of this reaction in the overall bromate formation is not entirely clear yet, but preliminary experiments indicate that it might be of minor importance under conventional ozonation conditions.

The possible reactions of secondary oxidants with hypobromite are shown by Reactions 24 to 26 in Table 3. CO_3^- radicals are formed through Reactions 27 to 30. In contrast to CO_3^-, no kinetic data are available for the reaction of HCO_3 radicals with $HOBr/OBr^-$. By including the latter reactions into the model it could be shown, however, that they do not alter the rate of formation of bromate since Reaction 26 seems fast and selective enough to guarantee an oxidation of OBr^- to BrO in the pH region above 7.

The common product of Reactions 23 to 26 is BrO. If this radical is not consumed by other solutes, it undergoes a disproportionation to form bromite and hypobromite (Equation 27). According to Reaction 4, bromite is then readily oxidized by molecular ozone to bromate. In waters with high OH radical concentrations (advanced oxidation) we can also assume that bromite can be oxidized to bromate by the reaction sequence 32 to 35.

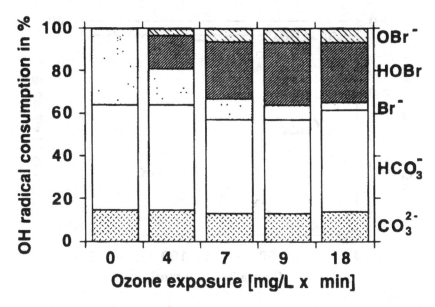

Figure 4 Calculated fate of OH radicals expressed as percentage of OH radicals reacting with different solutes. Ozone dose 2 mg/l, bromide concentration 1 mg/l, 2.5 mM $[HCO_3^-]_{tot}$, pH 8.

Under conventional drinking water treatment conditions molecular ozone still plays an important role for two oxidation (steps also within the OH radical mechanism): (1) formation of hypobromite (Equation 1) which then reacts with Br_2^-, OH, and CO_3^- to form BrO and (2) oxidation of bromite (Equation 4) which leads to bromate. A schematic representation of bromate formation through the interaction of the molecular ozone mechanism and the OH radical mechanism is shown in Figure 1.

To evaluate the major pathways for OH radical consumption during ozonation, competition kinetics for parallel reactions can be employed. Figure 4 shows an example of calculated OH radical consumption by different reactants as a function of the ozone exposure at pH 8 in the absence of DOM. Of the OH radicals not consumed by ozone, approximately 60% react with carbonate and bicarbonate, and the remaining 40% are distributed between bromide, hypobromous acid, and hypobromite. Based on the effective reaction rate constant (Equation 14), bromide only plays a role as an OH radical scavenger at the very beginning of the ozonation process.

RESULTS AND DISCUSSION

Experimental and analytical methods have been described elsewhere.[24] To plan experiments and interpret data, kinetic simulations based on the rate

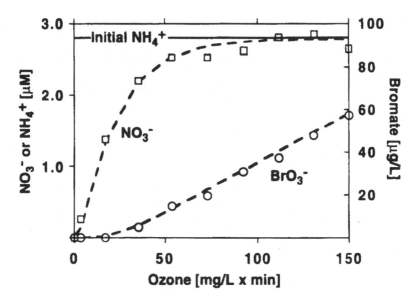

Figure 5 Bromate and nitrate formation as a function of the ozone exposure. Continuous ozonation in the presence of 0.22 mg/l bromide ($2.75 \cdot 10^{-6}$ *M*) and 0.05 mg/l NH_4^+ ($2.8 \cdot 10^{-6}$ *M*), an OH radical scavenger (2 mg/l octanol) at pH 7. Experimental data, lines: kinetic simulation with Reactions 1 to 9.[24]

constants presented in Reactions 1 to 9 and 19 to 35 were performed with the computer program LARKIN.[32] This program can solve systems of nonlinear differential equations and gives a time-dependent behavior of the species under question.

The model presented here accounts for reactions with molecular ozone as well as for reactions with OH radicals. Apart from the decisive factors described above, it is necessary to have information about ozone concentration vs. time and the rate of OH radical production to apply our model correctly. The OH radical production was taken from the experimentally measured ozone decomposition curve by assuming that the OH radicals produced amount to 50% of the ΔO_3.[33] Experimental testing of our model did not reveal any significant discrepancies.

OXIDATION BY MOLECULAR OZONE

To further test and demonstrate the kinetic model for oxidation by molecular ozone (Reactions 1 to 9), we performed a first set of experiments at pH 7 in the presence of bicarbonate and octanol. For such conditions transformation of ozone into OH radicals is slow and OH radicals are efficiently scavenged. An example of an ozonation of a solution containing bromide and ammonium is presented in Figure 5. Experimental and simulated bromate and nitrate formation are shown as a function of the ozone exposure. Virtually

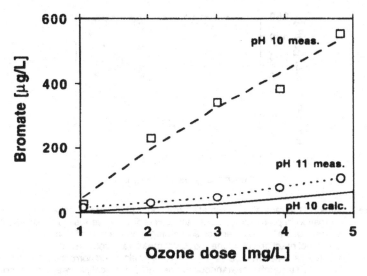

Figure 6 Bromate formation as a function of the ozone dose. Measured data for complete decomposition of ozone are shown for Br⁻ as starting compound (1 mg/l) both for pH 10 (squares) and 11 (circles). The lowest line represents calculated bromate formation if only ozone reactions are considered (Equations 1 to 9).

complete oxidation of ammonium to nitrate occurs (Equations 6 and 7). The ozone exposure for which the formation of nitrate levels off corresponds to the end point of monobromamine oxidation. Beyond this region the ozonation process can be described by Equations 1 to 5. There is an excellent agreement between measured and simulated data indicating that the processes in the bromide-ammonium-ozone system in which OH radicals are scavenged (by octanol) can be accurately described by the above mechanism (Equations 1 to 9). Therefore, Reaction 18, accounting for the decomposition of bromamines, can be neglected under these conditions.

OXIDATIONS BY HYDROXYL RADICALS

OH Radical Reactions at pH 10 and 11

In the bromate formation reactions presented in Figure 1, there is a synergistic effect of ozone and OH radicals, that is, bromide has to be oxidized to hypobromite first to become further oxidized by OH radicals. In experiments performed at pH 10 and 11 such a synergism could be affirmed. If measured after complete ozone decomposition, bromate formation was much higher at pH 10 than at pH 11 (Figure 6). The first reason for this difference is the longer half-life of ozone at pH 10 (about 20 s) still allowing for a relevant ozone exposure compared to pH 11 (2 s). Higher bromate concentrations are

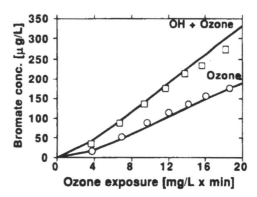

Figure 7   Effect of OH radicals on bromate formation at pH 8. Circles: measured bromate formation in presence of an organic OH radical scavenger (0.1 mM t-BuOH). Curve "ozone": calculated corresponding bromate formation only considering the direct reactions of ozone. Squares: measured bromate formation including OH radical reactions. Curve "OH + ozone": calculated corresponding bromate formation through simultaneous ozone and OH radical pathways. Initial conditions: 1 mg/l bromide, 2.5 mM carbonate at pH 8, and an ozone dose of 2 mg/l.[24]

a direct consequence of this fact. Figure 6 also shows bromate calculated for Reactions 1 to 5 (ozone reactions only) at pH 10. If we assume two separate pathways through ozone or OH radicals, bromate formation at pH 10 should be the sum of calculated bromate formation at pH 10 (contribution of ozone) and measured bromate formation at pH 11 (contribution of OH radicals). However, bromate formed at pH 10 is significantly higher than this superposition (Figure 6). This shows that there is indeed a synergistic effect of ozone and OH radicals. The results presented here were further confirmed by using hypobromite instead of bromide as a starting compound.[24]

Role of OH Radical Mechanism at pH 7.5 to 8

To test the contribution by the OH radical mechanism, batch-type experiments with a single ozone dose (2 mg/l) were performed at pH 8 (alkalinity 2.5 mM). Bromate formation as a function of the ozone exposure in the absence and in the presence of t-BuOH (0.1 mM) are shown in Figure 7. The difference in bromate formed is due to the inhibition of OH radical reactions. The curves included in Figure 7 show bromate formation calculated for oxidation by molecular ozone and the simultaneous reaction of ozone and OH radicals. The calculations are based on the above kinetic data, assuming an OH radical yield of 50% with respect to decayed ozone.[33] A first-order rate constant $k = 1.5 \cdot 10^{-4} \text{ s}^{-1}$ for the conversion of O_3 into OH radicals was calculated from a fit of observed ozone decay curves. There is a good agreement between experimental and calculated data.

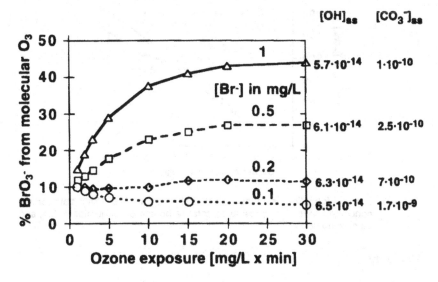

Figure 8 Calculated relative importance of the ozone and the OH radical mechanism: effect of initial bromide concentration at pH 7.5 (2.5 mM HCO$_3^-$). Calculated bromate formation vs. ozone exposure is shown for 0.1, 0.2, 0.5, and 1 mg/l bromide. The calculated steady-state concentrations for OH and carbonate radicals (reaction with OBr$^-$ as main sink) are also shown.

The relative importance of the overall sequence of reactions leading to the oxidation by ozone or OH radicals is dependent on the bromide concentration. The percentage of bromate formed through ozone reactions as a function of the ozone exposure and the initial bromide concentration is shown in Figure 8 for kinetic calculations at pH 7.5 with a carbonate concentration of 2.5 mM. The molecular ozone mechanism is only predominant for high bromide levels, whereas for small bromide concentrations the OH radical pathway is by far more important. In carbonate-containing waters, Reactions 24 and 25 no longer play a decisive role for the oxidation of hypobromite and hypobromous acid, because the steady-state concentrations of carbonate radicals are more than three orders of magnitude higher than the steady-state concentration of OH radicals. The increase of the steady-state concentration of carbonate radicals with decreasing bromide concentrations is due to a competition of bromide and carbonate for OH radicals. However, the pathway through the formation of Br$_2^-$ (Reactions 19 to 21) seems to be less efficient than the formation of HOBr by ozone followed by and oxidation with carbonate radicals.

Figure 8 might also explain some of the differences found by different authors with respect to the importance of either molecular ozone or OH radicals if working with different initial bromide concentrations.

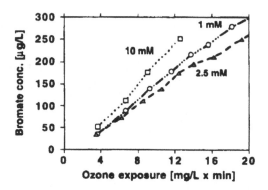

Figure 9   Bromate formation as a function of the alkalinity at pH 8. The initial bromide concentration is 1 mg/l; results for total carbonate concentration 1, 2.5, and 10 mM are shown.[36]

Alkalinity

Carbonate and bicarbonate act as OH radical scavenger and normally affect ozonation processes in two ways: (1) inhibition of the radical-type chain reaction leading to an enhanced decomposition of ozone through OH radicals, and (2) inhibition of reactions of OH radicals as secondary oxidants. To investigate the influence of carbonate on bromate formation, a series of experiments was performed at pH 10.2.[24] At this pH, the rate of ozone decomposition and the rate of OH radical formation is high. Reduction of bromate formation was found if an organic OH radical scavenger was added to the solution. This is a direct result of the inhibition effect of OH radical reactions. If carbonate was added to the same solution bromate formation increased again indicating the possibility of carbonate radicals acting as secondary oxidants (Equation 26). The higher bromate formation in the case of carbonate addition is not a consequence of a longer half-life of ozone and thus a higher ozone exposure. In the range of scavenger concentrations applied here, the half-life of ozone is fairly constant.[34]

This hypothesis was further tested through ozonation experiments on bromide-containing solutions at pH 8 with carbonate concentrations ranging from 1 to 10 mM (Figure 9). The calculated fraction of OH radicals scavenged by $[CO_3^{2-}]_{tot}$ varied between 50% (1 mM $[CO_3^{2-}]_{tot}$) and 90% (10 mM $[CO_3^{2-}]_{tot}$). No systematic variation was observed in the bromate formed for differing alkalinities. These results again support the model assumptions in Equations 24 to 26, where $[HOBr]_{tot}$ can either react with OH or carbonate radicals. Correspondingly, kinetic simulations based on Reactions 1 to 9 and 19 to 35 have shown that the bromate formed through the OH radical mechanism is nearly independent of total carbonate concentrations. This is the first case reported in which carbonate radicals are observed as transient species that act as secondary oxidants in ozonated water.

Organic Matter

Both the transformation of ozone into OH radicals and the rate of OH radical scavenging are influenced by the pH and the type and concentration of dissolved organic matter (DOM) present in natural waters. Experiments were performed with two sources (mesotrophic and eutrophic lake) and various concentrations of DOM. Within certain limits, bromate formation as a function of the ozone exposure (c·t) did not depend on type and concentration of DOM.[24] Based on this finding it is possible to simulate bromate formation in drinking water treatment, knowing the characteristics of a representative model water. An example of such a calculation is given by von Gunten and Hoigné.[24] For these waters bromate formed through reactions with molecular ozone contributes in the range of 30 to 80% to the overall bromate formation. Since no reactions with DOM are included in our model, the agreement of experimental and calculated data suggests that bromate formation per c·t is independent of DOM. Because the contribution of the molecular ozone mechanism is well defined for a given ozone exposure, this means that the effect of OH radicals has to be similar for the different waters. This is notable because in our experiments the second half-lives of ozone varied between 2 and 10 min. It could be expected that a factor of 5 difference in the second half-lives of ozone would lead to significant differences in the OH radical concentrations. However, it has been shown that the steady-state ratio of [OH]/[O_3] during continuous ozonation of several DOM-containing surface waters has a fairly constant value, averaging $1.3 \cdot 10^{-7}$.[36] This implies that the rate of OH radical consumption by DOM is proportional to the ozone depletion rate which is also controlled by the DOM.

The small influence of type and concentration of DOM on bromate formation is also a consequence of the relatively slow reaction rate for the reaction of HOBr and other bromine species with DOM. The overall yield of bromate remains fairly constant for similar waters containing DOM or being DOM-free. The steady-state concentration of HOBr thus remains almost unaffected by DOM. Knowing the characteristics of a DOM-free system it is possible to estimate the total bromate formed (within 20 to 30%) as a function of ozone exposure for any natural water.

ROLE OF HYDROGEN PEROXIDE

In advanced oxidation processes often a combination of ozone and hydrogen peroxide are applied. Deprotonated hydrogen peroxide (HO_2^-) acts as initiator for the chain reactions that transform ozone into hydroxyl radicals ($k_{O3,HO2^-} = 5.5 \cdot 10^6\ M^{-1}s^{-1}$).[34]

Semi-batch-type experiments at pH 7.5 (4 mM[CO_3^{2-}]$_{tot}$) were performed with an initial concentration of 0.2 mg/l bromide. Ozone and hydrogen peroxide were added continuously to achieve w/w ratios of H_2O_2/O_3 between 0 and

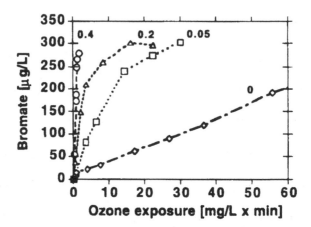

Figure 10 Bromate formation in advanced oxidation as a function of the ozone exposure. Results for different ratios H_2O_2/O_3 (0, 0.05, 0.2, 0.4) are shown. Experimental conditions: 0.2 mg/l bromide, pH 7.5, 4 mM HCO_3^-.[35]

0.4. However, the ratios applied here do not reflect drinking water conditions because a semi-batch system was used. From comparative studies with full-scale treatment a typical advanced oxidation process is reflected by a ratio H_2O_2/O_3 of 0.05 in this study. The ratio H_2O_2/O_3 was calculated according to

$$\frac{H_2O_2}{O_3} = \frac{[H_2O_2]_{aq} \times (\text{flow rate})_{H_2O_2}}{[O_3]_g \times (\text{flow rate})_{O_3}} \quad (36)$$

Figure 10 shows the bromate formation as a function of the ratio H_2O_2/O_3. Ratios as small as 0.05 already lead to a substantial increase of the bromate formation compared to the system free from hydrogen peroxide. This is due to the fact that both molecular ozone and OH radicals are kept at steady-state concentrations which allow bromate production through both the molecular ozone mechanism and the OH radical mechanism. In these combined H_2O_2/O_3 systems the conditions for a synergistic effect of ozone and OH radicals are almost ideal.

Figure 10 shows that, under the treatment conditions applied here, the addition of hydrogen peroxide always leads to faster bromate production if plotted against the ozone exposure. However, there is also a reaction that can delay bromate formation:

$$H_2O_2 + HOBr \rightarrow H^+ + Br^- + H_2O + O_2$$
$$k_{37} = 2 \cdot 10^4 \, M^{-1}s^{-1} \quad (37)$$

The effect of this reaction could only be observed if an efficient organic OH radical scavenger was added to a solution with a ratio $H_2O_2/O_3 > 0$. The bromate produced as a function of the ozone exposure was far below the case in which ozone alone was applied with the same scavenger concentration. This indicates that the bromate production is slowed down or even hindered in the extreme case, because Reaction 37 becomes dominant and consumes HOBr before it can be further oxidized to bromate.[37]

For H_2O_2/O_3 ratios larger than about 0.1, Reaction 22a becomes an important pathway for bromate formation. In the case of a ratio of 0.4 in Figure 10, the steady-state ozone concentration is below 0.1 mg/l indicating very high OH radical fluxes. The bromate production in this experiment can only be explained by a direct oxidation of bromide to bromate through a pure OH radical mechanism. However, as discussed earlier, these conditions do not reflect conventional drinking water ozonation.

If a single dose of ozone and hydrogen peroxide are applied the bromate formation is controlled by the half-life of ozone and the rate of OH radical production. A more detailed discussion of the influence of advanced oxidation processes on bromate formation is given elsewhere.[38]

CONCLUSIONS

Bromate formation by ozonation of bromide-containing waters can be described by a limited set of chemical equations including direct reactions with molecular ozone and reactions with OH radicals.

In systems where OH radicals are scavenged by organic compounds, bromate formation can be accurately predicted as a function of pH and ammonia concentrations using kinetic calculations based on published reaction-rate constants. An excellent agreement between experimental data and model predictions was achieved by slightly adjusting the reported pK_a value of HOBr.

In systems where OH radicals are not scavenged by organic compounds, considerably higher amounts of bromate can be formed. The resulting bromate concentration in systems where a substantial part of the ozone is transformed into OH radicals is often 30 to 70% greater than the bromate concentration found in systems where only reactions with molecular ozone occur. In drinking water ozonation, an ozone exposure of 4 mg/l·min at a pH of 8.5 and an initial bromide concentration of 200 µg/l will result in a final bromate concentration of approximately 25 µg/l, the present standard used by the WHO. By lowering either one of the three decisive input parameters (ozone exposure, pH, or bromide concentration) the bromate concentration can be reduced below the current standard.

The effect on bromate formation of the following parameters was studied:

- Alkalinity — carbonate radicals, which are formed through OH radical scavenging, react as secondary oxidants for hypobromite.
- Ammonium — in the presence of organic OH radical scavengers, the pres-

ence of ammonium results in a lag time for bromate formation. At pH 8, the bromate standard of 25 µg/l is reached for a c·t-value of 4 mg/l·min and an initial bromide concentration of 1 mg/l. The effect of ammonium on bromate formation in the presence of OH radicals is not entirely understood.

- Dissolved organic matter — DOM accelerates the transformation of ozone into OH radicals, but also accelerates the scavenging. This might be the reason why in our experiments the bromate yield and thus the OH radical yield per ozone exposure (c·t value) appears to be independent of DOM.
- Hydrogen peroxide — Hydrogen peroxide highly accelerates the production of OH radicals through ozone decomposition. In semi-batch experiments a continuous addition of ozone and hydrogen peroxide leads to a substantial increase in bromate formation if expressed as a function of the ozone exposure. For small ratios $H_2O_2/O_3 < 0.05$ this is due to a synergistic effect between ozone and OH radicals. For increasing H_2O_2/O_3 ratios larger than 0.1 a direct pathway from bromide to bromate via OH radicals gets more and more important; ozone concentrations become smaller while OH radical steady-state concentrations are high. However, this pathway is not important in conventional drinking water treatment because the OH radical steady-state concentrations are much smaller.

ACKNOWLEDGMENTS

The authors thank H. Bader, M. Boller, S. Canonica, and D. Sedlak for fruitful discussions and corrections in the manuscript. The useful comments of two anonymous reviewers are also acknowledged.

REFERENCES

1. Kurokawa, Y., Hagashi, Y., Maekawa, Y., Takahashi, M., and Kokubo, T., Induction of renal cell tumors in F-344 rats by oral administration of potassium bromate, a food additive, *Gann*, 73, 335, 1982.
2. Means, E. G., III and Krasner, S. W., D-DBP Regulations: issues and ramifications. *J. AWWA*, 85, 68, 1993.
3. World Health Organization, Guidelines for Drinking Water Quality, Geneva, 1993.
4. Krasner, S. W., Gramith, J. T., Coffey, B. M., and Yates, R. S., Impact of water quality and operational parameters on the formation and control of bromate during ozonation, preprints of International Workshop on Bromate and Water Treatment, Paris, 1993, 157.
5. Legube, B., Bourbigot, M. M., Bruchet, A., Deguin, A., Montiel, A., and Matia, L., Bromide/bromate survey on different water utilities, preprints of International Workshop on Bromate and Water Treatment, Paris, 1993, 135.
6. Hautmann, D. P., Analysis of trace bromate in drinking water using selective anion concentration and ion chromatography, Water Quality Technology Conference, Toronto, 1992.

7. von Gunten, U. and Hoigné, J., Factors controlling the formation of bromate during ozonation of bromide-containing waters, *Aqua*, 41, 299, 1992.

8. Home, R. A., *Marine Chemistry*, Wiley-Interscience, New York, 1969, 153.

9. Heller-Grossman, L., Manka, J., Limoni-Relis, B., and Rebhun, M., Formation and distribution of haloacetic acids, THM and TOX in chlorination of bromide-rich lake water, *Water Res.*, 27, 1323, 1993.

10. Klein, M. J., personal communication, 1993.

11. Lundström, U. and Olin, Å, Bromide concentration in Swedish precipitation, surface and ground waters, *Water Res.*, 6, 751, 1986.

12. Baudepartement Basel-Stadt, "Qualität der Oberflächengewässer im Kanton Basel-Stadt," Basel, 1992.

13. Rook, J. J., Bromierung organischer Wasserinhaltsstoffe als Nebenreaktion der Chlorung, *Vom Wasser*, 44, 57, 1975.

14. Greve, P. A., personal communication, 1982.

15. Bean, R. M., Mann, D. C., and Riley, R. G., Analysis of Organohalogen Products from Chlorination of Natural Waters under Simulated Biofouling Control Conditions, U.S. NRC Report NUREG/CR-1301, Washington, DC, 1980.

16. Amy, G. L., Chadik, P. A., King, P. H., and Cooper, W. J., Chlorine utilization during trihalomethane formation in the presence of ammonia and bromide, *Environ. Sci. Technol.*, 18, 781, 1984.

17. Högl, O., *Die Mineral- und Heilquellen der Schweiz*, Verlag Paul Haupt, Bern, 1980.

18. Crecelius, E. A., The production of bromine and bromate in seawater by ozonation, *Ozone News*, 5 (part 2), 1, 1978.

19. Richardson, L. B., Burton, D. T., Helz, G. R., and Roderick, J. C., Residual oxidant decay and bromate formation in chlorinated and ozonated sea-water, *Water Res.*, 15, 1067, 1981.

20. Wajon, J. E. and Morris, J. C., Rates of formation of N-bromo amines in aqueous solution, *Inorg. Chem.*, 21, 4258, 1982.

21. Haag, W. R. and Hoigné, J., Ozonation of bromide-containing waters: kinetics of formation of hypobromous acid and bromate, *Environ. Sci. Technol.*, 17, 261, 1983.

22. Haag, W. R., Hoigné, J., and Bader, H., Improved ammonia oxidation by ozone in the presence of bromide ion during water treatment, *Water Res.*, 18, 1125, 1984.

23. Siddiqui, M. S. and Amy, G. L., Factors affecting DBP formation during ozone-bromide reactions, *J. AWWA*, 84, 63, 1993.

24. von Gunten, U. and Hoigné, J., Bromate formation during ozonation of bromide-containing: interaction of ozone and hydroxyl radical reactions, *Environ. Sci. Technol.*, 28, 1234, 1994.

25. Zehavi, D. and Rabani, J., The oxidation of aqueous bromide ions by hydroxyl radicals. A pulse radiolytic investigation, *J. Phys. Chem.*, 76, 312, 1972.

26. Sutton, H. C., Adams, G. E., Boag, J. W., and Michael, B. D., Radiolysis yields and kinetics in the pulse radiolysis of potassium bromide solutions in *Pulse Radiolysis*, Ebert, M., Keene, J. P., Swallow, A. J., and Baxendale, J. H., Eds., Academic Press, London, 1965, 61.

27. Buxton, G. V. and Dainton, F. S., The radiolysis of aqueous solutions of

oxybromine compounds; the spectra and reactions of BrO and BrO_2, *Proc. R. Soc. A*, 304, 427, 1968.

28. Kläning, U. K. and Wolff, T., Laser Flash Photolysis of HClO, ClO^-, HBrO, and BrO^- in aqueous solution, *Ber. Bunsenges. Phys. Chem.*, 89, 243, 1985.

29. Buxton, G. V., Greenstock, C. L., Helman, W. P., and Ross, A. B., Critical review of rate constants for reactions of hydrated electrons, hydrogen atoms and hydroxyl radicals in aqueous solution, *J. Phys. Chem. Ref. Data*, 17, 677, 1988.

30. Buxton, G. V. and Elliot, A. J., Rate constants for reaction of hydroxyl radicals with bicarbonate ions, *Radiat. Phys. Chem.*, 27, 241, 1986.

31. Chen, S., Cope, V. W., and Hoffman, M. Z., Behavior of CO_3^- radicals generated in the flash photolysis of carbonatoamine complexes of cobalt(III) in aqueous solution, *J. Phys. Chem.*, 77, 1111, 1973.

32. Deuflhard, P., Bader, G., and Nowak, U., Larkin—a software package for the simulation of large systems arising in chemical reaction kinetics, in *Modelling of Chemical Reaction Systems*, Ebert, K. H., Deuflhard, P., and Jäger, W., Eds., (Springer Series Chem. Phys.), Springer-Verlag, New York, 1981, 18.

33. Hoigné, J. and Bader, H., Ozonation of water: "oxidation-competition values" of different types of waters used in Switzerland, *Ozone Sci. Eng.*, 1, 357, 1979.

34. Staehelin, J. and Hoigné, J., Decomposition of ozone in water: rate of initiation by hydroxide ions and hydrogen peroxide, *Environ. Sci. Technol.*, 16, 676, 1982.

35. von Gunten, U., Hoigné, J., and Bruchet, A., Bromate formation during ozonation of bromide-containing waters, preprints of International Workshop on Bromate and Water Treatment, Paris, 1993, 51.

36. Haag, W. R. and Yao, C. C. D., Ozonation of U. S. drinking water sources: OH radical concentration and oxidation-competition values. *Proc. 11th Ozone World Congress*, San Francisco, Vol. 2, Port City Press, New York, 1993, 119.

37. Taube, H., Reactions of solutions containing ozone, H_2O_2, H^+ and Br^-, *J. Am. Chem. Soc.*, 64, 2468, 1942.

38. von Gunten, U., Bruchet, A., and Costentin, E., Bromate formation in conventional and advanced oxidation processes (O_3/H_2O_2): Experiments on full-scale drinking water plants, JAWWA, 1995, submitted.

CHAPTER 9

Removal of Bromate after Ozonation during Drinking Water Treatment

Mohamed Siddiqui, Gary Amy, Wenyi Zhai, and Larry McCollum

ABSTRACT

As mandated by the 1986 Safe Drinking Water Act amendments, the USEPA is currently developing regulations for disinfection by-products (DBPs). In addition to potentially lowering the existing THMs standard, the USEPA may regulate other DBPs. One ozone by-product under consideration is bromate. The surface water treatment rule (SWTR) has established disinfection requirements for utilities based on minimum CT values. Utilities must control pathogenic microorganisms while minimizing the fomation of other harmful DBPs. Proposed drinking water regulations in the U.S. will specify a maximum contaminant level (MCL) of 10 $\mu g/l$ for bromate ion (BrO_3^-), an ozonation by-product produced during drinking water treatment. Several different options to remove BrO_3^- after its formation, applicable to conventional surface water treatment plants contemplating the use of ozone at various points of application, have been evaluated: ferrous iron reduction (Fe^{2+}), activated carbon surface reduction (GAC), and ultraviolet irradiation (UV). In all processes, chemical analysis of the treated water showed the formation of bromide ion (Br^-) indicating that chemical reduction of BrO_3^- is the significant mechanism; adsorption of BrO_3^- onto the surface of iron floc is insignificant. BrO_3^- removal is activated carbon specific, and not all carbons have shown the ability to reduce BrO_3^- to Br^-.

INTRODUCTION

During the chemical oxidation and/or disinfection of natural waters containing Br^- with ozone, BrO_3^- is formed at concentrations ranging from 0 to 50 $\mu g/l$ under normal drinking water treatment conditions.[8,11,17] Br^- itself

1-56670-136-8/96/$0.00+$.50

occurs ubiquitously, with an average concentration in the U.S. of almost 100 $\mu g/l$.[1] BrO_3^- can form through both a molecular ozone (O_3) pathway and a hydroxyl radical (OH·) pathway depending on the dissolved organic carbon (DOC), Br^- content, and pH of the source water. By the molecular ozone pathway, Br^- is first oxidized by dissolved ozone to hypobromite ion (OBr^-) which is then further oxidized to BrO_3^-. Molecular ozone-bromide ion reactions leading to the formation of bromate are shown below (values of rate constants can be obtained from Reference 14).

$$Br^- \xrightarrow{O_3} BrO^- \xrightarrow{O_3} BrO_2^- \xrightarrow{O_3} BrO_3^-$$

This reaction is pH dependent since OBr^- is in equilibrium with hypobromous acid (HOBr). The molecular ozone theory suggests that BrO_3^- formation is directly driven by dissolved ozone (DO_3) and the OH radical theory indicates that dissolved ozone plays only an indirect role by decomposing to produce radicals that further react with bromine species to produce BrO_3^-.[6,21]

$$Br^- \xrightarrow{OH·} Br· \xrightarrow{O_3} BrO· \xrightarrow{OH·} BrO_2^-$$
$$BrO_3^- \xleftarrow[O_3]{OH·} BrO_2· \xleftarrow{OH·}$$

von Gunten and Hoigné[20] have reported that 30 to 80% of BrO_3^- formation occurs through the molecular ozone pathway in contrast to Yates and Stenstrom[21] who report less than 10% bromate formation through the molecular ozone pathway in natural source waters.

Proposed drinking water regulations in the U.S. will specify an MCL of 10 $\mu g/l$ for BrO_3^- and a best available treatment (BAT) of pH adjustment. To date, most BrO_3^- control strategies have involved inhibiting (minimizing) BrO_3^- formation through acid or ammonia addition.[9] Adjustment to pH 6.0 prior to ozonation will significantly reduce BrO_3^- formation; however, acid addition may not be viable or cost effective for high-alkalinity waters. Ammonia addition can theoretically tie up HOBr/OBr$^-$ as bromamines and scavenges free radicals thereby reducing the formation of BrO_3^-; however, the complexity of bromamine chemistry has yielded inconsistent BrO_3^- formation results ranging from reduction to no reduction in lab and pilot studies.[11] Also according to von Gunten et al.,[20] ammonia addition only results in a time-lag minimization of BrO_3^- formation. The work reported herein involves removing BrO_3^- after its formation, when minimization strategies are not cost effective and/ or reliable. If the draft BrO_3^- MCL in the U.S. is lowered further, a combination of minimized formation and subsequent removal may be required as a BrO_3^- control strategy.

This work has evaluated several different options to remove BrO_3^- applicable to conventional surface treatment plants contemplating the use of ozone at various points of application. Several means of removing BrO_3^- after its formation have been evaluated: use of a chemical reducing agent/coagulant ($FeSO_4$), activated carbon, and ultraviolet irradiation.

EXPERIMENTAL AND ANALYTICAL METHODS

Surface waters evaluated included California State Project Water, CA (SPW), Contra Costa Water, CA (CCSW), Colorado River Water, CA (CRW), Silver Lake Water, CO (SLW), and DOC-free (≤ 0.2 mg/l) Milli-Q water (MQW). Samples of SPW and CCSW was taken from the Sacramento-San Joaquin River Delta, which is subject to saltwater intrusion. SLW is a subalpine reservoir in the Rocky Mountains. CRW bromide levels may be influenced by salt deposits (connate sea water) on the western slope of Colorado. Important characteristics of the source waters studied are summarized in Table 1.

FE(II) REDUCTION

Fe(II) reduction was evaluated in conventional jar test experiments. After adjusting samples to a desired pH, aliquots of these samples were placed under a six-place stirrer. With the stirrer operating at 100 rpm, varying amounts of reductant/coagulant ($FeSO_4 \cdot 7H_2O$) were added to the samples and the pH was allowed to drift from an initial pH to a final lower pH. Each sample was flocculated for 30 min at 30 rpm after 1 min of rapid mixing, settled for 1 h, decanted, and filtered through a prewashed 0.45-μm membrane filter. The presence of residual Fe^{2+} did not interfere with measurements of BrO_3^-.

BATCH PAC EXPERIMENTS

BrO_3^- reduction experiments using powdered activated carbon (PAC) (<325 mesh) were performed by the batch shaker test. The appropriate PAC dose (milligrams per liter) was introduced into 100-ml amber bottles. The bottles were put on a shaker table for a specified time and filtered through

Table 1   Source Water Characteristics

Water source	DOC (mg/l)	pH	Alkalinity (mg/l CaCO₃)	Br⁻ (µg/l)
MQW	0.3	6.0	<10	<5
SLW	3.2	7.0	18	5
SPW	4.4	8.0	86	160
CCSW	2.8	7.1	80	155
CRW	3.4	8.0	100	70

Table 2   Characteristics of Different Carbons Employed

Carbon	D-10	D-10[a]	HDB	PK	F-400
Origin	Peat	Peat	Lignite	Peat	Bituminous
Surface area (m^2/g)	550	620	525	850	1000
Density (g/ml)	0.57	0.61	0.5	0.29	0.44
Ash (%)	6	5	5	7	6
Isoelectric point	8.0	8.2	3.5	8.0	6.3
Manufacturer	Norit	Norit	Norit	Norit	Calgon

[a] HCl washed carbon.

prewashed 0.45-μm filters. A complete list of activated carbons used along with their total surface area is given in Table 2.

RSSCT TESTS

Rapid small-scale column testing (RSSCT) experiments were performed to evaluate bromate removal by granular activated carbon (GAC) under dynamic, continuous-flow conditions. Typical operating conditions for RSSCT evaluation are given in Table 3. Design parameters and scale-up have been addressed through similitude relationship between RSSCTs and pilot-scale plants, based on a constant-diffusivity modeling approach.[4] The carbon beds were secured between glass wool in 0.9-cm glass column. Solutions were pumped downwardly from 100-liter plastic tank using a peristaltic pump at a rate of 10 to 20 ml/min.

UV IRRADIATION

UV irradiation of BrO_3^- was evaluated in a batch reactor, with an initial BrO_3^- concentration in a higher range than might be encountered in water treatment (50 to 100 μg/l). Solutions of BrO_3^- were prepared in MQW and source waters, and irradiated in a cylindrical reactor equipped with a variable-intensity low-pressure mercury lamp. The irradiance of the UV lamp (180 to 254 nm) was measured by a UV intensity meter (850 to 2000 μW/cm^2).

Table 3   Rapid Small-Scale Column Testing (RSSCT) Operating Conditions

Design parameters	RSSCT	Pilot scale[a]
Bulk density (g/ml)[b]	0.5	0.5
Particle radius (cm)	0.01 (50 × 80)	0.051 (12 × 40)
EBCT (min)	0.32	8–10
Flow rate (ml/min)	12–30	170
Column diameter (cm)	1.1	5.0
Time of operation (d)	2–5	50–100
Water volume required (l)	120	12,258
Glass column length (cm)	14	80
Mass of carbon (g)	7	—

[a] Based on approximate equivalent pilot plant scale.
[b] Calgon Filtrasorb-400.

Table 4   Properties of Bromate, Bromite, and Hypobromite Ions in Aqueous
           Solution

Ion	λ_{max} (nm)	ΔH_f° (kcal mol^{-1})	ΔG_f° (kcal mol^{-1})
BrO_3^-	195	-15.95	$+4.55$
BrO_2^-	260	-9.9	$+2.1$
BrO^-	330	—	—

Note: ΔH_f° = heat of formation; ΔG_f° = free energy change.

Relevant spectrophotometric and thermodynamic properties of bromate and
other bromate decomposition products are given in Table 4.

ACID AND BASIC GROUPS OF ACTIVATED CARBON

Acid-base titrations were used to evaluate the number of acidic and basic
groups on the surface of activated carbon samples. A 1-g carbon sample was
placed in 40 ml of 0.05 N solutions of sodium hydroxide (NaOH), sodium
carbonate (Na_2CO_3), sodium bicarbonate ($NaHCO_3$), and hydrochloric acid
(HCl). The polythene vials were sealed and agitated for 24 h. The number of
acidic sites of carbon was calculated under the assumption that NaOH neutral-
izes carboxyl, phenolic, and lactonic groups. The number of surface basic sites
was calculated from the amount of HCl which reacted with the carbon sample.

ANALYTICAL METHODS

Br^- and BrO_3^- measurements were accomplished by ion chromatography
(IC) using a Dionex 4500i series system (Dionex Corp., Sunnyvale, CA) with
an IonPac AS9-SC column. A 2-mM Na_2CO_3/0.75 mM $NaHCO_3$ eluent was
used for Br^- measurement and a 40-mM H_3BO_3/20-mM NaOH eluent was
employed for BrO_3^- determination. Minimum detection limit (PQL) for BrO_3^-
was 2 µg/l. For samples with high chloride ion (Cl^-) content, a syringe silver
cartridge filter (silver coating on high-capacity strong acid cation exchanger)
was used to precipitate Cl^- as silver chloride.

Ferrous iron (Fe^{2+}) concentration was measured with TPTZ (2,4,6-tripyri-
dyltriazine) with a detection limit of 0.01 mg Fe^{2+}/l.[6]

Isoelectric points of different carbons were evaluated by measuring zeta
potential at various pH levels with a Zeta Plus instrument equipped with a
laser Doppler velocimetry unit (Brookhaven Instrument Corporation, NY).

RESULTS AND DISCUSSION

In this study, the experimental matrix was designed to study commonly
used operating conditions, and generally encompassed the ranges of conditions
as indicated in Table 5.

Table 5   Range of Experimental Variables

BrO_3^- ($\mu g/l$)	pH	DOC (mg/l)	Fe^{2+} (mg/l)	Fe^{2+} time (min)	PAC dose (mg/l)	PAC time (h)	UV intensity ($\mu W/cm^2$)
25–100	6.5–8.5	0.2–4	25–100	5–30	25–100	0.5–24	800–2000

Waters subjected to BrO_3^- reduction were characterized according to Br^-, BrO_3^-, pH, UV absorbance (at 254 nm), and DOC. Observed levels of BrO_3^- formed in same source waters upon ozonation in bench-scale studies were used as a basis to select BrO_3^- concentrations which were then spiked into these source waters and MQW prior to investigation of the various BrO_3^- removal processes.

REMOVAL OF BrO_3^- AFTER PRE-OZONATION

Pre-ozonation before coagulation is being considered by various utilities in the U.S., with treatment objectives including CT credit, ozone-enhanced coagulation, and taste and odor removal. If Br^- is present, BrO_3^- can form during pre-ozonation and will pass without removal through a conventional water treatment plant with a traditional configuration. If BrO_3^- is present in (pre)ozonated water entering the coagulation process, several options exist for its removal. An aqueous-phase reducing agent (e.g., Fe^{2+}) can be added at the rapid mix step. Powdered activated carbon can likewise be added as a solid-phase reductant to remove BrO_3^- and DBP precursors.

Several commonly used coagulant chemicals have been investigated to quantify the extent of adsorption of the BrO_3^- to the surface of the precipitate phase, or its reduction to Br^- (Figure 1). Trivalent metal hydroxides, such as $Al(OH)_3(s)$ and $Fe(OH)_3(s)$, exhibit a pH-dependent amphoteric behavior. At a pH below the pH_{zpc} (isolectric point), the floc surface will be positive, a condition favorable for the adsorption of BrO_3^-. Alum (Al^{3+} = 60 mg/l) removed up to 5% BrO_3^- as opposed to up to 20% removal by ferric chloride (Fe^{3+} = 60 mg/l), both presumably due to adsorption of BrO_3^- to the $Al(OH)_3$ (pH_{zpc} = 5.2) or $Fe(OH)_3$ (pH_{zpc} = 8.5) floc surface. These experiments illustrate the poor capabilities of coagulating agents and the apparent inability of metal hydroxide adsorption to significantly reduce BrO_3^- concentrations in natural waters.

Chemical reduction of BrO_3^- to Br^- is feasible through the use of ferrous iron (Fe^{2+}) as a reducing agent which itself is then converted to ferric iron (Fe^{3+}). Also, in an ozonated water, there will be ample dissolved ozone (DO_3) or dissolved oxygen (DO) to convert any remaining Fe(II) to Fe(III). Relevant reactions are summarized below:

$$BrO_3^- + 6Fe^{2+} + 6H^+ \rightarrow Br^- + 6Fe^{3+} + 3H_2O \qquad (1)$$

$$4Fe^{2+} + O_2 + 4H^+ \rightarrow 4Fe^{3+} + 2H_2O \qquad (2)$$

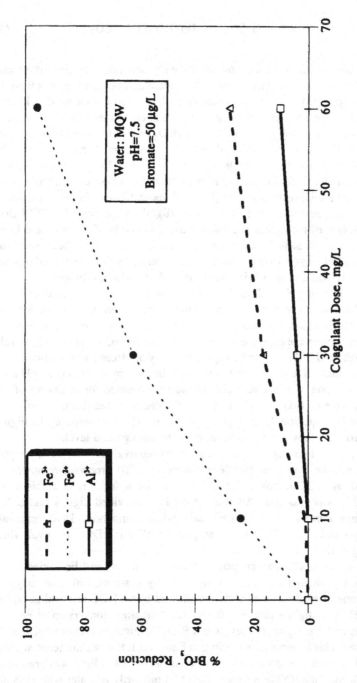

Figure 1. Bromate reduction with common coagulants.

$$Fe^{3+} + 3HCO_3^- \rightarrow Fe(OH)_3(s) + 3CO_2 \qquad (3)$$

The rates of BrO_3^- reduction are strongly dependent on Fe^{2+} dose and pH. In general, the majority of the BrO_3^- reduction occurred in less than 10 min and appeared to reach equilibrium within 15 min (During this kinetic study, the coagulation protocol explained in the experimental section was not followed to investigate the reduction of BrO_3^- for 0 to 30 min). Also over 65% of Fe^{2+} added was converted to Fe^{3+} in less than 30 min in both SPW and CRW source waters at ambient pH levels (DO $\sim$ 7 mg/l). These kinetics demonstrate that BrO_3^- reduction is feasible within a flocculation basin, with a typical hydraulic residence time (HRT) of about 30 min. First-order kinetics for Fe^{2+} disappearance were found to be slightly more rapid for CRW than SPW because of the higher alkalinity (same DO levels). Fe^{2+} persisted in its dissolved state in SLW for a longer time than SPW and CRW because of the very low alkalinity (<15 mg/l $CaCO_3$) and relatively similar dissolved oxygen level (BrO_3^- reduction kinetics were favored at lower alkalinities).

The effect of pH on BrO_3^- reduction by Fe^{2+} was evaluated over the range of 6.5 to 8.5 in SPW with an initial bromate concentration of 30 μg/l (Figure 2). Equation 1 would predict better BrO_3^- reduction at lower pH, a trend observed in the experiments. Upon Fe^{2+} addition, the pH of MQW (alk = 40 mg/l) and SLW dropped to approximately 5.5 from an initial pH of 7.5, while the final pH levels of CRW and SPW dropped by less than one pH unit.

Br^- analysis of coagulated filtered water accounted for nearly all of the disappearance of BrO_3^- as Br, verifying the chemical reduction of BrO_3^- as predicted by Equation 1, as opposed to surface (floc) adsorption. In Figure 2, the bromide shown is the amount over the background level.

The effect of adding comparable amounts (equivalent in terms of weight) of Fe^{2+} and Fe^{3+} salts to the SPW source on DBP precursor removals as indicated by DOC removal and UV absorbance reductions at ambient pH levels (7.5) was evaluated. Addition of the Fe^{2+} provided slightly less UV_{254} absorbance removal than the Fe^{3+} salt, while comparable DOC removals were observed. Thus, Fe^{2+} salts can provide effective DOC removal while reducing BrO_3^-.

The effect of DO concentration on BrO_3^- removal and Fe^{2+} kinetics in CRW was evaluated at 7 and 10 mg/l DO by sparging with pure oxygen. Measurements made after coagulation showed better DOC removal at higher DO levels and slightly impaired BrO_3^- removal since conversion of Fe^{2+} to Fe^{3+} was faster at higher DO levels and BrO_3^- reduction kinetics were largely unaffected. This is because, according to Equation 1, the stochiometric amount of Fe^{2+} required to reduce BrO_3^- to Br^- is only 6:1 but Fe^{2+} was present in excess of this ratio (50 times more than that primarily in water with ambient DO levels of approximately 7 mg/l). However, after ozonation, DO levels could be as high as 15 to 20 mg/l.

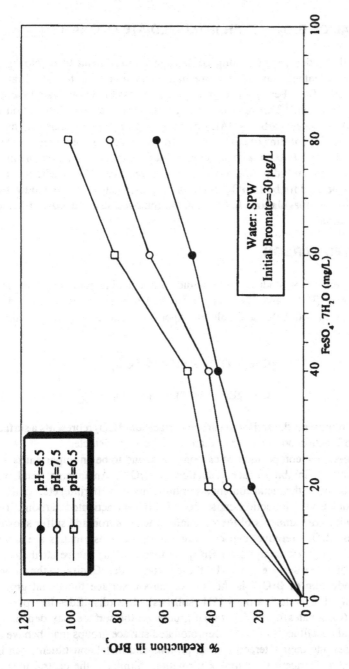

Figure 2 Effect of pH and reductant dose on BrO$_3^-$ reduction.

REMOVAL OF BrO$_3^-$ AFTER INTERMEDIATE OZONATION

Not all utilities contemplating ozone application intend to employ pre-ozonation but rather may use intermediate ozonation prior to the filtration process. Apart from Fe^{2+} reduction, removal is possible by activated carbon and UV irradiation.[17] This approach has potential relevance to integration of granulated activated carbon (GAC) columns into a process train or, more realistically, to retrofitting of rapid sand filters with GAC-capped filters (replacing anthracite with GAC in a conventional filter). Alternatively, after chemical pretreatment for turbidity and organic matter removal, UV irradiation can potentially be used for both BrO$_3^-$ removal and disinfection. For groundwaters that require no coagulation, BrO$_3^-$ can be removed using a GAC filter or UV irradiation.

ACTIVATED CARBON

Activated carbon studies were conducted to examine both powdered activated carbon (PAC) and GAC applications. Bromate ion was reduced to bromide ion in both batch and column experiments, according to the following reaction:

$$*C + BrO_3^- \rightarrow BrO^- + *CO_2$$

$$*C + 2BrO^- \rightarrow 2Br^- + *CO_2$$

where *C represents the activated carbon surface and *CO$_2$ represents a surface oxide. No reaction products other than bromide were detected.

The inorganic composition of carbon was found to be an important factor in its ability to inhibit surface reduction of BrO$_3^-$. An inverse trend was observed between the metal content (strontium, titanium, barium, etc.) of these carbons and BrO$_3^-$ reduction capacities of different activated carbons. The carbons with least amounts of these metals as well as iron and sulfur showed the highest BrO$_3^-$ removal capacity suggesting that these metals are acting as inhibitors and not catalysts for BrO$_3^-$ reduction. This can be attributed to blockage of active surface groups by these metal oxides. Central to the carbon surface reduction of BrO$_3^-$ is the role of carbon surface functional groups present on all activated carbons [SH (sulfides), -S-S (disulfides like pyrite), -CO$_2$OH (carbonates)]. The functional groups on the surface may depend on the mineral constituents of PAC. Deprotonated surface groups may behave as Lewis bases and their interaction with protons and metal constituents can be understood as competitive complex formation. Similarly, the central ions of the carbon surface can exchange functional groups for other ligands.

The chemical characteristics of the surface of the activated carbon was analyzed by using the acid-base titration method and the results are presented

Table 6   Acid and Basic Groups of Different Activated Carbons

Carbon	D-10	D-10[a]	HDB	F-400	F-400[b]	PK
Acid groups (meq/100 g carbon)	23	17	80	30	34	24
Basic groups (meq/100 g carbon)	64	84	12	60	45	60
Carboxyl + lactonic groups (meq/100 g carbon)	15	12	45	26	31	13
Carboxyl groups (meq/100 g carbon)	5	4	30	15	20	4

[a] 5 N hydrochloric acid washed carbon.
[b] Pregenerated carbon.

for virgin carbons and other treated carbons in Table 6. The indicated categories of surface chemical group are classified according to their acid-base character. Groups that are neutral in the acid-base reactions cannot be detected using this method. Carbon D-10 was found to possess a much higher density of basic groups than HDB carbon. This is consistent with bromate removal results. Acid washing by 5 N nitric acid did not change the number of total basic sites. On the other hand acid washing by 5 N HCl increased the number of basic groups and decreased the total number of acidic and carboxylic sites. The ability of carbon to remove bromate correlated with the number of basic groups on the surface of the activated carbon.

PAC RESULTS

The extent of BrO_3^- reduction was found to vary significantly from carbon to carbon, ranging from 5% to almost complete removal (Figure 3). BrO_3^- removal capacities for different activated carbons were found to vary from 0.02 to 0.45 mg BrO_3^-/g of carbon under PAC application conditions. The number of basic groups on the surface of activated carbons varied from carbon to carbon and possessed significantly different isoelectric points. The HDB carbon was found to be the least effective in removing bromate since this carbon had the lowest number of basic groups and a very low zero point charge ($pH_{zpc} \sim 3.5$). Activated carbon HDB is more negatively charged and repulsive to bromate than carbon D-10.

Waters containing DOC concentrations ranging from 0.15 to 4 mg/l were evaluated for BrO_3^- reduction to see if natural organic matter (NOM) would compete with BrO_3^- for active sites on the carbon or whether adsorbed NOM would mask potential surface reduction sites. MQW water with a DOC of 0.15 mg/l showed the highest BrO_3^- removal when compared with other DOC-containing waters. Thus, NOM appears to compete for or mask active sites.

The removal of BrO_3^- was found to increase upon lowering pH from 8.0 to 6.0 (Figure 4). Similar to the coagulation experiments, Br^- analysis showed that surface reduction of BrO_3^- to Br^- was occurring, presumably through a preliminary transient adsorption step (Figure 5).

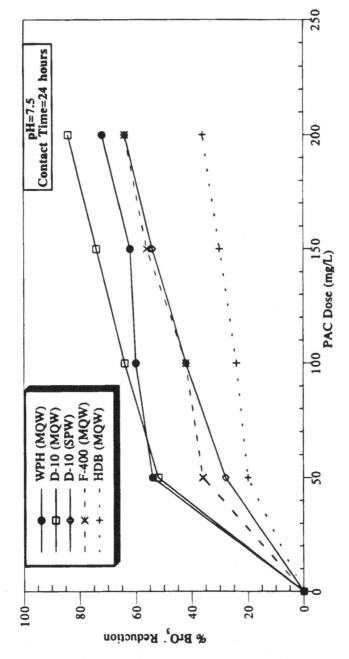

Figure 3 Effects of PAC dose and type on bromate reduction.

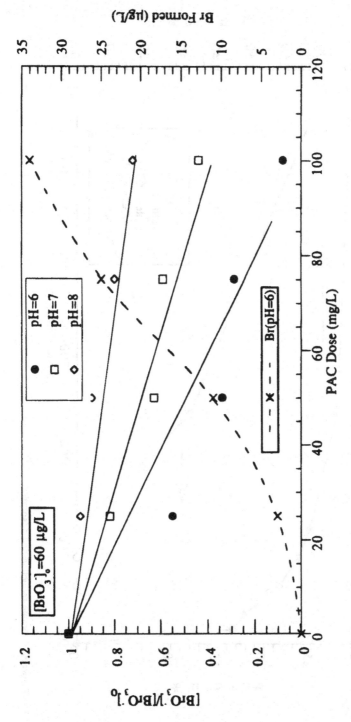

Figure 4 Effects of pH on bromate reduction by PAC (D-10, SPW).

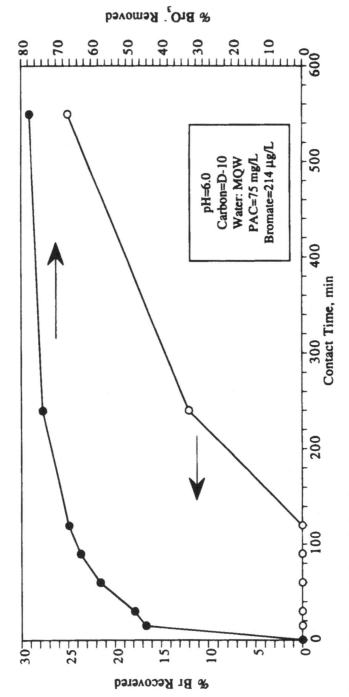

Figure 5 Bromate reduction and adsorption.

BrO_3^- removal efficiencies were compared using PAC with and without acid washing using 5 N HCl. BrO_3^- removal was significantly enhanced by acid washing PAC, presumably due to activation of inactive sites and removal of surface impurities; this is shown at two different pH levels (8.0 and 6.0) for carbon D-10 (Figure 6). Acid washing affects the surface chemistry of the carbon. The number of basic groups after acid washing increased from 64 to 84 meq/100 g of carbon. The analysis of PAC after extraction with acid showed almost 50% reduction in ash content. BrO_3^- reductions by regenerated F-400 activated carbon and virgin F-400 activated carbon were similar, indicating that BrO_3^- reduction capacity could be regenerated and/or regeneration does not adversely affect the surface chemistry responsible for BrO_3^- reduction (nor does it improve it).

Some carbons, in PAC form, have provided effective but kinetically slow removal of BrO_3^- (Figure 5). Because of time limitation, PAC does not seem to be a viable technique to remove BrO_3^- unless an appropriate PAC is employed. Moreover, the amount of PAC required is much larger than that needed for treatment objectives.

RSSCT COLUMN STUDIES

A rapid method for design of a large-scale fixed bed column from small column studies, known as the rapid small-scale column test (RSSCT), has been used. This test procedure dramatically reduces the time and effort necessary to select the conditions necessary for GAC reduction of BrO_3^-. The RSSCT test may be conducted in a fraction of the time required to conduct a pilot study. The RSSCT similitude methodology for constant diffusion is employed to simulate a pilot plant with an EBCT of 10 to 15 min (Table 3). Although the RSSCT column testing scale-up methodology employed may be valid only for the removal of DOC, it can still be used for preliminary evaluations of BrO_3^- removal as long as one keeps this objective in mind.

EFFECT OF SOURCE WATER

Figure 7 shows BrO_3^- breakthrough curves for SLW and SPW waters. Since the SPW source water is subjected to saltwater intrusion and contained sulfate, chloride, and other anions at much higher levels than SLW which may compete with BrO_3^- for adsorption sites, BrO_3^- breakthrough occurred earlier for SPW than SLW (1000 BV (bed volumes) vs. 7000 BV). In these experiments, raw chemically untreated source waters were used and the low capacity for BrO_3^- may be attributed to preferential adsorption by the GAC of DOC rather than BrO_3^- and competition from other anions such as sulfate. The results suggest that a GAC filter cap would not be suitable for removing BrO_3^- from chemically untreated SPW water but, since a filter cap is used after coagulation, the filter cap would remove BrO_3^- from chemically pre-

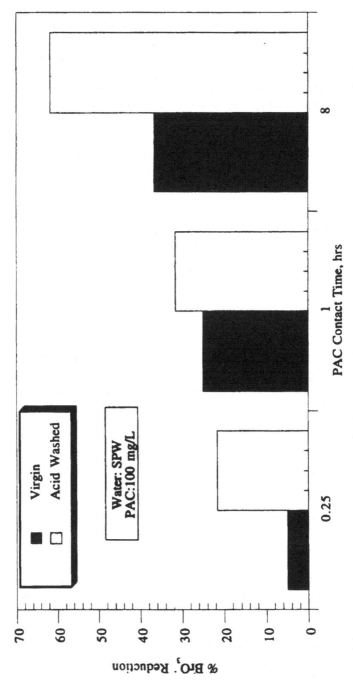

Figure 6 Effect of acid washing on bromate reduction.

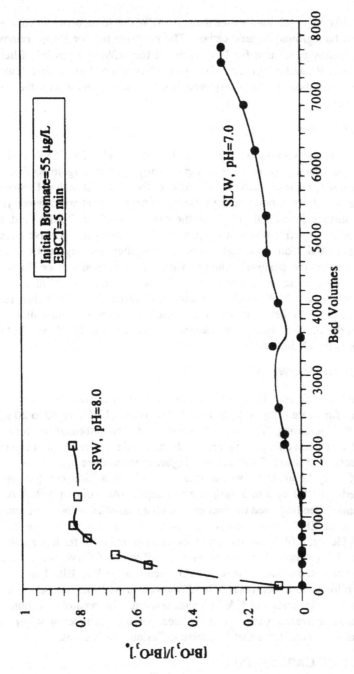

Figure 7 RSSCT breakthrough curves for SLW and SPW.

treated SPW for much longer time than raw SPW due to reduction in competition from background organic carbon. The effective life for BrO_3^- removal
could be many times that for DOC removal for SLW, and possibly filtered
SPW water. Breakthrough for BrO_3^- was never reached even after 1 week
for SLW, even though DOC breakthrough was reached after 4 d for the same
source water.

EFFECT OF pH

The effect of varying pH was studied at pH 6 and 8 for CCSW and the
removal of bromate at pH 6 was significantly greater (Figure 8). This is
possible since, for most carbons investigated, the charge at high pH values is
negative, which corresponds to the presence of negatively charged carboxylate
anionic surface functional groups on the carbon. As the pH is lowered, the
weakly acidic functional groups are protonated, resulting in a positive change
in zeta potential. Continued addition of H^+ ions after reaching the isoelectric
point increases the positively charged sites on the carbon surface. In water
treatment decreasing pH to 8 from 6 to enhance bromate removal may not be
cost effective. For high-alkalinity waters, decreasing the pH and then read-
justing to pH 7 before distribution can result in increase in total dissolved
solids in the product water. The reaction of bromate with GAC can alter the
characteristics of the carbon.

EFFECT OF FLOW RATE

Effect of different EBCTs on bromate reduction was evaluated by varying
the water flow rate from 10 to 20 ml/min (equivalent EBCT of 10 to 20 min,
Figure 9). Increasing flow rates accelerated the breakthrough of bromate
through the carbon column. In other words, more effective removal of bromate
was observed at lower flow rates and higher contact times.

GAC may be an effective treatment scenario as a filter cap (on top of
sand media: EBCT of 3 to 5 min) or a full depth GAC column (EBCT of 10
to 12 min) depending upon the background characteristics of the source water.
Variations of EBCT from 3 to 20 min were found to equally reduce BrO_3^-
in SLW. Hence, a filter cap scenario may be more effective for low alkalinity
and low TDS waters than waters such as SPW and CCSW. Marhaba and
Medler[12] reported greater than 90% BrO_3^- removal with an EBCT of 5 to 10
min in pilot-scale experiments. The difference in performance between the
bench-scale PAC tests and GAC column tests may be possibly attributed to
the overall increased surface contact area per unit volume of water, and
turbulent flow conditions that minimize diffusional limitations.

EFFECT OF CARBON TYPE

Three different activated carbons (F-400, F-300, and PK) with different
surface and origin were evaluated and the results are shown in Figure 10 for

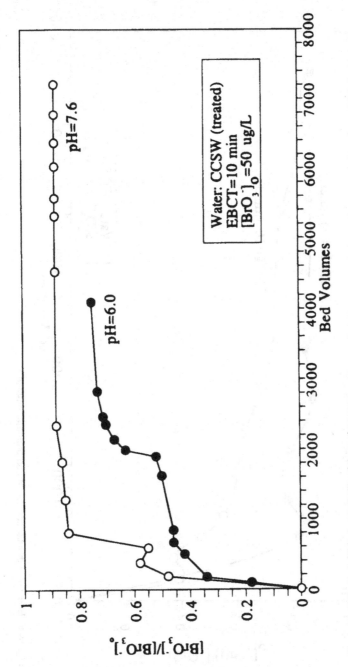

Figure 8 Effect of pH on bromate breakthrough curves.

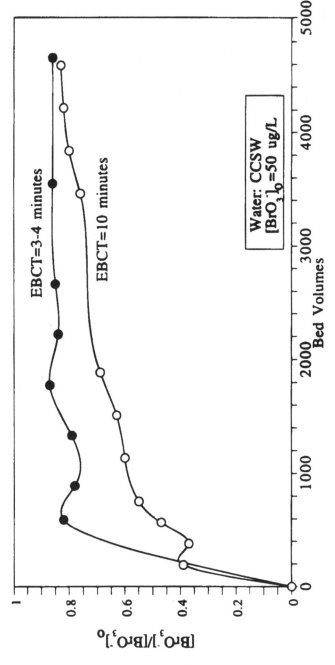

Figure 9 Effect of hydraulic loading rate on bromate removal.

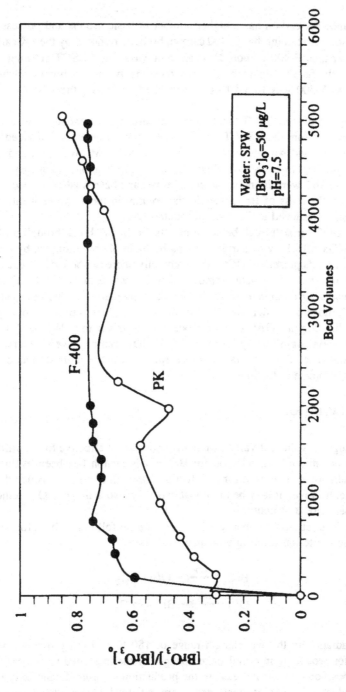

Figure 10 Effect of carbon type on bromate removal.

SPW. Since PK carbon has a higher number of basic groups and possesses a higher isoelectric point than F-400 carbon, bromate removal by the PK carbon was better than F-400 carbon. Bromate removal using RSSCT columns was significantly faster than in batch tests presumably due to higher diffusion capability. F-300 was found to be more effective than either F-400 or PK carbons.

After one of the RSSCT breakthrough experiments, spent activated carbon was removed from the RSSCT column (8000 bed volumes) and divided into three aliquots and washed with acid (0.1 M H_2SO_4), MQW water, and base (0.1 M NaOH). No BrO_3^- was found in any of these aliquots but increasing amounts of Br^- were found in the acid, MQW, and base washes, respectively. Measuring the mass of Br species in the column influent and effluent, 95% of the BrO_3^- reduced to Br^- was accounted for.

The DOC breakthrough behavior affects the BrO_3^- breakthrough profiles for RSSCTs. Blocking of surface groups by hydrophobic moieties by humic and fulvic acids present in DOC will eventually decrease the BrO_3^- reduction capacity of GAC. Much better removal of BrO_3^- was achieved by GAC using rapid small scale column tests (RSSCT), as compared to using the same carbon in a PAC mode of application. The kinetics of BrO_3^- reduction do not appear to be as limiting in a GAC column mode in comparison to a PAC batch mode of application; significantly higher rates of BrO_3^- removal were observed in the former. The effect of total dissolved salts was much more dramatic than DOC on bromate reduction.

UV IRRADIATION

Recognizing that activated carbon may not be cost effective for all utilities, the use of ultraviolet irradiation for BrO_3^- destruction has been evaluated. UV irradiation can be used for disinfection although it is not considered cost effective; however, it may be more viable if they also achieve BrO_3^- removal and other DBP destruction.

The destruction of bromate can be represented as follows with the formation of hypobromite ion as an intermediate:

$$BrO_3^- \xrightarrow{UV} OBr^- + O_2$$
$$OBr^- \xrightarrow{UV} Br^- + 1/2\,O_2$$

UV irradiation in the wavelength range of 180 to 300 nm provides energy ample for producing intermolecular changes of the irradiated molecules. The decomposition of bromate leads to the production of bromide and oxygen as end products via complex reactions that are initiated by the products generated by the primary reactions of photolysis. The reaction proceeds through internal transition of bromate and does not involve a charge-transfer-to-solvent (CTTS) intermolecular process.[22]

In Figure 11 the concentration changes of these three species during a typical UV irradiation run are shown. pH had no significant effect on BrO_3^- reduction by UV irradiation since bromate exists in solution in ionic form over the pH range investigated. The presence of organic matter (DOC) lowered the reduction of BrO_3^- and decreased the production of Br^- in natural waters, presumably due to the consumption of free radical species and UV absorbance by organic carbon. The presence of bromine can lead to the formation of organic bromine because of secondary reactions between humic material and bromine in irradiated solutions of BrO_3^-. Upon UV irradiation of MQW containing high levels of Br^-, no BrO_3^- was formed indicating that UV irradiation can be used irreversibly to convert BrO_3^- to Br^-. The effect of UV intensity (900 to 1900 $\mu W/cm^2$) and contact time on bromate reduction was investigated and the results are shown in Figures 12 and 13. The contact time necessary to reduce bromate can be reduced by increasing the intensity of the UV lamp or by employing a medium-pressure arc lamp. Percentage of BrO_3^- reductions for contact time ranging from 0.6 sec to 6.0 sec using a medium pressure lamp are summarized in Table 7.

CONCLUSIONS

Several methods have shown promise in removing BrO_3^- applicable to conventional surface treatment plants contemplating the use of ozone at various points of application.

1. In all processes evaluated, Br^- was found after treatment ($\geq 90\%$ recovery of influent Br associated with BrO_3^-), indicating that chemical reduction (aqueous or surface) is dominating as opposed to adsorption (to floc or activated carbon).
2. The presence of background DOC had a significant effect on BrO_3^- reduction using Fe^{2+}, GAC, and UV irradiation. pH variation had a strong effect on BrO_3^- reduction using Fe^{2+} and activated carbon but had an insignificant effect on UV irradiation.
3. Chemical reduction of BrO_3^- to Br^- is feasible through the use of ferrous iron (Fe^{2+}) as a reducing agent which itself is then converted to ferric iron (Fe^{3+}). In reacting with BrO_3^- and the dissolved oxygen present in an ozonated water, Fe(II) is oxidized to Fe(III) which can then function as a coagulant for DBP precursors. The ferrous reduction studies were performed primarily in water with ambient dissolved oxygen (DO) levels. However, after ozonation, DO could be as high as 15 to 20 mg/l. Additional work is needed to evaluate the effect of higher DO.

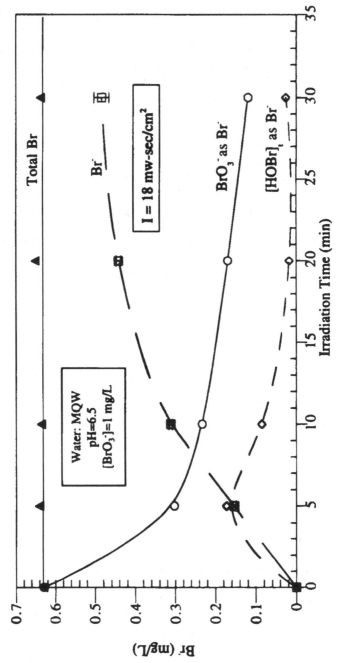

Figure 11   UV irradiation of BrO_3^- solution; formation of hypobromous acid and Br^-.

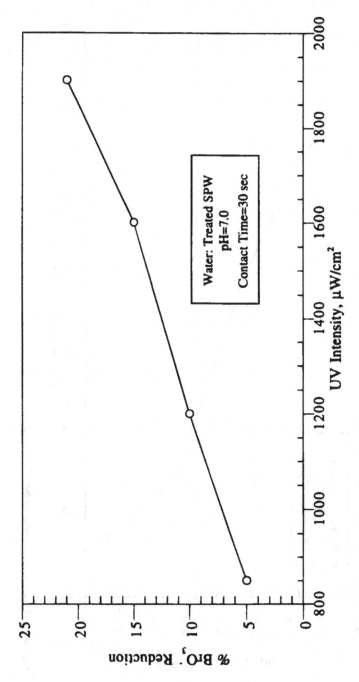

Figure 12   Effect of UV intensity on bromate reduction.

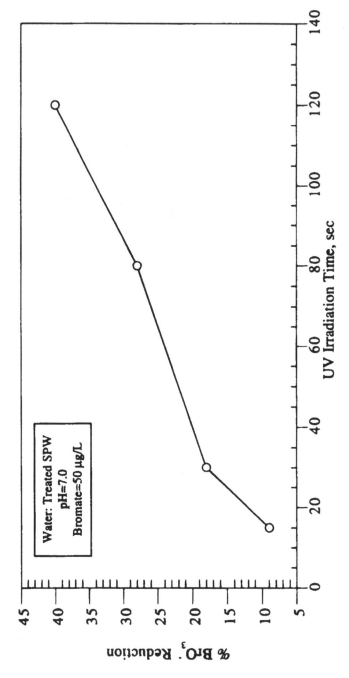

Figure 13   Effect of contact time on BrO$_3^-$ destruction.

Table 7 Bromate Destruction by UV Irradiation (Medium Pressure Lamp): Pilot-Scale Studies (CCSW)[a]

Sample	Flow (gpm)	pH	Time (s)	Temp In (°C)	Temp Out (°C)	UV Dose[b] ml/cm^2	Influent Measured [BrO$_3^-$]$_o$, (μg/l)	% BrO$_3^-$ Reduction
1	16	8.7	6.0	19	20	600	11	46
5	32	8.7	3.0	19	19	300	13	30
9	45	8.7	1.5	19	19	150	16	7
13	140	8.7	0.6	19	19	60	23	5
2	16	7.7	6.0	19	19	600	11	27
6	32	7.7	3.0	19	20	300	12	25
10	45	7.7	1.5	19	19	150	18	12
14	140	7.7	0.6	19	19	60	22	5
4	16	8.7	6.0	19	19	600	19	31
8	32	8.7	3.0	19	19	300	16	19
12	45	8.7	1.5	19	20	150	28	15
16	140	8.7	0.6	19	19	60	29	7
3	16	7.7	6.0	19	19	600	16	31
7	32	7.7	3.0	19	19	300	16	31
11	45	7.7	1.5	19	19	150	26	16
15	140	7.7	0.6	19	19	60	38	24

[a] Lamp details (Aquionics UV-500; Reactor volume = 1.5 gal; Length/Diameter = 2.4; Power = 2 kW).

[b] 1 mJ/cm^2 = 1 mW-s/cm^2.

4. Use of GAC (RSSCT) columns showed almost complete removal of BrO$_3^-$. BrO$_3^-$ breakthrough was much slower than DOC breakthrough (SLW) indicating that GAC columns may be used economically to remove BrO$_3^-$ in low TDS waters.
5. UV irradiation can also reduce BrO$_3^-$ but its cost of operation has not been adequately evaluated for this purpose.

ACKNOWLEDGMENTS

This work has been supported by a consortium of California drinking water utilities (MWD of Southern California, Alameda County Water District, Contra Costa Water District, LA Dept. of Water & Power) through the California Urban Water Agencies, CA, with Lyle Hoag as the Project Manager. The authors thank Chandra Mysore for measuring zeta potentials of different activated carbons.

REFERENCES

1. Amy, G., Siddiqui, M., Zhai, W., and Debroux, J., National Survey of Br$^-$ in Drinking Waters, 1993 AWWA Annual Conference Proceedings, San Antonio, TX, 1993.
2. Buxton, G. V. and Dainton, F. S., The radiolysis of aqueous solutions of oxybromine compounds; the spectra and reactions of BrO$^-$ and BrO$_2^-$, *Proc. R. Soc.*, 304, 1968.

3. Buxton, G. and Greenstock, C., Critical review of rate constants for reactions of hydrated electrons, hydrogen atom and hydroxyl radicals in aqueous solutions, *J. Phys. Chem. Ref. Data*, 17(2), 1988.
4. Crittenden, J. C., Hand, D. W., Reddy, P. S., Arora, H., Trynoski, J. M., and Notthakun, S., Prediction of GAC performance using rapid small-scale column tests, AWWA Annual Conference Proceedings, Los Angeles, 1989.
5. Dixon, K. L. and Lee, R. G., The effect of sulfur-based reducing agents and GAC filtration on chlorine dioxide by-products, *JAWWA*, 83, 5, 1991.
6. Dougan, W. K. and Wilson, A. L., Absorptiometric determination of iron with T.P.T.Z., *Water Treat. Exam.*, 22, 1973.
7. USEPA, Draft D/DBP Rule, USEPA Office of Groundwater and Drinking Water, Washington, DC, February 4, 1994.
8. Glaze, W. H., Weinberg, H., and Cavanagh, J., Evaluating the formation of brominated DBPs during ozonation, *JAWWA*, 85, 96, 1993.
9. Gramith, J. T., Coffey, B. M., Krasner, S. W., Kuo, C., and Means, E. G., Demonstration-Scale Evaluation of Bromate Formation and Control Strategies, AWWA Annual Conference Proceedings, San Antonio, TX, 1993.
10. Haag, W. R. and Hoigne, J., Ozonation of bromide containing waters: kinetics of formation of hypobromous acid and bromate, *Environ. Sci. Technol.*, 17, 261, 1983.
11. Krasner, S. W., Glaze, W., Weinberg, H., Daniel, P., and Najm, I., Formation and control of BrO_3^- during ozonation of waters containing bromide, *JAWWA*, 85, 96, 1993.
12. Marhaba, T. and Medler, S., Removing bromate from water, *Ozonews*, 21, 3, 1993.
13. Miller, J., Snoeyink, V. L., and Harrell, S., The Effect of GAC Surface Chemistry on Bromate Reduction, preprints, Environmental Chemistry Division, 206th ACS National Meeting, Chicago, 1993, 237; also Chapter 13 in this text.
14. Neta, P., Huie, R., and Ross, A., Rate constants for reactions of inorganic radicals in aqueous solutions, *J. Phys. Chem. Ref. Data*, 17, 3, 1988.
15. Prados-Ramirez, M. J., Ciba, N., and Bourbigot, M. M., Available Techniques for Reducing Bromate in Drinking Water, Proceedings, IWSA Workshop on Bromate and Water Treatment, Paris, 1993.
16. Richardson, L., Burton, D. T., and Helz, G. R., Residual oxidant decay and bromate formation in chlorinated and ozonated sea-water, *Water Res.*, 15, 1067, 1981
17. Siddiqui, M., Westerhoff, P., Ozekin, K., and Amy, G., Removal of Bromate from Drinking Water, preprints, Environmental Chemistry Division, 206th ACS National Meeting, Chicago, 1993, 221.
18. Siddiqui, M. and Amy, G., DBPs formed during ozone-bromide reactions in drinking water treatment, *JAWWA*, 85, 63, 1993.
19. von Gunten, U., Hoigne, J., and Bruchet, A., Bromate Formation during Ozonation of Bromide Containing Waters, IWSA Workshop on Bromate and Water Treatment, Paris, 1993, 51.
20. von Gunten, U. and Hoigne, J., Factors affecting the formation of BrO_3^- during ozonation of bromide-containing waters, *J. Water SRT-Aqua* 41, 5, 1992.
21. Yates, R. S. and Stenstrom, M. K., Bromate Production in Ozone Contactors, AWWA Annual Proceedings, San Antonio, 1993.
22. Amichai, O., Czapski, G., and Treiniu, A., Flash Photolysis of the Oxybromine Anions, *Israel J. Chem.*, 7, 351, 1969.

CHAPTER 10

Ozone-Bromide Interactions with NOM Separated by XAD-8 Resin and UF/RO Membrane Methods

Rengao Song, Roger Minear, Paul Westerhoff, and Gary Amy

ABSTRACT

Ozone-bromide interactions in the presence of selectively isolated natural organic matter (NOM) fractions were studied under batch conditions. XAD-8 resin and ultrafiltration and reverse osmosis (UF/RO) membrane isolation methods were used for concentration of NOM from the Teays Aquifer, Champaign, IL. Experimental results show that both decomposition rates of aqueous ozone and disinfection by-product (DBP) formation are the functions of raw water characteristics, treatment process parameters, NOM isolate molecular weights, and isolation procedures. Bromate (BrO_3^-) formation is favored at high pH, high ozone dosage, and high initial bromide concentrations, but low dissolved organic carbon (DOC) levels. On the other hand, total organic bromine (TOBr) formation is favored at low pH, high bromide levels, and high O_3/DOC ratios. The addition of alkalinity promotes BrO_3^- formation. An inhibitory effect on bromate formation is observed upon ammonia addition. The addition of hydrogen peroxide increases bromate formation for membrane isolates, while TOBr production is significantly reduced. Finally, the highest bromate yields are associated with NOM isolated by the XAD-8 resin technique, whereas the highest TOBr yields are found with the RO isolate.

INTRODUCTION

Since ozone was first used as a disinfectant for drinking water treatment in 1893 at Oudshoorn, Netherlands, water ozonation has been extended to the oxidation of ferrous and manganous ions,[1] control of taste and odor,[2]

1-56670-136-8/96/$0.00+$.50

enhancement of coagulation-flocculation, reduction in the formation of chlorinated DBPs,[3] and removal of color.[4] Interest in the use of ozone has been
increasing steadily, primarily as a result of both health concerns associated
with the formation of chlorinated DBPs[5] such as trihalomethanes (THMs) and
in light of the fact that ozone is a much stronger oxidant ($E^o = +2.07$ V)
than chlorine ($E^o = +1.36$ V). Thus, ozonation is one of the most powerful
treatment options available for the enhancement water quality under some
circumstances in which other oxidants are not able to meet certain
requirements.

   In contrast to these benefits of water ozonation, there are three basic
disadvantages of ozone utilization in water treatment.[6] First, ozone reacts with
natural organic matter to form assimilable organic carbon (AOC) that is more
biodegradable than its precursor in promoting possible biological growth in
the distribution system. Second, as a disinfectant, the half-life ($t_{1/2}$) of aqueous
ozone is too short to provide a sufficient disinfectant residual within distribution systems. The final disadvantage is that ozonation of bromide-containing
water can produce brominated DBPs including bromate, bromoform, and
bromoacetic acids, and some of these inorganic and organic by-products are
potential carcinogens. While the first two disadvantages may be overcome and/
or beneficially converted to advantages by the utilization of specific treatment
strategies such as biologically activated carbon (BAC) and postdisinfection,[7,8]
respectively, brominated by-products formation, especially bromate production, can be a challenging obstacle for water utilities. The concern for bromate
regulation occurred primarily as a result of both improvement of analytical
capabilities and the recognition of its adverse human health impact. Analytical
developments, especially the use of ion chromatography (IC), make it possible
to measure part-per-billion (ppb) bromate levels.[9] The carcinogenicity of bromate in animal experiments led to the regulation of this compound in potable
water.[10] The guideline level for bromate proposed by the World Health Organization (WHO) is 0.5 μg/l at a risk level of 10^{-5}.[11] Moreover, a proposed
maximum contaminant level (MCL) for bromate has been set at 10 μg/l during
the regulation-negotiation (reg-neg) agreements between federal regulators
and water utility experts, based on the practical quantitative level.[12] Thus,
considering the average raw water bromide level of about 0.1 mg/l in the
U.S., moderate ozone doses applied to a water may lead to violation of the
proposed MCL.

   Brominated organic by-products can also be produced during ozonation
of bromide-containing water.[13] Although very few species are identifiable, the
formation of total organic bromine could be significant. Furthermore, some
potential strategies used to control bromate formation such as pH depression
can, on the other hand, promote TOBr formation.[14] Bromoform, dibromoacetic
acid, and some other identifiable TOBr species are toxic compounds and
could be potential constraints for water treatment after stricter disinfectants-
disinfection by-products (D-DBPs) rules are put into effect. Therefore, while

some other research focus was on bromate formation, TOBr formation was also investigated in this work.

BACKGROUND

Bromate (BrO_3^-) formation during ozonation of water containing bromide ions (Br^-) has been well documented. Haag and Hoigné[15] proposed the following kinetic relationships, based on experiments in NOM-free deionized waters:

$$O_3 + Br^- \rightarrow O_2 + OBr^- \qquad k = 1.6 \times 10^2 \ M^{-1}s^{-1} \qquad (1)$$

$$O_3 + OBr^- \rightarrow 2O_2 + Br^- \qquad k = 3.3 \times 10^2 \ M^{-1}s^{-1} \qquad (2)$$

$$O_3 + OBr^- \rightarrow O_2 + BrO_2^- \qquad k = 1.0 \times 10^2 \ M^{-1}s^{-1} \qquad (3)$$

$$O_3 + BrO_2^- \rightarrow O_2 + BrO_3^- \qquad k > 10^5 \ M^{-1}s^{-1} \qquad (4)$$

These mechanisms involve only molecular ozone (O_3) as an oxidant. Thus, they are referred to as the direct ozonation/oxidation and/or molecular ozone pathway (mechanism).

Since the Haag-Hoigné model was developed, no attempt has been made to develop a model to predict bromate formation in natural water systems. Recently, Yates and Stenstrom[16] conducted a pilot-scale study of surface waters to investigate BrO_3^- formation and stated that hydroxyl radical may play a dominant role in bromate formation, and that direct ozonation can be considered insignificant. The skeleton of their model is

$$3O_3 \rightarrow 2 \ HO\cdot \qquad (5)$$

$$Br^- + HO\cdot \rightarrow BrOH^-\cdot \qquad k = 1.1 \times 10^{10} \ M^{-1}s^{-1} \qquad (6)$$

$$BrOH^-\cdot \rightarrow Br\cdot + OH^- \qquad k = 4.2 \times 10^6 \ s^{-1} \qquad (7)$$

$$Br\cdot + O_3 \rightarrow BrO\cdot + O_2 \qquad k = 1.0 \times 10^{10} \ M^{-1}s^{-1} \qquad (8)$$

$$HOBr + HO\cdot \rightarrow BrO\cdot + H_2O \qquad k = 2.0 \times 10^9 \ M^{-1}s^{-1} \qquad (9)$$

$$2BrO\cdot + H_2O \rightarrow BrO_2^- + HOBr + H^+ \qquad k = 2.6 \times 10^9 \ M^{-1}s^{-1} \qquad (10)$$

$$BrO_2^- + O_3 \rightarrow BrO_2\cdot + O_3^-\cdot \qquad k = 1.0 \times 10^5 \ M^{-1}s^{-1} \qquad (11)$$

$$2BrO_2\cdot + H_2O \rightarrow BrO_3^- + BrO_2^- + 2H^+ \qquad k = 4.0 \times 10^7 \ M^{-1}s^{-1} \qquad (12)$$

$$BrO_2 \cdot + HO \cdot \rightarrow BrO_3^- + H^+ \qquad k = 2.0 \times 10^9 \, M^{-1} \, s^{-1} \qquad (13)$$

The entire model postulated by Yates and Stenstrom includes 65 species and 190 reactions. They believe that HO radical and its daughter-radicals dominate bromate formation. This kind of model is referred to as a indirect oxidation/ozonation and/or radical pathway (mechanism).

Brominated organic DBP formation during ozonation has also been intensively investigated by Glaze et al.[13] and Amy et al.[17] Both Glaze and Amy pointed out that TOBr formation is favored at low pH values. The generally accepted pathway for TOBr formation is the reaction between HOBr and NOM as indicated below.

$$O_3 + Br^- \rightarrow O_2 + OBr^- \qquad k = 1.6 \times 10^2 \, M^{-1} s^{-1} \qquad (14)$$

$$OBr^- + H^+ \rightleftharpoons HOBr \qquad pK_a = 8.8 \text{ at } 20°C \qquad (15)$$

$$HOBr + NOM \rightarrow TOBr(CHBr_3, CHBr_2CN, CHBr_2COOH, \text{etc.}) \qquad (16)$$

Von Gunten and Hoigné,[18] based on their own and other researchers' work, concluded that the two different mechanisms might be associated with the formation of bromate and the brominated organics. Gordon[19] further proposed that TOBr is favored below pH 7 and bromate is favored above pH 7.

Work using tertiary-butanol as a radical scavenger confirms that at least two pathways are responsible for bromate formation and suggests that the radical pathway appears to be much more influential in the presence of NOM.[20] The experimental results indicate that NOM is perhaps the determining factor, along with pH, in dictating bromate formation. In light of this fact and since the investigation of the role of different NOM isolates separated by different methods in initiating and/or promoting ozone decomposition and their effects on DBP formation are not well illustrated, the primary goal of this study was to assess, in the presence of different NOM fractions, the influence of raw water characteristics and treatment process parameters on bromate and TOBr formation.

EXPERIMENTAL SECTION

NOM ISOLATION

Raw groundwater was collected from the Teays Aquifer, the drinking water source for Champaign, IL, and transported back to storage at a cold room (4°C) and immediately filtered with a 0.22-μm tangential flow filter to remove particulate matter. One portion of the permeate was passed through an Na^+ cation exchange column to remove Ca^{2+} and Mg^{2+}, then applied to a continu-

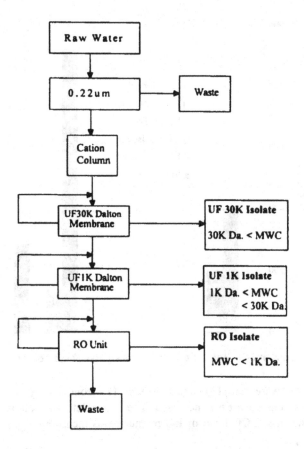

Figure 1   Ultrafiltration and reverse osmosis concentration system.

ous flow system consisting of ultrafiltration (UF) and reverse osmosis (RO) membranes in series with large pore-size membranes preceding smaller pore-sizes (Figure 1).[21] The remaining portion of the 0.22-µm filtrate was acidified pH to 2 with hydrochloric acid, and then applied to an XAD-8 resin (Rohm and Haas, Philadelphia, PA) column (2 l) to isolate humic substances.[22] The NOM recovery (measured as DOC) from the filtered water on XAD-8 was 52% whereas for the membrane system it was 95%. All DOC concentrations were determined with a Shimadzu TOC-500. Figure 2 shows the carbon mass of DOC isolated and concentrated from the Teays Aquifer.

RO ISOLATE DESALTING

A batch pressure UF system was used to desalt the RO retentate as much as possible by dialysis using a 500-Dalton membrane. During this processing, a fixed amount of deionized, DOC-free water was added to the system, and

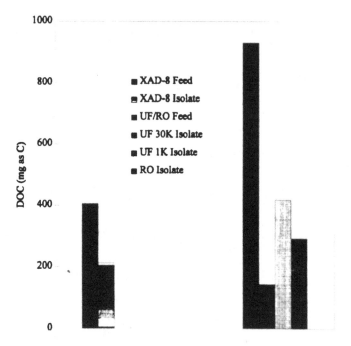

Figure 2   Mass of DOC isolated and concentrated from Teays Aquifer.

the filtrates were sampled and analyzed. The conductivity of the filtrates declined as the wash time increased. The RO isolate was totally washed 12 times. The final DOC level of RO retentate was measured.

OZONE STOCK SOLUTION

An OREC Model 03V10-0 ozone generator using extra-dry oxygen feed and a set of semi-batch contact columns was operated to reach an equilibrium of ozone dissolution. The output gas mixture passed through two 500-ml glass washing bottles aligned in series in which the first bottle contained 100 ml of 0.05 M phosphate buffer (pH = 6) serving as a sink for nonozone oxidants (nitrogen-containing compounds). The second bottle was cooled in an ice-water bath, and contained deionized water. The ozone stock concentration, produced by exhaustive ozonation, was typically 40 mg/l.

BATCH OZONATION

A batch reactor was used to study ozone-bromide-NOM reactions. The reactor was a 500-ml graduated cylinder with a side outlet 0.8 in. from the bottom to allow sample collection over time. A 0.25-in. thick Teflon cover that has the same diameter as the cylinder was kept on the solution surface to avoid the loss of dissolved ozone and any volatile DBPs through the water-air interface. A magnetic stirring bar was used to provide continuous mixing to distribute ozone and other reactants homogeneously throughout. The ozone stock solution was injected below the water surface of the reactor with a syringe. Following the addition of ozone stock solution, the Teflon® cover was immediately inserted.

ANALYTICAL METHODS

Dissolved ozone, pH, total HOBr (including OBr^-), bromide, and bromate concentrations were determined at 1, 5, 10, 20, and 30 min after the addition of ozone. TOBr analysis was performed only after 24 h of elapsed time, approaching its maximum yield.[23] For bromate and bromide measurements, the samples were quenched with one drop of 0.1 N $Na_2S_2O_3$ per 10-ml sample, and stored at 4°C until analysis. Dissolved ozone was determined by the indigo method.[24] A Hach Colorimeter, based on the DPD method, was used to measure the total HOBr concentration.[25] Residual ozone was removed from the sample by purging with argon. The individual OBr^- and HOBr concentrations were calculated according to the total concentration, the pKa (8.8 at 20°C), and the measured pH. Total organic bromine was measured with a Dohrmann TOX analyzer (DX-20A) after adsorption onto GAC. Both 2,4,6-trichlorophenol and bromoform were employed as standards to measure the adsorption recovery (92 and 90%, respectively). A Dionex (DX300) ion chromatograph coupled with an AS9-SC anion column and guard column was used to determine bromide and bromate concentrations.

RESULTS AND DISCUSSION

In this work, a variety of experiments have been performed to investigate the interactions of ozone-bromide in the presence of NOM using the model waters under batch conditions. The focus has been on comparison of DBP formation in the presence of different NOM fractions: an XAD-8 isolate corresponding to the humic fraction, and UF/RO isolates in which an RO isolate corresponds to membrane size of smaller than 1 K Dalton (NOM molecular weight cutoff <1 K Dalton) and two UF isolates correspond to 1 K Dalton < molecular weight cutoff (MWC) <30 K Dalton (UF30K isolate) and 30 K Dalton < MWC (UF30K isolate), respectively. The experimental variables embodied raw water characteristics (pH, alkalinity, and initial Br^- level) and water treatment process parameters (pH adjustment, hydrogen peroxide addition, ammonia addition, and ozone dose).

OZONE DECAY

Two-step ozone decomposition behavior was observed for all NOM fractions (Figure 3). A rapid ozone loss occurs during the first minute of reaction

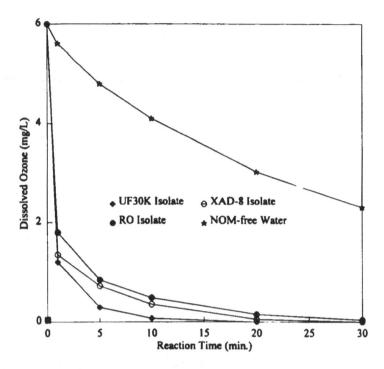

Figure 3   Ozone decay for different NOM isolates; pH = 7.5, ozone dose = 6 mg/l,
DOC = 3 mg/l, bromide = 0.4 mg/l.

time. This phenomenon has also been reported by other researchers.[26] It is
believed that this rapid dissolved ozone concentration drop is due to the
selective oxidation of NOM by ozone. Second, a slow ozone decomposition
followed which lasts at least 30 min. Table 1 shows these two different
decomposition rates in which $d[O_3]/dt = -k_1[O_3]$ from t = 0 to t = 1 min
and $d[O_3]/dt = -k_2[O_3]$ from t = 1 min to t = 30 min.

This two-step ozone decomposition behavior directly influences bromate
formation. For example, for all isolates, a rapid period of bromate formation
was observed to coincide with the first rapid ozone decomposition step.

It was also found that aqueous ozone decomposition behavior varies as a
function of different fractions from a particular NOM. While using different

Table 1   Two-Step Dissolved Ozone Decomposition Rates

NOM Isolate	$k_1 \times 10^2$ (s^{-1})	$k_2 \times 10^3$ (s^{-1})	r^2 for k_2
XAD-8	2.41	1.26	0.99
UF30K	2.86	2.28	0.99
UF1K	2.29	1.22	0.92
RO	2.04	0.92	0.99

Note: pH = 7.5, O_3 dose = 6 mg/l, DOC = 3 mg/l, Br^- = 0.4 mg/l, temperature = 20°C.

Table 2   Influence of pH on DBPs Formation

DBPs	NOM Isolate	pH 6.5	pH 7.5	pH 8.5
BrO_3^- ($\mu g/l$)	UF30K	22	37	62
	UF1K	27	45	70
	RO	30	50	81
	XAD-8	43	71	89
TOBr ($\mu g/l$ as Br^-)	UF30K	15	9	4
	UF1K	28	17	11
	RO	37	23	17
	XAD-8	31	12	2

Note: O_3 dose = 6 mg/l, DOC = 3 mg/l, Br^- = 400 $\mu g/l$, temperature = 20°C.

NOM fractions and keeping all other experimental conditions identical, the aqueous ozone decomposition rates for different isolates are in the following order:

$$\text{UF30K isolate } (k_2 = 2.28 \times 10^{-3}\ s^{-1}) > \text{XAD-8 resin isolate } (k_2 =$$
$$1.26 \times 10^{-3}\ s^{-1}) = \text{UF1K isolate } (k_2 =$$
$$1.22 \times 10^{-3}\ s^{-1}) > \text{RO isolate } (k_2 =$$
$$0.92 \times 10^{-3}\ s^{-1}).$$

EFFECTS OF pH

The effect of pH on DBPs formation has been evaluated at pH value of 6.5, 7.5, and 8.5, held constant with a phosphate buffer (1 mM). The maximum values for bromate within the 30-min reaction time period for these constituents are presented in Table 2. Figure 4 illustrates the time series behavior for bromate formation plus the attendant O_3 decomposition for the RO isolate sample. The UF30K, UF1K, and XAD-8 isolates show similar trends. Increasing the pH from 6.5 to 8.5 leads to a decrease in the half-life of dissolved ozone, increases the OBr^-/HOBr ratio and promotes bromate formation, and decreases TOBr yields. The data demonstrate that the formation of bromate is favored at high pH, while TOBr is favored at low pH. This may be due to several factors. As pH increases, dissolved ozone reacts with the increased concentration of OH^- to generate more hydroxyl radicals, thus, promoting the radical mechanism for bromate formation. This suggests that the HO radical mechanism may play an important role. At the same time, from the Haag-Hoigné model, we know that only OBr^- can be oxidized to bromate. Thus, a high OBr^-/HOBr ratio at high pH would also contribute to BrO_3^- formation. At lower pH, higher concentrations of HOBr are observed, and the half-life of ozone is longer, resulting in continuous HOBr generation, and commensurate higher TOBr formation.

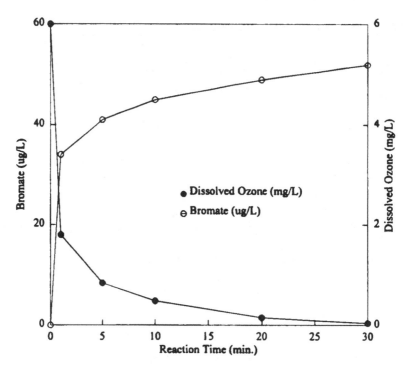

Figure 4 Bromate formation and ozone decay—RO isolate; pH = 7.5, ozone dose = 6 mg/l, DOC = 3 mg/l, bromide = 0.4 mg/l.

The extent of bromate formation observed for three different NOM fractions is related to both the characteristics of the fraction itself and the $t_{1/2}$ of dissolved ozone. While the UF 30 K isolate is consuming the most ozone and yielding the lowest bromate, the XAD-8 resin isolate exerts the an intermediate ozone demand relative to UF/RO fractions, but produces the highest bromate concentration. Thus, the results indicate that molecular ozone may only play a partial role as a reactant for bromate formation. Furthermore, BrO_3^- formation is sensitive to different NOM fractions.

There is a relationship between TOBr formation and ozone demand. The highest TOBr is found with the RO isolate which has the longest half-life of aqueous ozone. However, the pH effects on the TOBr formation are different for the three different isolates. This suggests that brominated organic DBP formation may be not only dependent on concentrations of reactants but also variation in NOM structure. Similar results were demonstrated for chlorination of humic substances.

EFFECTS OF O_3 DOSE/DOC RATIO

The O_3/DOC (mg/mg) ratio was observed to have effects on brominated DBP formation. These effects of O_3/DOC ratio on bromate and TOBr formation

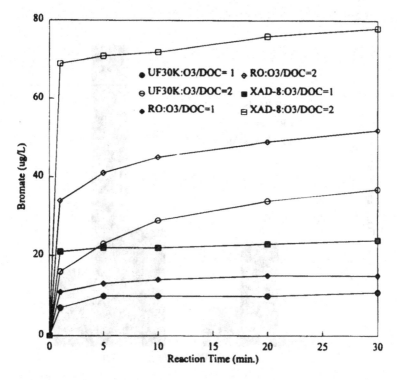

Figure 5 Bromate formation vs. ozone dose/DOC ratio; pH = 7.5, bromide = 0.4 mg/l, DOC = 3 mg/l.

are shown in Figures 5 and 6, respectively. For all experiments, increasing the O_3/DOC ratio while keeping DOC constant at 3 mg/l results in an increased bromate formation. This may indicate that at low O_3/DOC ratios NOM consumes most of the ozone and that low ozone residuals results in less OBr^-/HOBr formation and/or higher oxidation states of bromine species. Once the DO_3 residual dissipates, bromate is no longer formed, and the measured bromate concentration remains constant. This is simply due to the termination of both direct ozonation and radical oxidation once aqueous ozone is depleted. It has been reported that, at a high O_3/DOC ratio, the formation of brominated organics is reduced.[27] It is not intended to further increase the ozone dose above 6 mg/l in this study, so this phenomenon was not found.

EFFECTS OF BROMIDE CONCENTRATION

A series of experiments was conducted to examine the effects of bromide on bromate and TOBr formation. Bromide levels were varied from 0.1, 0.2, 0.4, to 1 mg/l to observe the influence of bromide while staying within analytical confines, although these values are relatively high compared to the

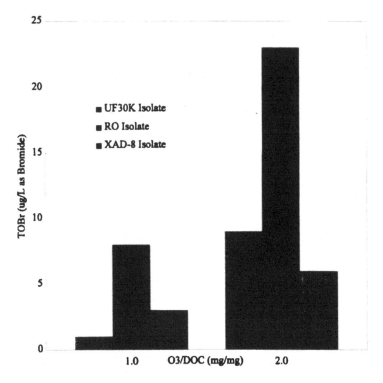

Figure 6   TOBr formation vs. ozone dose/DOC ratio; pH = 7.5, bromide = 0.4 mg/l, DOC = 3 mg/l.

average freshwater bromide background (almost 0.1 mg/l) over the U.S.[28] Table 3 shows the influences of bromide concentration on bromate and TOBr formation.

As expected, for all isolates, increasing bromide from 0.1 to 1 mg/l with all other parameters constant resulted in increased BrO_3^- and TOBr formation, probably due to increased OBr^- and/or HOBr concentrations. The half-life of DO_3 (Figure 7) was also increased. The increase in $t_{1/2}$ of dissolve ozone with the higher spiked bromide level may be good evidence for the production

Table 3   Influences of Bromide Concentration on DBP Formation

DBPs	NOM Isolate	[Br⁻] = 0.1 (mg/l)	[Br⁻] = 0.4 (mg/l)	[Br⁻] = 1.0 (mg/l)
BrO_3^- (μg/l)	UF30K	9	37	54
	RO	11	50	121
	XAD-8	18	71	163
TOBr (μg/l as Br⁻)	UF30K	4	9	62
	RO	10	23	73
	XAD-8	2	12	57

Note: pH = 7.5, O_3 dose = 6 mg/l, DOC = 3 mg/l, temperature = 20°C.

and involvement of bromide species radicals. As the initial bromide ion concentration increased, the bromide ions could act as hydroxyl radical scavengers, thus leading to a relatively longer half-life of dissolved ozone (e.g., Equation 6).

Bromate concentrations increased roughly in proportion to the increase in spiked bromide concentrations, whereas TOBr underwent a larger increase as initial bromide ion concentration was varied from 0.4 to 1 mg/l but not so with 0.1 to 0.4 mg/l. This may result from the higher concentration of HOBr at 1 mg/l of bromide level, which means longer contact time under higher HOBr concentration between HOBr and NOM. It is also possible that other mechanisms[29] could be involved in the formation of organo-bromine species, such as bromide radical involvement through the following reaction:

$$NOM(H) + HO\cdot \rightarrow NOM\cdot + H_2O \tag{17}$$

$$Br\cdot(Br_2^-\cdot) + NOM\cdot \rightarrow TOBr + (Br^-) \tag{18}$$

$$Br\cdot(Br_2^-\cdot) + NOM \rightarrow TOBr + (Br^-) \tag{19}$$

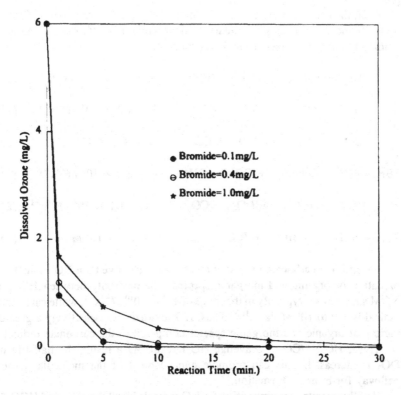

Figure 7 Ozone decay vs. bromide concentration—UF 30 K isolate; pH = 7.5, ozone dose = 6 mg/l, DOC = 3 mg/l.

Table 4   TIC Effects on Ozone Decomposition Rates

NOM Isolate	TIC (mg/l)	$k_1 \times 10^2 (s^{-1})$	$k_2 \times 10^3 (s^{-1})$	r^2 for k_2
	0	2.86	2.28	0.99
UF30K	24	2.19	0.93	0.98
	216	2.01	0.35	0.98

Note: pH = 7.5, O_3 dose = 6 mg/l, DOC = 3 mg/l, Br^- = 0.4 mg/l, temperature = 20°C.

EFFECTS OF ALKALINITY

The effects of alkalinity have been investigated at two levels, 2 and 18 mM of bicarbonate which correspond to 24 and 216 mg/l as total inorganic carbon (TIC). TIC addition has been found to increase the half-life of dissolved ozone (Table 4). It is also observed that, regardless of the NOM fraction examined, bromate formation increases with increasing alkalinity concentration. Von Gunten and Hoigne[18] recently reported that bromate formation is also affected with addition of alkalinity under true batch conditions.[30] At least four possible reasons may account for the enhanced bromate formation in the presence of TIC.

First, the carbonate radical (CO_3^-) formed in the presence of ozone (Equations 20 and 21) acts as a secondary oxidant which like HO radical can also oxidize hypobromite ions (OBr^-) (Equation 22):

$$HO \cdot + HCO_3^- \rightarrow OH^- + HCO_3 \cdot \qquad k = 1.0 \times 10^7 \, M^{-1}s^{-1} \quad (20)$$

$$HO \cdot + CO_3^{2-} \rightarrow OH^- + CO_3^- \cdot \qquad k = 4.0 \times 10^8 \, M^{-1}s^{-1} \quad (21)$$

$$CO_3^- \cdot + OBr^- \rightarrow BrO \cdot + CO_3^{2-} \qquad k = 4.3 \times 10^7 \, M^{-1}s^{-1} \quad (22)$$

$$2BrO \cdot + H_2O \rightarrow BrO_2^- + OBr^- + 2H^+ \qquad k = 2.6 \times 10^9 \, M^{-1}s^{-1} \quad (23)$$

$$CO_3^- \cdot + BrO_2^- \rightarrow BrO_2 \cdot + CO_3^{2-} \qquad k = 1.1 \times 10^8 \, M^{-1}s^{-1} \quad (24)$$

$$2BrO_2 \cdot + H_2O \rightarrow BrO_3^- + BrO_2^- + 2H^+ \quad k = 4.0 \times 10^7 \, M^{-1}s^{-1} \quad (25)$$

Second, the carbonate radical is much more selective than HO radical for oxidation of organic and inorganic species. The reactivity between HO and NOM substrates is typically in the range of 10^8 to $10^{10} \, M^{-1}s^{-1}$, while carbonate radical is 10^4 to $10^8 \, M^{-1}s^{-1}$.[31,32] Thus, HO· oxidizes nonselectively a greater variety of organic and inorganic species compared to the carbonate radical.

Third, HCO_3^-/CO_3^{2-} scavenge HO radical. As a result, the half-life of DO_3 is increased. This can promote enhancement of the molecular ozone pathway for bromate formation.[33]

Finally, despite the competition for HO radicals by NOM, Br^-, and HCO_3^-/CO_3^{2-}, bromate formed via a radical oxidation pathway may still be significant

Table 5   Kinetic Analysis of the Reaction Pathways for HO Radicals

HO· scavenger	$[S_i]$ (M)	k_i $(M^{-1}s^{-1})$	k_i $[S_i]$ (s^{-1})	$\Omega_{HO,i}$ (after TIC addition)	$\Omega_{HO,i}$ (before TIC addition)
Br^-	5.0×10^{-6}	2.0×10^9	1.0×10^4	0.04	0.12
NOM	2.5×10^{-4}	3.0×10^8	7.5×10^4	0.28	0.88
HCO_3^-	1.7×10^{-2}	1.0×10^7	1.7×10^5	0.64	0
CO_3^{2-}	2.2×10^{-5}	4.0×10^8	8.8×10^3	0.03	0

Note: pH = 7.5, O_3 dose = 6 mg/l, DOC = 3 mg/l, Br^- = 400 μg/l, temperature = 20°C.

(Table 5). The fraction of HO radical ($\Omega_{HO,i}$) that reacts with a system component, S_i, with second-order rate constant k_i can be described as follows:

$$\Omega_{HO,i} = k_i[S_i]/\Sigma\, k_i[S_i]$$

where Ω is the "competitive-oxidation" value and $\Sigma\, k_i\,[S_i]$ accounts for all the important HO scavengers. It can be seen from Table 5 that NOM exerts significant demand for HO radicals, before and after carbonate alkalinity addition.

TOBr formation was also enhanced in the presence of carbonate alkalinity. This may be due to the increased $C_{HOBr}T$ values.

EFFECTS OF AMMONIA

A 1-mol NH_3:1-mol Br^- ratio was used to examine the effects of ammonia. Table 6 shows the influence of ammonia addition on DBP formation.

After ammonia addition, an obvious decrease of bromate yield was observed. However, it is noteworthy that ammonia addition did not stop the formation of target DBPs through the conversion of HOBr to bromoamines.[34]

$$NH_3 + HOBr \rightarrow NH_2Br + H_2O \qquad k = 7.4 \times 10^7\ M^{-1}\ s^{-1} \qquad (26)$$

Table 6   Influence of Ammonia on DBP Formation

DBPs	NOM isolate	$NH_3 = 0$ (mM)	$NH_3 = 1$ mol/ mol Br	Percent decrease[a]
BrO_3^- (μg/l)	UF30K	37	32	14
	RO	45	36	20
	XAD-8	71	48	32
TOBr (μg/l as Br^-)	UF30K	9	12	−33
	RO	23	13	43
	XAD-8	12	20	−67

Note: pH = 7.5, O_3 dose = 6 mg/l, DOC = 3 mg/l, Br^- = 0.4 mg/l, temperature = 20°C.
[a] A negative decrease equates to an increase in DBP formation.

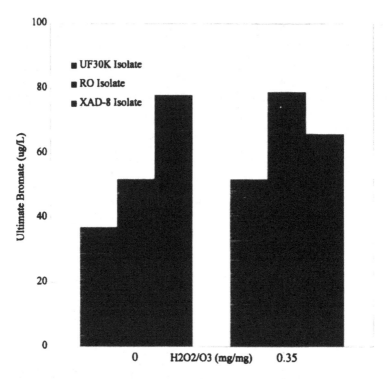

Figure 8 Bromate formation in the presence of hydrogen peroxide; pH = 7.5, ozone
dose = 6 mg/l, DOC = 3 mg/l, bromide = 0.4 mg/l.

This may be attributable to: (1) the addition of this stoichiometric amount of
ammonia is not high enough to promote the predominance of the ammonia
and HOBr reaction; (2) NH_2Br may be converted to bromide upon an excess
of dissolved ozone;[35] (3) the continuation of cycling of $Br^- \rightarrow HOBr \rightarrow$
$NH_2Br \rightarrow Br^-$ in the presence of O_3 reduces the overall amount of bromide
available for oxidation in the system; and (4) some bromate is not formed
from the direct oxidation of $HOBr/OBr^-$.

Experimental results for brominated organics formation were interesting.
Increased TOBr yields for the UF30K and the XAD-8 isolates after ammonia
addition may be interpreted, in addition to the above first two reasons, as
NH_2Br reacting with NOM to form TOBr. This may occur at relatively slower
reaction rates compared to HOBr alone.

EFFECTS OF HYDROGEN PEROXIDE

The role of hydrogen peroxide (H_2O_2) was studied at the level of 0.35 mg
H_2O_2/mg O_3. H_2O_2 addition eliminated the ozone residual immediately (i.e.,
DO_3 = 0 after only 1 min), but still produced significant bromate (Figure 8).

This phenomenon suggests that the peroxone (O_3 + H_2O_2) has a higher oxidation ability for bromine species than ozone alone and may originate from the generation of HO radicals:[36]

$$H_2O_2 \rightarrow HO_2^- + H^+ \qquad pK_a = 11.6 \qquad (27)$$

$$O_3 + HO_2^- \rightarrow HO\cdot + O_2^-\cdot + O_2 \qquad k = 2.8 \times 10^6 \, M^{-1}s^{-1} \quad (28)$$

$$O_2^-\cdot + O_3 \rightarrow O_3^-\cdot + O_2 \qquad k = 1.6 \times 10^9 \, M^{-1}s^{-1} \qquad (29)$$

$$O_3^-\cdot + H_2O \rightarrow HO\cdot + O_2 + OH^- \qquad k = 25 \, M^{-1}s^{-1} \qquad (30)$$

The effects of hydrogen peroxide on bromate formation were different for the various NOM isolates. It can be seen from **Figure 8** that bromate production in peroxone system was higher than in ozone alone for UF30K and RO isolates, but a small decrease in bromate formation was observed in the presence of H_2O_2 for the XAD-8 resin isolate. The promoted bromate formation in the presence of H_2O_2 may be due to a synergistic effect of ozone and HO radicals.

TOBr formation rate was slower than bromate formation (Figure 9). Thus, the quick loss of dissolved ozone led to low TOBr yield. Another possible reason for the low TOBr production may be the quenching effect of hydrogen peroxide on hypobromous acid.

$$H_2O_2 + HOBr \rightarrow Br^- + O_2 + H_2O + H^+ \quad k = 6.0 \times 10^4 \, M^{-1}s^{-1} \qquad (31)$$

SUMMARY

In the presence of NOM, aqueous ozone decomposition can be described by a two-step behavior, first involving a short-term rapid decomposition followed by a second slow decomposition. Different NOM isolation procedures result in NOM fractions that exert different dissolved ozone demands. The RO isolate is the least reactive with ozone, followed by the XAD-8 resin fraction and UF1K isolate, and finally the UF30K isolate. The NOM structure, composition, and size likely play important roles in influencing the decomposition rates of dissolved ozone.

Both isolation procedures and the resultant isolates have effects on bromate and TOBr formation. The highest bromate formation is observed with the XAD-8 isolate experiments, while the least bromate is found with the UF30K isolate experiments. These experimental results indicate that molecular ozone is only one of the oxidants responsible for bromate formation. On the other hand, other oxidant mechanisms can be responsible for DBP formation. It

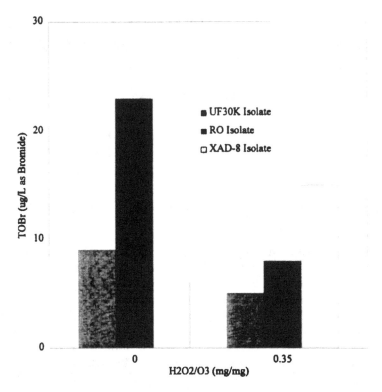

Figure 9   TOBr formation in the presence of hydrogen peroxide; pH = 7.5, ozone dose = 6 mg/l, DOC = 3 mg/l, bromide = 0.4 mg/l.

was also observed that TOBr formation may be a function of NOM structure and composition, and hence dependent on the specific NOM isolation method.

Water quality characteristics have significant influences on brominated DBP formation. The formation of bromate is favored at high pH, ozone doses, and bromide levels. Increasing DOC concentration increased the ozone demand by NOM, leading to low bromate yields. TOBr formation is favored at low pH, high bromide concentration, and high O_3/DOC ratios. Carbonate and bicarbonate species stabilizes aqueous ozone decomposition and promotes both TOBr and bromate formation.

Water treatment process parameters also affect aqueous ozone decomposition behavior and DBP formation. Addition of ammonia increases decomposition rates of aqueous ozone and decreases bromate formation. This is presumably due to the scavenging effect of ammonia on HOBr/OBr⁻. Hydrogen peroxide addition eliminated the ozone residual immediately, but still produced significant amounts of bromate due to the high oxidation potential of hydroxyl radical. Because of the quick loss of aqueous ozone after peroxide addition, TOBr production is heavily reduced.

ACKNOWLEDGMENTS

The authors wish to express their appreciation to the American Water Works Research Foundation (AWWARF) for providing financial support. Special gratitude is extended to Jeff Oxenford (AWWARF Senior Project Manager) and the rest of the Project Advisory Committee (P. Singer, J. Symons, W. Haag, and R. Powell). Dr. Richard Larson and Dr. Vernon Snoeyink of the University of Illinois and Gary Peyton of the Illinois State Water Survey are also gratefully acknowledged for their valuable advice.

REFERENCES

1. Stoebner, R. A. and Rollag, D. A., *Aqua*, 291, 1981.
2. Lalezary, S., *J. Am. Water Works Assoc.*, 78(3), 62, 1986.
3. Rice, R. G., *Ozone Sci. Eng.*, 2(1), 75, 1980.
4. Flogstad, D. M. and Odegaard, H. O., *Ozone Sci. Eng.*, 7(2), 121, 1985.
5. Rook, J. J., *J. Am. Water Works Assoc.*, 23, 168, 1976.
6. Cavanagh, J. E., Master's Dissertation, University of Carolina at Chapel Hill, 1991.
7. Jekel, M. R., in *Oxidation Techniques in Drinking Water Treatment*, Kuhn, W. and Sontheimer, H., Eds., U.S. EPA-570/9-79-020, 1979.
8. Young, J. S. and Singer, P. C., *J. Am. Water Works Assoc.*, 71, 87, 1979.
9. Van der Jagt, H., Noij, Th. H. M., and Ooms, P. C. A., Proceedings of the International Water Supply Association International Conference: Bromate and Water Treatment, Paris, November 22–24, 1993, 25.
10. Kurokawa, Y., Hagashi, Y., Maekawa, Y., Takahashi, M., and Kokubo, T., Gann, 73, 335, 1982.
11. World Health Organization, *Report of the Review Group on Disinfectants and Disinfection By-Products*, Geneva, 1991.
12. Krasner, W., Glaze, W., Weinberg, W., Daniel, P., and Najm, I. N., *J. Am. Water Works Assoc.*, 85(1), 73, 1993.
13. Glaze, W. H., Weinberg, H. S., and Cavanagh, J. E., *J. Am. Water Works Assoc.*, 85(1), 96, 1993.
14. Song, R., Minear, R. A., Amy, G., and Westerhoff, P., Preprints of the Environmental Chemistry Division Papers (206th ACS Meeting), 33(2), 225, 1993.
15. Haag, W. R. and Hoigne, J., *Environ. Sci. Technol.*, 17(5), 261, 1983.
16. Yates, R. S. and Stenstrom, M. K., American Water Works Association Annual Conference Proceedings, San Antonio, TX, 1993.
17. Siddiqui, M. S. and Amy, G. L., *J. Am. Water Works Assoc.*, 85(1), 63, 1993.
18. Von Gunten, U. and Hoigne, J., Proceedings of the International Water Supply Association International Conference: Bromate and Water Treatment, Paris, November 22–24, 1993, 51.
19. Gordon, G., Proceedings of the International Water Supply Association International Conference: Bromate and Water Treatment, Paris, November 22–24, 1993, 41.

20. Westerhoff, P., Amy, G., Ozekin, K., Siddiqui, M., Minear, R., and Song, R., Fifth Quarterly Report: Bromide-Ozone Interactions in Water Treatment, American Water Works Association Research Foundation, 1993.
21. Nanny, M. A., Kim, S., Gadomski, J. E., and Minear, R., *Water Res.*, 28(6), 1355, 1994.
22. Leenheer, J. A., in *Water Analysis*, Vol. 3, Minear, R. A. and Keith, L. H., Eds., Academic Press, San Diego, 1984, chap. 3.
23. Westerhoff, P., Amy, G., Ozekin, K., Siddiqui, M., Minear, R., and Song, R., Third Quarterly Report: Bromide-Ozone Interactions in Water Treatment, American Water Works Association Research Foundation, 1993.
24. *Standard Methods for the Examination of Water and Wastewater*, 17th ed., APHA, AWWA, and WPCF, Washington, D.C., 1989.
25. Bader, H. and Hoigné, J., *Water Res.*, 15, 449, 1981.
26. Xiong, F., Croue, J.-P, and Legube, B., *Environ. Sci. and Technol.*, 26, 1059, 1992.
27. Siddiqui, M. S. and Amy, G. L., *J. Am. Water Works Assoc.*, 85(1), 63, 1993.
28. Amy, G. L. and Odem, W., Bromide Occurrence: National Bromide Survey, 1994.
29. Song, R., Minear, R. A., and Eyring, A., Investigations of potential radical interaction in the formation of brominated organic compounds, National Conference on Environmental Engineering, Boulder, CO, July 11–13, 1994.
30. Von Gunten, U. and Hoigne, J., Preprints of the Environmental Chemistry Division Papers (206th ACS Meeting), 33(2), 217, 1993.
31. Buxton, G. V., Greenstock, C. L., Helman, W. P., and Ross, A. B., *J. Phys. and Chem. Ref. Data*, 17(2), 677, 1988.

CHAPTER 11

Evaluation of Bromate Formation and Ozone Decomposition Kinetics

Paul Westerhoff, Gary Amy, Rengao Song, and Roger Minear

ABSTRACT

The rate of ozone decomposition controls the yield of bromate and is influenced by water quality characteristics and water treatment parameters. Hypobromite ion (OBr⁻) and hydroxyl (HO) radical concentrations affect the extent and rate of bromate formation as well. Through the use of t-butanol as an HO radical scavenger, delineation of indirect HO radical and direct molecular ozone oxidation pathways are inferred. Approximately 75% of the bromate formation is controlled by HO radical mechanisms, while the remainder forms via molecular ozone mechanisms alone. Oxidation of bromine appears to be a critical oxidation step in either pathway.

INTRODUCTION

BACKGROUND

Since the early 1900s the drinking water industry has relied primarily upon chlorine to inactivate microbiological constituents of health concern (e.g., *Giardia lamblia* cysts, fecal coliforms, and most recently *Cryptosporidium* oocysts). Additionally, chlorine can oxidize reduced metals (e.g., Fe^{+2}) into hydroxide precipitates [e.g., $Fe(OH)_{3(s)}$], reduce unaesthetic color attributes in water by oxidizing natural organic matter, and/or reduce unaesthetic odors in water by oxidizing algae. Rook observed haloform formation in natural waters upon chlorination and attributed these by-products to the reaction between chlorine or bromine with natural organic matter (NOM) precursor material (mainly the aromatic constituents) in water.[28,29] The U.S. Environmental Protection Agency (USEPA) later regulated the haloform and other chlorinated organic disinfection by-product (DBP) concentrations due to health concerns.

These regulations require researchers and drinking water utilities to reevaluate techniques to disinfect/oxidize water without forming chlorinated DBPs. Aside from altering the chlorine dosages or increasing DBP precursor material (i.e., NOM) removal through enhanced coagulation practices,[12,35,36] alternative disinfection scenarios may provide an avenue to comply with health concerns. Ozone is recognized as a more powerful disinfectant/oxidant than chlorine and has been gaining attention in the U.S., while European countries widely employ ozone technologies in water treatment. As new regulations require higher finished water quality, ozone has emerged as a viable alternative disinfectant/oxidant, yet unanswered health-related concerns of ozonation DBPs have emerged. Organic ozone DBPs such as formaldehyde have been reported as well as inorganic DBPs such as bromate.[30,39]

Ozonation of bromide-containing waters forms bromate and brominated organic DBPs.[19-21,30] Ozonation of bromide, and subsequent formation of bromine ($HOBr/OBr^-$), can lead to 0 to 30% of the bromide becoming substituted into NOM, resulting in brominated organic DBPs. Techniques are available to quantify the total brominated organic DBP concentration, but only a small fraction is specifically identifiable.[9] Under typical ozonation conditions, the concentrations of identifiable brominated organic DBP concentrations do not exceed regulated maximum contaminant levels (MCLs). On the other hand, inorganic bromate concentrations in many drinking water supplies may exceed the proposed USEPA MCL of 10 μg/l (78 nM) for bromate.[1] This MCL has been established based upon a practical quantifiable limit (PQL) (e.g., analytical limitation) and not on a typical USEPA excess health risk of 10^{-6} or 10^{-5} (corresponding to bromate concentrations of 0.05 and 0.5 μg/l respectively).[19] The risk assessment studies that form the basis of health risk evaluation have drawn criticism that excessively large potassium bromate concentrations may have biased animal study results, and requires further assessment prior to lowering the bromate MCL.[24]

A recent survey of 100 U.S. drinking water sources shows an average national bromide concentration of almost 100 μg/l (1.25 μM), with concentrations ranging from 0 to 2.3 mg/l (28 μM).[3] The probability frequency distribution of bromide concentrations in random and targeted water sources is shown in Figure 1. As illustrated in Figure 1, river and ground waters exhibit higher bromide concentrations than lake waters in the survey. Batch ozonation at 62.5 μM (3 mg/l) of 31 of those natural waters (DOC = 0.04 to 1 mM; pH = 6 to 9) resulted in bromate concentrations ranging from nondetectable (<0.015 μM) to 1.950 μM, with an average of 0.235 μM.[1] Other studies have observed bromate concentrations in excess of the proposed MCL in pilot-, full-, and laboratory-scale ozone reactors.[19-21,30] Although organic ozonation DBPs pose concerns, regulation of bromate may potentially pose significant limitations on the use of ozone. Understanding bromate formation mechanisms

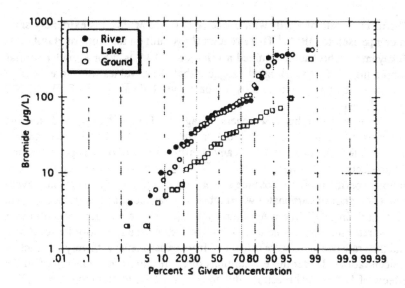

Figure 1 Distribution of nationwide bromide occurrence in U.S. drinking water supplies.

will aid in assessing potential bromate formation and implementing bromate control strategies.

PATHWAYS FOR BROMATE FORMATION

Chlorination in the presence of sunlight or ozonation of seawater (bromide = 0.8 mM) was observed to form bromate.[23,27] Chlorine oxidizes bromide to bromine and sunlight is capable of producing excited state oxidants, such as hydroxyl (HO) radicals, which can oxidize bromine to bromate.[7] Haag and Hoigne[11] concluded that direct oxidation of bromide by molecular ozone (O_3) leads to bromate during ozonation of drinking water, and proposed a series of kinetic reactions depicting the bromate formation pathway:[11]

$$Br^- + O_3 \rightleftarrows OBr^- + O_2 \qquad k_1 = 160 \pm 20\ M^{-1}s^{-1} \qquad (1)$$

$$OBr^- + O_3 \rightleftarrows Br^- + 2O_2 \qquad k_2 = 330 \pm 60\ M^{-1}s^{-1} \qquad (2)$$

$$OBr^- + O_3 \rightleftarrows BrO_2^- + O_2 \qquad k_{3a} = 100 \pm 20\ M^{-1}s^{-1} \qquad (3)$$

$$BrO_2^- + O_3 \rightleftarrows BrO_3^- + O_2 \qquad k_{3b} > 10^5\ M^{-1}s^{-1} \qquad (4)$$

$$H^+ + OBr^- \rightleftarrows HOBr \qquad pK_a = 8.7 \qquad (5)$$

The reaction rate of HOBr with ozone ($k < 0.013$ $M^{-1}s^{-1}$) is extremely slow in comparison to OBr$^-$.[11] These relationships built on the work of Haruta and Takeyama[13] who had determined a value of 170 $M^{-1}s^{-1}$ for k_1 under similar conditions. Haag and Hoigne[11] suggest that these values are more reliable than previous values reported by Taube[33] and Garland et al.[8] of 90 and 230 $M^{-1}s^{-1}$, respectively.

In contrast to other researchers, Haag and Hoigne[11] suggested that the direct oxidation reactions with bromide species are the primary mechanisms for bromate formation rather than any significant role of ozone decomposition by-products such as HO radicals.[5-7,13,27] These conclusions evolved from ozonation experiments in the absence and presence of different HO radical scavengers (t-butanol or carbonate) where observed rate constants were unchanged. This work in DOC-free water employed large stoichiometric concentrations of one reactant (e.g., ozone or bromide) to create pseudo-first-order reactions, and then rate constants were calculated from experiments with varying reactant concentrations. Under these conditions direct reactions may have masked the effects of HO radical reactions due to a high ozone to HO radical ratio.[10]

Yates and Stenstrom[41] conducted pilot-plant experiments with a natural water containing bromide with and without an HO radical scavenger (isopropyl alcohol) and concluded that HO radicals are solely responsible for initiating the oxidation of bromide to bromate.[41] The selection of IPA as a radical scavenger may not have been judicious since their experiments clearly demonstrate that this alcohol decreases the half-life of ozone due to direct reactions. They postulated a model including 65 species and 190 reactions based primarily on radiolysis work to simulate bromate formation. The complete model accounts for some dissolved organic matter (DOM) reactions, but in general assumes that molecular ozone and HO radicals produced during ozonation oxidize bromide species, and postulate that disproportionation is the final step in bromate formation:

$$O_3 + H_2O \rightarrow OH + OH^- + [products] \tag{6}$$

$$OH + Br^- \rightarrow BrOH^- \rightarrow Br + OH^- \tag{7}$$

$$Br + O_3 \rightarrow BrO + O_2 \tag{8}$$

$$2BrO + H_2O \rightarrow BrO_2^- + HOBr + H^+ \tag{9}$$

$$O_3 + BrO_2^- \rightarrow O_3^- + BrO_2 \tag{10}$$

$$2BrO_2 + H_2O \rightarrow BrO_3^- + BrO_2^- + 2H^+ \tag{11}$$

Although disproportionation is favored at high pH and the high rate constant for molecular ozone with bromite is ignored, there is evidence that oxidation of bromite to bromate may occur via disproportionation at near neutral pH levels. Others observed that bromite reacted too rapidly with ozone to be measured directly, and proposed the rate was greater than $10^5 \ M^{-1}s^{-1}$.[16-18] A stoichiometric yield factor for $\Delta BrO_2^-/\Delta O_3$ was estimated at 0.3, but became twice as high in the absence of radical scavengers. This later observation suggests that radicals may be oxidizing or reducing bromite. Yates and Stenstrom concluded that bromate production resulted from rapid free radical processes, and that the slower, recursive reactions involving the direct oxidation of bromide and hypobromite by molecular ozone are insignificant.[41]

Von Gunten and Hoigne proposed a combined OH radical and molecular ozone pathway for bromate formation.[37,38] Four key elements should be noted. First, a constant ozone concentration (mg/l) is maintained for a set duration (minutes) to represent the ozone exposure (min-mg/l). Ozone exposure is the graphically integrated area under an ozone residual vs. time plot. Thus, although the ozone concentration decreases over time during the actual experiment, the model assumes a constant concentration. Second, the model assumes an HO radical yield of 0.5 mol/mol of ozone decomposed at near neutral pH ranges. This value is based at pH 11.[14] Third, molecular ozone is assumed to be responsible for oxidation of bromide to bromine and bromite to bromate, while HO radicals oxidize bromine. Fourth, DOM is only considered as a slowly reacting sink for hypobromous acid. They concluded that the OH radical mechanism can contribute significantly (>60%) to bromate formation.[37,38]

The above models differ, but two key elements permeate throughout. First, molecular ozone is required either to directly oxidize bromide species or to provide OH radicals via free radical ozone decomposition. Second, the pH dependency of bromine plays a dominate factor in regulating the equilibrium hypobromite ion (OBr^-) concentration which strongly influences the formation of bromate.

OBJECTIVES

This paper presents batch experimental ozonation results to address the following hypotheses:

1. The kinetics of bromate formation are related to ozone decomposition kinetics, and depend upon water quality characteristics and treatment parameters.
2. Bromate formation occurs via combined indirect HO radical reaction and direct molecular ozone reaction pathways. The oxidation of bromide to bromine is controlled by direct reactions, while HO radicals contribute significantly to oxidizing bromine to bromate.
3. Ozone decomposition rates and solute concentrations regulate HO radical concentrations, and the rate of reactions by specific solutes for molecular ozone and HO radicals equates to a series of competitive kinetic expressions.

Table 1   Source Description and Specific Absorbance of Organic Matter Isolates

Identification name	Source description	Specific absorbance at 254 nm (l-cm/mgC)
Milli-Q	Laboratory clean water	0.0
TEAYS	Shallow aquifer serving as water supply for Urbana, IL	0.0457
SPW	Sacramento River Delta, with freshwater tributaries (San Joaquin and Sacramento Rivers), and saltwater intrusion via San Francisco Bay system	0.0430
SRHA	Black water river in Georgia	0.0917

EXPERIMENTAL METHODS AND PROCEDURES

The waters examined herein are model solutions consisting of organic matter isolated from NOM, bromide, phosphate, inorganic carbon, t-butanol, and/or hydrogen peroxide. All chemicals were reagent grade or better. Organic matter (humic and fulvic acids) was isolated and concentrated from Teays Aquifer, Illinois (TEAYS), and State Project Water, California (SPW) by XAD-8 resin adsorption (Rohm and Haas), following the procedures of Thurman and Malcolm.[34] Suwannee River humic acid (SRHA) was obtained from the International Humic Substances Society (Golden, CO). Descriptions of the sources from which the water was obtained and the specific absorbance at 254 nm of each isolate are presented in Table 1.

Phosphate buffers (0.001 M) were added to all solutions to minimize pH fluctuations to ± 0.1 units (Fisher Accumet pH meter Model 230A). Solutions were prepared using deionized (Millipore Water Purification System) laboratory clean water (Milli-Q) which has a dissolved organic carbon (DOC) concentration of approximately 1.67 μM (0.2 mgC/l).

Batch ozonation experiments were conducted in a continuously mixed reactor (magnetic stirrer) built from a glass graduated cylinder (500 ml or 1 l) adapted with a sample port at the base. Time series samples were collected from a sample port at the bottom of the reactor. An adjustable Teflon disc was suspended on the water surface within the reactor to minimize the loss of ozone to the atmosphere. A concentrated stock ozone solution (~ 0.8 mM at 4°C) was prepared daily by continuously bubbling ozone through Milli-Q water within a gas washing bottle.[4] Ozone was generated from pure oxygen by an Orec ozone generator (Model O3V5-O). The aqueous ozone concentration in the absence of organic matter was determined by UV absorbance (Shimadzu UV-160U spectrophotometer) at 254 nm, based on an extinction coefficient of 2950 $M^{-1}cm^{-1}$. This extinction coefficient is equivalent to 3100 $M^{-1}cm^{-1}$ at 258 nm; it correlated with ozone concentration measurements using the indigo method. The indigo method was used to measure ozone concentrations in the presence of organic matter.

Ozone was transferred from the concentrated stock solution to the reactor via a wide-mouth pipette (rinsed with stock solution to prevent ozone demand from the glassware). The pipette was placed beneath the water surface in the completely mixed reactor. The adjustable Teflon disc was quickly suspended on the water surface. Volumes of Milli-Q water and reagent and ozone stock solutions were precalculated to have a final volume of either 500 or 1000 ml. Ozone was stripped by sparging the sample for 45 s with inert argon gas; control experiments indicated that at least 90% of the ozone was stripped from Milli-Q water at pH 6.5 within 15 s.

A Dionex (DX300) ion chromatograph (IC) measured bromide and bromate using an Ionpac column (AS9-SC) and Dionex guard columns (AG9-3C and NGI). Alkaline eluent (2 mM Na$_2$CO$_3$/0.75 mM NaHCO$_3$) was employed for bromide analysis and provided a detection limit of 13 nM (10 μg/l). A 20 mM NaOH/40.3 mM H$_3$BO$_3$ eluent provided a lower detection limit for bromate of approximately 16 nM (2 μg/l).

A colorimetric analysis was used to determine the bromine concentration (HOBr/OBr$^-$) based on the DPD method (lower detection limit of 0.13 μM). Residual ozone can interfere with the DPD method; therefore, bromine was measured after stripping ozone.

These analyses facilitated a complete Br mass balance, accounting for all the major bromide-containing species. Waters containing NOM were also analyzed for total organic bromine (TOBr) after 24 h with a TOX analyzer (Dohrman DX-20) according to standard methods,[32] and typically accounted for less than 5% of the total bromide. Typical bromide recoveries ranged from 98 to 101% of the initial spiked bromide.

RESULTS AND DISCUSSION

The hypotheses set fourth will be addressed in three parts. First, bromate formation and ozone decomposition kinetics in the presence and absence of organic matter will be presented to illustrate the apparent linear relationship between the moles of ozone consumed and the moles of bromate formed. Second, a conceptual model is developed for understanding the relative contribution of two dominant bromate formation pathways. Finally, kinetic trends in bromate concentrations are discussed with respect to the influence of water quality characteristics.

CORRELATION BETWEEN OZONE CONSUMPTION AND BROMATE FORMATION

Bromate Formation Kinetics

Upon ozonation (all ozone applied at time zero) of Milli-Q water containing bromide, bromide is oxidized to measurable quantities of bromine and bromate

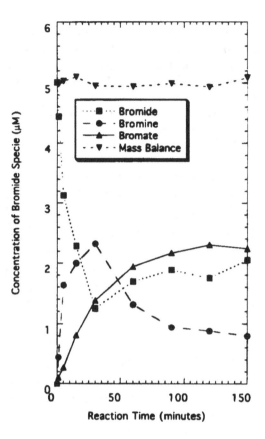

Figure 2 Mass balance on bromide containing species after ozonation of Milli-Q; pH = 7.5, $[Br^-]_o$ = 5 μM, $[O_3]_o$ = 62.5 μM.

(BrO_3^-) (Figure 2). Initially bromide rapidly oxidizes to bromine. Bromine is then observed to oxidize to bromate, or be converted back to bromide. The aggregate sum of the bromide, bromine, and bromate concentrations at any time closes the mass balance on the initial bromide concentration. Closure of the mass balance signifies that measurable amounts of other intermediate oxidation products (e.g., bromite) do not accumulate and implies they probably react quickly once formed.

The presence of TEAYS organic matter isolate significantly decreases bromine and bromate concentrations over time after ozonation, in comparison to Milli-Q (Figure 3). The mass balance is not complete without consideration of TOBr, reported as *ultimate TOBr*, and represents TOBr formation after 24 h. In comparison to a lower ozone dose in Milli-Q water (Figure 2), lower bromine and bromate concentrations form in the presence of TEAYS isolate matter (Figure 3).

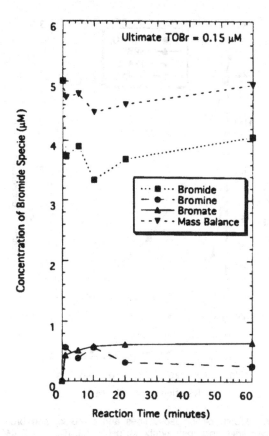

Figure 3   Mass balance on bromide containing species after ozonation of TEAYS isolate; pH = 7.5, $[Br^-]_o = 5 \ \mu M$, $[O_3]_o = 125 \ \mu M$, $[DOC]_o = 0.25 \ mM$.

Different sources of organic matter result in dissimilar bromate concentrations (Figure 4). These differences will be examined later in the context of composition of solutes to be oxidized by ozone and ozone decomposition by-products such as HO radicals. Ozone decomposition curves presented in Figure 4 show that ozone decomposes slowest in Milli-Q water, while the bromate formation curve illustrates that the Milli-Q water formed the most bromate. It should be noted that bromate ceases to form after the ozone is completely consumed (i.e., in the absence of an ozone residual). Overall, the three waters with organic matter form different ultimate bromate concentrations after all the ozone decomposes, and the kinetics of formation also differ.

Ozone Decomposition Kinetics

Understanding ozone decomposition in the presence of organic matter and other solutes will assist in developing a relationship between ozone consump-

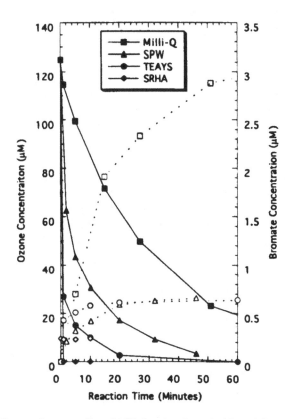

Figure 4   Ozone decomposition (solid lines and symbols) and bromate formation (dashed lines and open symbols); pH = 7.5, $[Br^-]_o$ = 5 μM, $[O_3]_o$ = 125 μM, and $[DOC]_o$ = 0.25 mM for SPW, TEAYS, and SRHA isolates.

tion and bromate formation. Ozone is rapidly consumed in waters containing organic matter compared to Milli-Q water (Figure 4). SRHA organic matter resulted in the most rapid consumption of ozone; in fact, it is completely consumed within the first minute of reaction. The specific absorbance of SRHA organic matter is greatest and this measurement may be an approximate indicator of ozone demand. UV absorbance at 254 nm is a measure of conjugated double bonds (e.g., aromatics and alkenes) and represent electron-rich molecular bonds susceptible to electrophilic attack by ozone.

Initial bromide concentration in Milli-Q water has an impact on the rate of ozone decomposition and subsequent bromate formation (Figure 5). Comparison of the curves can be facilitated by obtaining overall kinetic expression describing ozone decomposition. First-order ozone decomposition rate constants can be calculated by analytically or graphically solving the following equation(s):

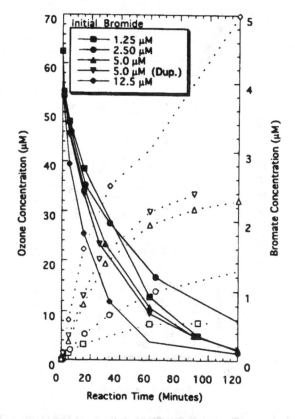

Figure 5 Effect of initial bromide concentration on ozone decomposition (solid lines and symbols) and bromate formation (dashed lines and open symbols); pH = 7.5, $[Br^-]_o$ = 5 μM, $[O_3]_o$ = 62.5 μM.

$$\frac{d[O_3]}{dt} = -k_{ov}[O_3] \tag{12}$$

or

$$\frac{[O_3]_t}{[O_3]_o} = \exp[-k_{ov}t] \tag{13}$$

where k_{ov} is the overall first-order ozone decomposition rate constant, $[O_3]_o$ is the initial ozone dose, and t is time. Based on the high r^2 values obtained, first-order kinetics accurately represent the ozone decomposition curves depicted in Figure 5 for Milli-Q water with varying initial bromide concentrations (Figure 6).

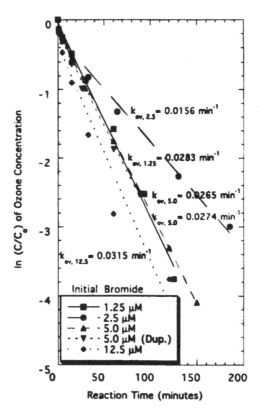

Figure 6   Effect of initial bromide concentration on k_{ov} in Milli-Q; pH = 7.5, $[O_3]_o$ = 62.5 μM.

A simple one-stage ozone decomposition expression, such as Equation 12, does not accurately describe the observed measurements in the presence of organic matter (Figure 4), since low r^2 values were obtained for k_{ov}. However, k_{ov} is representative of the general trend (fastest/slowest) in ozone decomposition. A two-stage ozone decomposition approach more closely simulates experimental data. The fast-stage ozone decomposition parameter (Δ_{01}) is the concentration of ozone rapidly consumed by the organic matter during the first minute of reaction and represents a pseudo zero-order approximation (based on two data points, 0 and 1 min) over a 1-min interval:

$$\Delta_{01} = [O_3]_o - [O_3]_{t=1} \tag{14}$$

The second-stage can be fit with a first-order kinetic expression:

Table 2   First- and Second-Stage Ozone Decomposition Kinetic Values for Solutions Containing Organic Matter

Source	Δ_{01} (mol)	k_2 (min^{-1})	r^2 value for k_2
SRHA	1.25×10^{-4}	Reaction too fast	Reaction too fast
TEAYS	9.79×10^{-5}	0.1227	0.991
SPW	3.13×10^{-5}	0.0634	0.994

Note: pH = 7.5, $[O_3]_o$ = 125 μM, $[Br^-]_o$ = 5 μM.

$$\frac{d[O_3]}{dt} = -k_2[O_3] \tag{15}$$

or

$$\frac{[O_3]_{t>1}}{[O_3]_{t=1}} = \exp[-k_2 t] \tag{16}$$

where k_2 is the rate constant calculated for ozone decomposition after the first minute (t > 1 min). The starting ozone concentration ($[O_3]_{t=1}$) for these calculations is the measured ozone concentration after 1 min. Based on the data presented in Figure 4, the value of Δ_{01} for SRHA exceeds that for TEAYS or SPW and illustrates that SRHA organic matter consumes ozone extremely fast (Table 2). A high r^2 value for k_2 signifies a good fit (Table 2). No k_2 value was calculated for SRHA since all the ozone was consumed during the first stage.

Bromate Yield

Observation of the exponential ozone decomposition and the apparently "inverse exponential" bromate formation curves depicted in Figure 5 suggest that there may be a relationship between the two. The existence of a simple *bromate yield*, or linear relationship between the total moles of bromate formed and the cumulative moles of ozone consumed, would imply an apparent stoichiometry. Such relationships could assist in evaluating pathways for or prediction of bromate formation.

Bromate yields correspond to slope between the moles of bromate formed and the moles of ozone consumed, determined by linear regression. An example calculation is presented in Table 3. The bromate yields for several experiments are summarized in Table 4, and demonstrate several important concepts. First, a high statistical relationship exists between the moles of bromate formed and the moles of ozone consumed, defined as *bromate yield*. The relationship implies an apparent stoichiometry between the two species, i.e., that bromate formation is a function of, or dependent upon, ozone consumption. Second,

Table 3   Example Calculation of Bromate Yield in Milli-Q Water

Reaction time (min)	Measured ozone (μM)	Cumulative ozone consumed (μM)	Measured bromate (μM)
0	62.5	0	0
1	54	8.5	0.10
5	46	16.5	0.27
15	35	27.5	0.81
30	23	39.5	1.38
60	11	51.5	1.94
90	5.0	57.5	2.17
120	2.3	60.2	2.30
150	1.0	61.2	2.30

Linear regression: $Y = mX + b$
m = slope or bromate yield
Y = moles bromate formed
X = moles ozone consumed
b = intercept
Results: $Y = 0.0406\ X - (2.2 \times 10^{-7})$
$r^2 = 0.99$
Bromate yield = 0.0406 mol BrO_3^- per mol O_3
Threshold ozone dose = $(2.2 \times 10^{-7}) \div (0.0406) = 5.42\ \mu M$

Note: pH = 7.5, $[O_3]_o$ = 62.5 μM, $[Br^-]_o$ = 5 μM.

Table 4   Bromate Yields (pH = 7.5)

Source	$[O_3]_o$ (μM)	$[Br^-]_o$ (μM)	Bromate yield (mol/mol)	r^2 for bromate yield	Intercept (b, nM)	Ozone threshold (μM)
Milli-Q	125	5.0	0.0279	0.98	65	−2.3
Milli-Q	62.5	1.25	0.0101	0.98	−25	2.5
Milli-Q	62.5	2.5	0.0253	0.97	−170	6.8
Milli-Q	62.5	5.0	0.0406	0.99	−230	5.4
Milli-Q[a]	62.5	5.0	0.0440	0.98	−240	5.5
Milli-Q	62.5	12.5	0.0701	0.86	−590	8.4
SPW	125	5.0	0.0056	0.94	−63	11.3
SPW[a]	125	5.0	0.0050	0.92	−63	12.5
TEAYS	125	5.0	0.0049	0.98	−11	2.3
SRHA[b]	125	5.0	0.0019	—[c]	—[c]	—[c]

Note: pH = 7.5.
[a] Indicates duplicate samples.
[b] Bromate yield for SRHA = $[BrO_3^-]_{t=1} - [BrO_3^-]_{t=0}$ since ozone was completely consumed during the first stage (t = 0 to 1 min).
[c] No statistical value is obtained from only two points.

the bromate yield in not uniform over ozone doses, bromide concentrations, or organic matter sources. This suggests that the apparent stoichiometry depends upon the specific reaction conditions and is at best a semi-empirical understanding or a *conditional* bromate yield, as opposed to a more *universal* bromate yield. Finally, a negative intercept (b) for bromate formation yield calculations commonly prevails, as depicted in Table 4, and can be used to infer a threshold ozone dose for bromate formation. For the example calculation presented in Table 3, solving the linear regression equation in Table 3 for zero bromate formation (Y) equates to 5.42 μM of ozone consumption before any bromate will form, defined as an *ozone threshold*. The ozone threshold is generally larger in waters containing organic matter (Table 4); TEAYS has a low ozone threshold due to greater than 80% of the ultimate bromate forming during the first minute, consequentially skewing the intercept (b) and ozone threshold calculations. In summary, a *conditional* bromate yield correlates with ozone consumption and implies that ozone oxidizes bromide species constantly throughout the reaction period, but depends upon water quality characteristics and treatment variables.

ROLE OF HO RADICAL IN BROMATE FORMATION

Estimating HO Radical Concentration

Hydroxyl (HO) radicals occur at much lower concentrations than molecular ozone residuals, but have extremely high rates of reaction compared to more selective molecular ozone reactions. Estimating HO radical concentrations can be useful in understanding the mechanisms of bromate formation.

Ozone decomposition by free radical reactions theoretically results in 0.67 mol of HO radicals per mole of ozone decomposed. For initiation of ozone decomposition by hydroxide, an HO radical yield can be represented by a set of free radical reactions:[22]

$$O_3 + OH^- \rightarrow HO_2 + O_2^- \tag{17}$$

$$HO_2 \rightleftarrows O_2^- + H^+ \tag{18}$$

$$2O_2^- + 2O_3 \rightarrow 2O_3^- + 2O_2 \tag{19}$$

$$2H^+ + 2O_3^- \rightarrow 2HO + 2O_2 \tag{20}$$

$$H_2O \rightleftarrows H^+ + HO^- \tag{21}$$

Cumulative addition (combination) of Equations 17 through 21 results in the following expression:

$$3O_3 + H_2O \rightarrow 2HO + 4O_2 \tag{22}$$

Equation 22 represents the HO radical yield (η) and corresponds to a stoichiometric ratio of HO to O_3 of 2/3. Hoigne and Bader[14] experimentally estimated

Table 5   Rates of Reactions between Solutes and
             HO Radicals

Solute	k_i $(M^{-1}s^{-1})$	Ref.
O_3	1.1×10^8	7
HPO_4^{-2}	1.5×10^5	7
$H_2PO_4^-$	1.5×10^5	7
Br^-	2×10^9	7
DOC	3.6×10^8	10

η to equal 0.5, and Peyton and Bell[25] suggest that η is greater than 0.5 based on experimental work.

During ozonation, the concentration of HO radicals is related to the rate of formation of HO radicals divided by the rate of HO radical consumption by aqueous solutes.[15] The rate of formation (k_f) of HO radicals can be represented as

$$k_f = \eta \, \frac{d[O_3]}{dt} \qquad (23)$$

The rate of HO radical consumption (k_d) by aqueous solutes (S_i) can be represented as

$$k_d = \Sigma \, k_i[S_i] \qquad (24)$$

where k_i is the rate of reaction between the solute and HO radicals. Thus, an expression for the HO radical concentration can be represented as

$$[HO] = \frac{k_f}{k_d} = \frac{\eta \, \dfrac{d[O_3]}{dt}}{\Sigma \, k_i[S_i]} \qquad (25)$$

Using a first-order expression for ozone decomposition and published k_{ov} values from pulse radiolysis work, HO radical concentrations can be predicted. Values of k_i for ozone, bromide, phosphate, and NOM are presented in Table 5. Simulations of HO radical concentrations and measured ozone concentrations in Milli-Q water and SPW organic matter are presented in Figure 7. In Milli-Q water, ozone decomposition was modeled with Equation 12 and k_{ov}, while ozone decomposition was modeled with a two-step function for SPW organic matter. Comparison of these two representative simulations with bromate formation graphs can be useful in gaining an understanding of oxidation pathways and bromate yields. The simulated trends presented in Figure 7 show that high HO radical concentrations are produced when ozone decom-

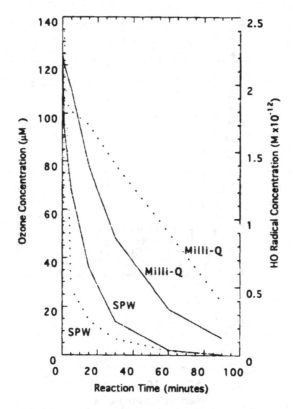

Figure 7   Simulated ozone (solid lines) and HO radical (dashed lines) concentrations in SPW isolate ($[DOC]_o$ = 0.25 mM) and Milli-Q; pH = 7.5, $[O_3]_o$ = 125 µM.

poses rapidly (SPW) in comparison to slower ozone decomposition Milli-Q water. After the initial rapid formation of HO radicals, the rate of formation decreases. Organic matter rapidly consumes HO radicals thus leading to lower HO radical concentrations in the presence of organic matter than in Milli-Q waters. Throughout the entire reaction the ratio of HO radicals to ozone residual (HO/O_3) is greater in Milli-Q than SPW water due to the presence of organic matter in the latter. The elevated HO/O_3 ratio in Milli-Q water may imply a more efficient overall oxidation of bromide to bromate, and corroborates with the higher bromate yields observed in the Milli-Q compared to SPW, TEAYS, or SRHA waters. However, once all the ozone is consumed HO radicals are no longer produced.

Because ozone is directly responsible for HO radical production, there is a relationship between the total moles of ozone consumed and the moles of HO radicals produced (η = 0.67 from Equation 22). Thus, with a linear yield between the moles of bromate formed and the moles of ozone consumed,

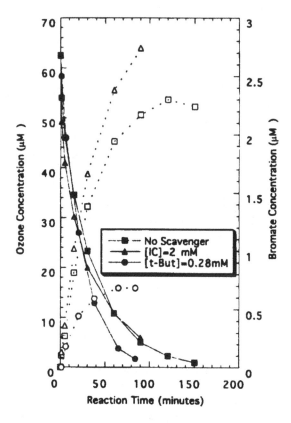

Figure 8 Effect of inorganic carbon (IC) and t-butanol addition on ozone decomposition (solid lines) and bromate formation (dashed lines and open symbols); pH = 7.5, $[Br^-]_o$ = 5 μM, $[O_3]_o$ = 125 μM.

there is a linear relationship between the moles of bromate formed and the moles of HO radicals produced. This conclusion suggests that an apparent stoichiometry exists between bromate and ozone, and bromate and HO radicals.

Delineating Direct and Indirect Oxidation Pathways

In the presence of HO radical scavengers a significant portion (>90%) of the HO radicals will react with the scavenger rather than with bromide species, thus aiding in delineating bromate formation pathways. Useful HO radical scavengers should have a low rate of reaction with molecular ozone, high rate of reaction with HO radicals and few reaction by-products; t-butanol is a short chain alcohol with these traits.[31] Carbonate and bicarbonate ions also efficiently scavenge HO radicals, but the subsequent carbonate or bicarbonate radicals may act as secondary oxidants for bromide species.[7] Scavenging HO radicals with either t-butanol or inorganic carbon (HCO_3^-/CO_3^{-2}) decreases the rate

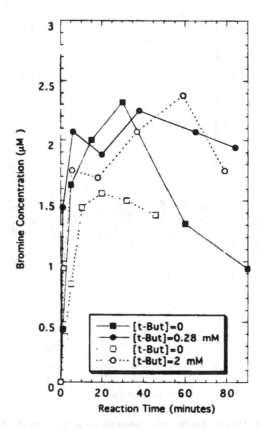

Figure 9 Effect of *t*-butanol addition (dashed lines) on bromine formation in Milli-Q (solid symbols) ($[O_3]_o$ = 62.5 μM) and SPW isolate (open symbols) ($[O_3]_o$ = 125 μM); pH = 7.5, $[Br^-]_o$ = 5 μM.

of ozone decomposition by inhibiting free radical reactions (Figure 8).[22,26] The addition of *t*-butanol results in a decrease in bromate formation by 66% while inorganic carbon addition results in an increase of 19%. The addition of *t*-butanol also results in a "lag" period prior to bromate formation. In these pure systems, inorganic carbon probably acts as a secondary oxidant, leading to higher bromate concentrations; *t*-butanol does not act as a secondary oxidant. These experiments suggest that, overall, the indirect HO radical reactions dominate over direct oxidation pathways for controlling bromate formation. Molecular ozone reactions are *important* in bromate formation via oxidation of some bromide species and formation of HO radicals, but HO radicals may actually *control* the majority of bromate formation.

The controlling reaction for bromate formation, involving HO radicals, may be elucidated by examining the influence of HO radical oxidation on bromine concentrations (Figure 9). Bromine rapidly forms upon ozone addition

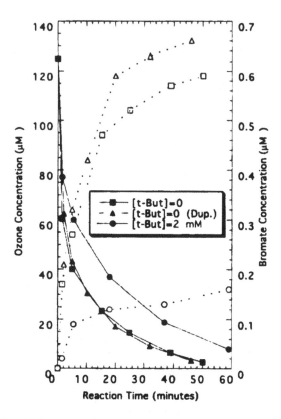

Figure 10 Effect of t-butanol addition on ozone decomposition (solid lines) and bromate formation (dashed lines and open symbols) in SPW isolate; pH = 7.5, $[Br^-]_o$ = 5 μM, $[O_3]_o$ = 125 μM, $[DOC]_o$ = 0.25 mM.

and then gradually decreases over time as bromate forms. In the presence of t-butanol in Milli-Q water, bromine still forms and higher bromine concentrations are observed than those in the absence of t-butanol. Bromine formation in the presence of t-butanol suggests that oxidation of bromide to bromine is not dependent upon HO radicals, and probably occurs via direct oxidation. Higher bromine concentrations in the presence of t-butanol suggest that oxidation of bromine is inhibited; therefore, this oxidation reaction probably depends on HO radicals. Oxidation of bromine probably forms bromine, but without bromite data conclusions cannot be made about complete oxidation of bromine to bromate. These simple experiments, along with many other experiments,[40] suggest that approximately 75% of the bromate forms through reactions controlled by indirect HO radical oxidation mechanisms.

In the presence of SPW organic matter, similar trends in the influence of indirect oxidation are observed. The addition of t-butanol increased the half-life of ozone and decreased bromate concentrations by 73% (Figure 10).

Table 6 Effect of Initial Bromide Concentration on k_{ov} and Ultimate Bromate Concentrations in Milli-Q Water (pH = 7.5, $[O_3]_o$ = 62.5 μM)

$[Br^-]_o$ (μM)	k_{ov} (min^{-1})	r^2 for k_{ov}	Ultimate bromate (μM)
1.25	0.0283	0.992	0.56
5.0	0.0265	0.998	2.38
5.0	0.0274	0.989	2.46
12.5	0.0315	0.942	4.99

Note: pH = 7.5, $[O_3]_o$ = 62.5 μM.

Bromine forms rapidly in the absence of t-butanol (Figure 9). Once formed in the presence of t-butanol, bromine concentrations did not decrease and had resulted in higher bromine concentrations when the ozone was completely consumed (Figure 9). Even in the presence of organic matter the indirect oxidation mechanism controls the majority of the bromate formation.

TRENDS IN BROMATE FORMATION

Effect of Bromide Concentration

Bromide reactions with molecular ozone and HO radicals influences ozone decomposition kinetics and bromate formation. In Milli-Q water, increasing initial bromide concentrations increases the rate of ozone decomposition and increases ultimate bromate concentrations (Figure 5 and Table 6). In the absence of organic matter, bromide reactions efficiently compete against hydroxide reactions for molecular ozone, thus leading to an increase in ozone consumption upon increasing initial bromide concentrations. In addition, as shown above, bromide species also kinetically compete for HO radicals and appear to act as an inhibitor. A bromate formation mechanism involving HO radicals that results in increasing ozone decomposition rates can also be made: (1) bromide consumes ozone during oxidation to bromine, (2) bromine scavenges HO radicals during oxidation to bromite, and then (3) bromite consumes ozone to form bromate. Either one of the above kinetic arguments may be valid for understanding the increase in ozone decomposition upon increasing initial bromide concentrations.

In the presence of TEAYS organic matter, increasing initial bromide concentrations decreases the rate of ozone consumption and increases the ultimate bromate concentrations (Table 7). The quantity of ozone consumed within the first minute (Δ_{01}) decreases, second-stage ozone decomposition rate constant (k_2) decreases, and an increase in ultimate bromate is observed. Rapid reaction rates between ozone and organic matter[18] lead to organic matter kinetically outcompeting bromide for ozone and controlling ozone consumption. Bromide species are oxidized to bromine in the presence of organic matter, although

Table 7   Effect of Initial Bromide Concentration on Δ_{o1}, k_2, and Ultimate Bromate Concentrations in TEAYS Isolate

[Br⁻]$_o$ (µM)	Δ_{o1} (µM)	k_2 (min⁻¹)	r^2 for k_2	Ultimate bromate (µM)
1.25	97.1	—[a]	—[a]	0.11
2.5	94.4	—[a]	—[a]	0.38
5.0	97.9	0.1227	0.991	0.64
12.5	66.7	0.1265	0.999	1.67

Note: pH = 7.5, $[O_3]_o$ = 125 µM, $[DOC]_o$ = 0.25 mM.

[a] Reaction occurred too rapidly and too few data points, if any, could be collected to calculate k_2 or r^2.

to a lesser extent than in Milli-Q water; bromine then appears to be capable of kinetically competing against organic matter for HO radicals since bromine oxidation progresses to form bromate. The overall stabilization of ozone decomposition with increasing initial bromide concentration is probably due to HO radical scavenging by bromine or other bromide species. This signifies that bromide species can and do react with HO radicals in water during ozonation. The effective kinetic competition for HO radicals by the bromide species reinforces the notion that HO radicals can play a dominant role in bromate formation.

Effects of pH

Increasing the pH from 4.5 to 8.5 in Milli-Q water results in more rapid ozone decomposition, more rapid bromate formation, and higher ultimate bromate concentrations (Figure 11). These trends represent the cumulative response to at least two shifts in important reactant concentrations. First, ozone oxidizes hypobromite ion significantly faster than hypobromous acid, and increasing the pH increases the hypobromite ion concentrations (from Equation 5 the pK_a = 8.7). Second, the molar quantity of ozone consumed at each pH is equal (ozone dose = 62.5 µM) and thus the molar quantity of HO radicals produced is also equal (η = 0.67). However, the concentration of HO radicals increases with pH since ozone decomposition occurs more rapidly (Equation 25). Overall increases in pH increase hypobromite ion and HO radical concentrations, leading to the observed increases in the extent and rate of bromate formation.

Effects of Ammonia

Ammonia addition (1 mol NH_3 per mole Br⁻) decreases the amount of bromate formed in either the absence or presence of organic matter (not shown). Theoretically ammonia can react with bromine rapidly to form bromamines such as monobromamine which can reform bromide upon ozonation.

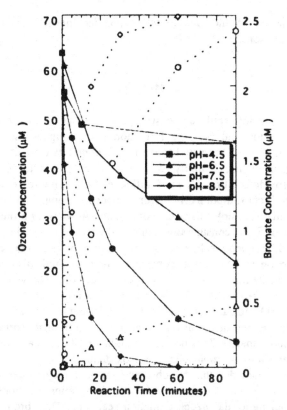

Figure 11 Effect of pH on ozone decomposition (solid lines) and bromate formation in Milli-Q (dashed lines and open symbols); pH = 7.5, $[Br^-]_o$ = 5 μM, $[O_3]_o$ = 62.5 μM.

Thus, ammonia can compete against HO radicals or molecular ozone for bromine and thus lead to a reduction in bromate. Another mechanism may also be considered to explain a decrease in bromate formation in the presence of ammonia. Ammonia, like bromine and t-butanol, can scavenge HO radicals and thus kinetically compete against bromine.

Effects of Hydrogen Peroxide Addition

Hydrogen peroxide (H_2O_2) addition (0.14, 0.50, and 0.71 mol H_2O_2 per mole O_3) increases bromate formation at pH 6.5 and decreases it at pH 8.5, both in the absence and presence of organic matter (not shown). H_2O_2 addition results in increases in HO radical concentrations through reaction of ozone with HO_2^-, and is highly dependent upon the equilibrium speciation of H_2O_2 and HO_2^- (pK_a = 11.6).[22] At pH 8.5 the HO radical concentration is much greater than at pH 6.5 and subsequently ozone decomposes very rapidly via free radical reactions.

SUMMARY

The results of this study demonstrate that a linear relationship between ozone consumption and bromate formation exists, suggesting that ozone decomposition "drives" bromate formation. Water quality characteristics (pH, bromide, organic matter, alkalinity, etc.) and treatment parameters (ozone dose, hydrogen peroxide addition, etc.) will influence the driving force, or bromate yield. The ozone consumption driving force concept encompasses both indirect HO radical and direct molecular ozone oxidation pathways for bromate formation. The rate of ozone consumption controls the rate of HO radical formation ($\eta = 0.67$), while aqueous solute concentrations control the HO radical concentrations. Increase in HO radical concentrations (e.g., high pH) can lead to increasing bromate formation. Bromide species have been shown to react with HO radicals during ozonation. Oxidation of bromine appears to be significantly impacted by HO radicals. Through the use of t-butanol as an HO radical scavenger, delineation of the overall contribution to bromate formation suggests that approximately 75% occurs via indirect HO radical oxidation and the remainder via direct molecular ozone oxidation.

This work concludes that hypobromite ion is a major intermediate for bromate formation, and that overall HO radical oxidation reactions are more significant that molecular ozone oxidation reactions. Thus bromate control strategies should involve decreasing hypobromite ion and/or HO radical concentrations or reactivities.

ACKNOWLEDGMENTS

This project is primarily funded by the American Water Works Association Research Foundation. We greatly appreciate the efforts of Jeff Oxenford (AWWARF Project Manager) and the rest of the AWWARF Project Advisory Committee (Philip Singer, James Symons, Werner Haag, and Robert Powell).

REFERENCES

1. Amy, G., Siddiqui, M., Ozekin, K., and Westerhoff, P., Threshold levels for bromate formation in drinking water, Proceedings of the International Water Supply Association International Conference: Bromate and Water Treatment, Paris, November 22–24, 1993, 169.
2. Amy, G., Siddiqui, M., Zhai, W., DeBroux, J., and Odem, W., Bromide occur-

rence: national bromide survey, AWWA Annual Conference: Bromide, Bromate, Brominated Disinfection By-Products ... Oh Brother!, San Antonio, TX, June 1993.

3. Amy, G., Siddiqui, M., Zhai, W., DeBroux, J., and Odem, W., Survey on Bromide in Drinking Water and Impacts on DBP Formation, American Water Works Research Foundation Report, Denver, 1994.

4. Bader, H. and Hoigne, J., Determination of ozone in water by the indigo method, *Water. Res.*, 15, 449, 1981.

5. Buxton, G. V. and Dainton, F. S., The radiolysis of aqueous solutions of oxybromine compounds; the spectra and reactions of BrO and BrO_2, *Proc. R. Soc.*, 304, 427, 1968.

6. Buxton, G. V. and Dainton, F. S., Radical and molecular yields in γ-radiolysis of water. V. The sodium hypobromite system, *Proc. R. Soc.*, 304, 441, 1968.

7. Buxton, G. V., Greenstock, C. L., Helman, W. P., and Ross, A. B., Critical review of rate constants for reactions of hydrated electrons, hydrogen atoms and hydroxyl radicals (OH^o/O^{o-}) in aqueous solution, *J. Phys. Ref. Data*, 17(2), 513, 1988.

8. Garland, J. A., Elzerman, A. W., and Penkett, S. A., The mechanism for dry deposition of ozone to seawater surfaces, *J. Geophys. Res.*, 85(C12), 7488, 1980.

9. Glaze, W. H., Weinberg, H. S., and Cavanagh, J. E., Evaluating the formation of brominated DBPs during ozonation, *J. Am. Water Works Assoc.*, 85(1), 96, 1993.

10. Haag, W. R. and David-Yao, C. C., Ozonation of U.S. drinking water: HO concentration and oxidation-competition values, International Ozone Association Conference Proceedings, San Francisco, September 1994, 119.

11. Haag, W. R. and Hoigne, J., Ozonation of bromide-containing waters: kinetics of formation of hypobromous acid and bromate. *Environ. Sci. Technol.*, 17(5), 261, 1983.

12. Hall, E. S. and Packham, R. F., Coagulation of organic color with hydrolyzing coagulants, *J. Am. Water Works Assoc.*, 57(9), 1149, 1965.

13. Haruta, K. and Takeyama, T., Kinetics of oxidation of aqueous bromide ion by ozone, *J. Phys. Chem.*, 85, 2383, 1981.

14. Hoigne, J. and Bader, H., The role of hydroxyl radical reactions in ozonation processes in aqueous solutions, *Water Res.*, 10, 377, 1976.

15. Hoigne, J. and Bader, H., Ozonation of water: oxidation competition values of different types of waters used in Switzerland, *Ozone Sci. Eng.*, 7, 357, 1979.

16. Hoigne, J. and Bader, H., Rate constants of reactions of ozone with organic and inorganic compounds in water. I. Non-dissociating organic compounds, *Water Res.*, 17, 173, 1985.

17. Hoigne, J. and Bader, H., Rate constants of reactions of ozone with organic and inorganic compounds in water. II. Dissociating organic compounds, *Water Res.*, 17, 185, 1985.

18. Hoigne, J., Bader, H., Haag, W. R., and Staehelin, J., Rate constants of reactions of ozone with organic and inorganic compounds in water. III. Inorganic compounds and radicals, *Water Res.*, 19(8), 993, 1985.

19. Krasner, W., Glaze, W., Weinberg, W., and Daniel, P., Bromate occurrence and control: pilot- and full-scale studies, AWWA Annual Conference: Bromide,

Bromate, Brominated Disinfection By-Products . . . Oh Brother! San Antonio, TX, June 1993.

20. Krasner, W., Glaze, W., Weinberg, W., Daniel, P., and Najm, I. N., Formation and control of bromate during ozonation of waters containing bromide, *J. Am. Water Works Assoc.*, 85(1), 73, 1993.

21. Krasner, S. W., Gramith, J. T., Coffey, B. M., and Yates, R. S., Impact of water quality and operational parameters on the formation and control of bromate during ozonation, Proceedings of the International Water Supply Association International Conference: Bromate and Water Treatment, Paris, November 22–24, 1993, 157.

22. Langlais, B., Reckhow, D. A., and Brink, D. R., *Ozone in Water Treatment: Applications and Engineering*, Lewis Publishers, Chelsea, MI, 1991.

23. Macalady, D. L., Carpenter, J. H., and Moore, C. A., Sunlight-induced bromate formation in chlorinated seawater, *Science*, 195, 1335, 1977.

24. Masschelein, W. J., On overloading in potassium as a factor of experimental carcinogenesis in animal tests with bromate, Proceedings of the International Water Supply Association International Conference: Bromate and Water Treatment, Paris, November 22–24, 1993, 13.

25. Peyton, G. R. and Bell, O. J., Practical Modeling of Contaminant Destruction by Hydroxyl Radicals Generated from Ozone/Peroxide/UV Combinations, International Ozone Association Conference Proceedings, San Francisco, September 1994, 1.

26. Reckhow, D. A., Legube, B., and Singer, P. C., The ozonation of organic halide precursors: effect of bicarbonate, *Water Res.*, 20(8), 987, 1986.

27. Richardson, L. B., Burton, D. T., Helz, G. R., and Rhoderick, J. C., Residual oxidant decay and bromate formation in chlorinated and ozonated sea-water, *Water Res.*, 15, 1067, 1981.

28. Rook, J. J., Formation of haloforms during chlorination of natural waters, *J. Water Treat. Exam.*, 23, 234, 1974.

29. Rook, J. J., Haloforms in drinking water, *J. Am. Water Works Assoc.*, 23, 168, 1976.

30. Siddiqui, M. S. and Amy, G. L., Factors affecting DBP formation during ozone-bromide reactions, *J. Am. Water Works Assoc.*, 85(1), 63, 1993.

31. Staehelin, J. and Hoigne, J., Decomposition of ozone in water in the presence of organic solutes acting as promoters and inhibitors of radical chain reactions, *Environ. Sci. Technol.*, 19(12), 1206, 1985.

32. *Standard Methods for the Examination of Water And Wastewater*, APHA, AWWA, and WPCF, Washington, DC, 1989.

33. Taube, H., Reactions in solutions containing O_3, H_2O_2, H^+, and Br^-. The specific rate of reaction $O_3 + Br^- \rightarrow$, *J. Am. Chem. Soc.*, 64, 2468, 1942.

34. Thurman, E. M. and Malcolm, R. L., Preparative isolation of aquatic humic substances, *Environ. Sci. Technol.*, 15(4), 463, 1981.

35. VanBenschoten, J. E. and Edzwald, J. K., Chemical aspects of coagulation using aluminum salts. I. Hydrolytic reactions of alum and polyaluminum chloride, *Water Res.*, 24(12), 1519, 1990.

36. VanBenschoten, J. E. and Edzwald, J. K., Chemical aspects of coagulation using aluminum salts. II. Coagulation of fulvic acid using alum and polyaluminum chloride, *Water Res.*, 24(12), 1527, 1990.

37. Von Gunten, U. and Hoigne, J., Bromate formation during ozonation of bromide-containing waters: a comparison of experiments with computer modelling, *Env. Sci. Tech.*, 28, 1234, 1994.

38. Von Gunten, U. and Hoigne, J., Bromate Formation During Ozonation of Bromide-Containing Waters, Proceedings of the International Water Supply Association International Conference: Bromate and Water Treatment, Paris, November 22–24, 1993, 51.

39. Weinberg, H. S., Glaze, W. H., Krasner, S. W., and Sclimenti, M. J., Formation and removal of aldehydes in plants that use ozonation, *J. Am. Water Works Assoc.*, 85(5), 72, 1993.

40. Westerhoff, P., Ozekin, K., Siddiqui, M., and Amy, G., Kinetic Modeling of Bromate Formation in Ozonated Waters: Molecular Ozone versus HO Radical Pathways, American Water Works Association National Conference Proceedings, New York, June 1994.

41. Yates, R. S. and Stenstrom, M. K., Bromate production in ozone contactors, AWWA Annual Conference Proceedings, San Antonio, TX, June 1993.

CHAPTER 12

Hydrolysis and Dehalogenation of Trihaloacetaldehydes

Yuefeng Xie and David A. Reckhow

ABSTRACT

The aqueous stability of three brominated trihaloacetaldehydes (THAs), bromodichloroacetaldehyde, chlorodibromoacetaldehyde, and tribromoacetaldehyde, was studied. All three brominated THAs were found to undergo rapid hydrolysis at high pHs forming trihalomethanes (THMs). The hydrolysis rate can be expressed as $-d[THA]/dt = k_{OH}[THA][OH^-]^{0.5}$. A rapid dehalogenation of these THAs was also found in the presence of sulfite over a range of pHs. The dehalogenation rate can be expressed as $-[THA]/dt = k_1[THA][HSO_3^-] + k_2[THA][SO_3^{2-}]$. These results indicate that THAs are unstable in alkaline drinking waters. The authors also recommend that high pH and sulfite quenching should be avoided during THA analysis.

INTRODUCTION

Since trichloroacetaldehyde (chloral hydrate) was first reported in chlorinated water, several studies have been conducted to investigate the occurrence and toxicity of this chlorination by-product.[1-4] Owing to its carcinogenicity and widespread occurrence in chlorinated drinking water, chloral hydrate has been included in the U.S. Environmental Protection Agency (USEPA) information collection rule (ICR) for possible regulation in the Stage II disinfectants and disinfection by-products (D-DBP) rule.[5] Like chloroform, chloroacetic acids, and cyanogen chloride, chloral hydrate has brominated analogs that have been reported in waters high in inorganic bromide.[6-8] The brominated

1-56670-136-8/96/$0.00+$.50

THMs are thought to be more toxic than chloroform, and the brominated THAs may pose human health concerns that equal or exceed those of chloral hydrate.

The chemical stability of chlorination by-products is of interest because (1) water samples need to be properly preserved for chemical analysis; (2) knowledge of degradation pathways may lead to new processes for DBP control; (3) DBP occurrence models must incorporate degradation rates; and (4) there may be a health concern over secondary degradation products. Chloral hydrate is known to undergo slow hydrolysis at neutral and high pHs.[2,3] It has been found to be relatively stable in the presence of another nucleophile; sulfite.[9] Very little is known about the chemical stability of chloral hydrate's brominated analogs. By analogy to other disinfection by-products[9,10] they may be more susceptible to decomposition than chloral hydrate.

The objectives of the present research are (1) to estimate the hydrolysis rate of brominated THAs at different pHs, (2) to study the degradation of brominated THAs in the presence of sulfite over a range of pHs, and (3) to evaluate the stability of THAs.

EXPERIMENTAL

SAMPLE PREPARATION

Two mixed bromochloro-THAs, bromodichloroacetaldehyde (BDCA) and chlorodibromoacetaldehyde (CDBA), were synthesized in the authors' laboratory. The method used was a modification of a published synthesis method for two mixed bromochloro-haloacetic acids.[11,12] All-glass repeated distillations were used to purify the BDCA and CDBA. Chloral hydrate* and tribromoacetaldehyde** (TBA) were obtained from commercial sources.

Aqueous solutions were prepared for each THA from methanolic stocks (or aqueous stock for chloral hydrate). These were buffered with phosphate (25 mM) at a pH from 3.4 to 10.7 in Super-Q water.*** All experiments were conducted with initial THA concentrations of 50 μg/l, a sulfite dose of 25 μM, and reaction times of up to 40 min at 20°C in the absence of light.

THAs ANALYSIS

At varying reaction times, 20 ml of sample was extracted with 4 ml of methyl *tert*-butyl ether (MTBE) (2-min agitation). The extracts were transferred to autosampler vials and submitted for gas chromatography (GC) analy-

* Columbia Organic Chemical Co., Inc.
** Aldrich Chemical Company, Inc.
*** Millipore Corporation.

sis. The THA analysis was carried out on an HP 5890A GC equipped with a 30 m × 0.32 mm, 0.25-μm film thickness SPB-5 capillary column. Nitrogen was used as a carrier gas at a flow rate of 2 ml/min. A split/splitless injector was operated at 200°C in splitless mode. Following an initial hold at 30°C for 2 min, the column oven temperature was increased to 40°C at a rate of 0.5°C/min. Then the temperature was ramped to 200°C at a rate of 25°C/min, and held for 2 min. An electron capture detector was employed in this analysis at a temperature of 300°C. All THA concentrations were expressed in units of micrograms per liter as unhydrated compounds.

RESULTS AND DISCUSSION

HYDROLYSIS OF BROMINATED THAs

Hydrolysis degradation of brominated THAs was found to occur over a range of pHs as shown in Figure 1. Application of gas chromatography-mass spectrometry (GC-MS) has shown that the principal halogenated by-products are the corresponding THMs. The straight lines in these semilogarithmic plots suggest the degradation of THAs was a pseudo-first-order reaction, as expressed in Equation 1.

$$-d[\text{THA}]/dt = k_h[\text{THA}] \tag{1}$$

The slope of each line represents the pseudo-first-order-reaction rate constant, k_h. These pH-dependent rate constants are shown in Table 1. The hydrolysis of chloral hydrate was also examined. At pH 10.0, the observed k_h for chloral hydrate was imperceptible (about $1 \pm 2 \; 10^{-5}\text{s}^{-1}$) over the timescale used in this study. Others have studied the hydrolysis of chloral hydrate and found it to be similarly slow.[2,3]

As shown in Table 1, the hydrolysis rate constants increase with increasing pH. By plotting the pseudo-first-order rate constants vs. the concentration of OH^-, one finds that all three THAs show a half-order dependence on OH^-. Figure 2 shows the pseudo-first-order rate constants along with a best fit line, imposing a slope of 0.5. The intercept of each line is then the pH-independent rate constant, k_{OH}.

$$-d[\text{THA}]/dt = k_{OH}[\text{OH}]^{0.5}[\text{THA}] \tag{2}$$

The best-fit values for k_{OH} are 0.0055 $M^{-0.5}\text{s}^{-1}$ for BDCA, 0.024 $M^{-0.5}\text{s}^{-1}$ for CDBA, and 0.12 $M^{-0.5}\text{s}^{-1}$ for TBA. These results show that the hydrolysis rate of THA increases with increasing bromine substitution.

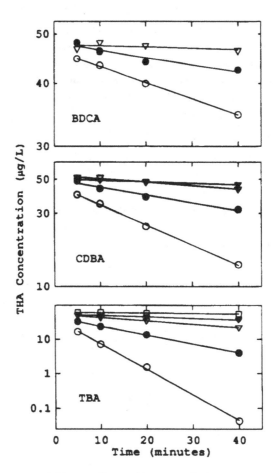

Figure 1   Hydrolysis of THAs at different pHs and 20°C. ○: pH 10.7, ●: pH 10.0, ▽: pH 9.0, ▼: pH 8.2, □: pH 7.0.

Table 1   Observed Pseudo-First-Order Reaction Rate Constants for the Loss of Trihaloacetaldehydes in the Presence and Absence of Sulfite at 20°C ($10^{-5}s^{-1}$)

	BDCA		CDBA		TBA	
pH	k_s	k_h	k_s	k_h	k_s	k_h
3.4					5 ± 3	
4.6			2.9 ± 0.5		8 ± 1	
5.8	12 ± 1		59 ± 1		205 ± 2	
7.0	54 ± 1		240 ± 20			
8.2	85 ± 1			3.3 ± 0.3		16.6 ± 0.7
9.0		0.9 ± 1.0		8.8 ± 0.9		37.6 ± 0.8
10.0		6 ± 1		20 ± 2		100 ± 1
10.7		12.4 ± 0.3		50 ± 1		283 ± 6

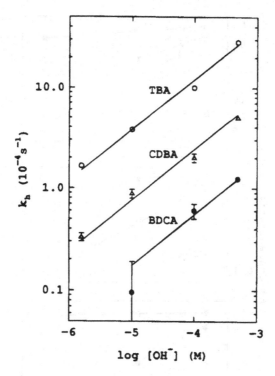

Figure 2 Relationship between k_h and [OH⁻] for the THAs.

THA DEHALOGENATION IN THE PRESENCE OF SULFITE

Rapid degradation of THAs was observed in the presence of sulfite. Semi-log degradation plots of the THAs in the presence of sulfite at different pHs are shown in Figure 3. The straight lines indicate that the loss of THA is through one or more pseudo-first-order reactions and the slope of the line is the observed aggregate pseudo-first-order rate constant; k_{obs}.

$$-d[THA]/dt = k_{obs}[THA] \qquad (3)$$

In this case, the observed rate constant, k_{obs}, is equal to the sum of hydrolysis rate, $k_{OH}[OH^-]^{-0.5}$, and the rate for the reaction with sulfites, k_s.

$$k_{obs} = k_{OH}[OH^-]^{-0.5} + k_s \qquad (4)$$

However, since $k_{OH}[OH^-]^{-0.5}$ is insignificant at the pHs used in the sulfite experiments, k_{obs} is simply taken to be k_s. For chloral hydrate, no sulfite reduction was observed in agreement with Croue and Reckhow.[9]

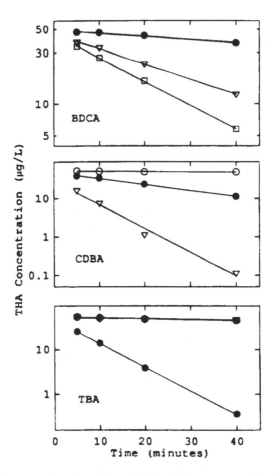

Figure 3 Dehalogenation degradation of THAs in the presence of sulfite at different pHs and 20°C. □: pH 8.2, ▽: pH 7.0, •: pH 5.8, ○: pH 4.6, ▼: pH 3.4.

The rate, k_s, represents the sum of the reactions with sulfite and bisulfite. Presuming that these reactions are first order in sulfite species concentration,[9] one obtains Equation 5. Combining this with the pH equilibria for sulfite species, one obtains a useful linear expression (Reference 9 and Equation 6).

$$k_s = k_1[HSO_3^-] + k_2[SO_3^{-2}] \qquad (5)$$

$$k_s(K_a + [H^+])/[S(IV)] = k_1[H^+] + k_2K_a \qquad (6)$$

The second-order rate constants, k_1 and k_2, can be obtained by plotting $k_s(K_a + [H^+])/[S(IV)]$ vs. $[H^+]$ (Figure 4). These are summarized in Table 2. The equilibrium constant, K_a, was adjusted for ionic strength using the extended Debye-Hückel equation. The values for k_1 and k_2 could not be determined for TBA, because only one precise value for k_s was available.

Table 2   Concentration Independent Rate Constants for the
          Reaction of Bisulfite, Sulfite, and Hydroxide
          with Trihaloacetaldehydes (20°C)

THAs	k_1 $(M^{-1}s^{-1})$	k_2 $(M^{-1}s^{-1})$	k_{OH} $(M^{-0.5}s^{-1})$
BDCA	2.5 ± 0.3	41 ± 2	0.0055
CDBA	14 ± 1[a]	200 ± 20[a]	0.024
TBA	—[b]	—[b]	0.12

[a] Based on pH 5.8 and 7.0 data only.
[b] Could not be determined.

Using GC-MS, the by-products of THA reduction have been found to be dibromoacetaldehyde and monobromoacetaldehyde for TBA, bromochloroacetaldehyde for CDBA, and dichloroacetaldehyde for BDCA. This indicates that three brominated THAs undergo a substitution of hydrogen for bromine in the presence of sulfite. As reported by Croue and Reckhow,[9] this sulfite reduction reaction is probably a one-step nucleophilic substitution reaction (S_N2). Since bromine is a better leaving group than chlorine, one would expect preferential loss of bromine. Reaction rates should also increase as the number of bromine atoms increases.

EVALUATION OF THE STABILITY OF THAs

As mentioned previously, an objective of this study was to evaluate the stability of THAs in drinking water. Using the estimated hydrolysis rate constants, k_{OH}, in Table 2, one may evaluate the stability of THAs at different

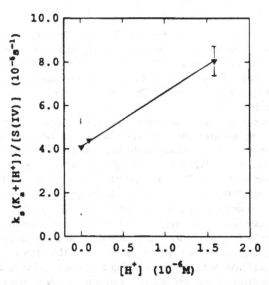

Figure 4   Decomposition of BDCA in the presence of 25 μM of S(IV) determination of k_1 and k_2 (20°C).

Table 3 Calculated Half-Lives of THAs at Different pH with or without Sulfite (20°)

THAs	pH 9.0 without sulfite (h)	pH 7.0 with 25 μM sulfite (min)
BDCA	11	20
CDBA	2.5	4
TBA	0.5	1[a]

Note: I = 0.01.

[a] Estimated assuming the ratio of k_1/k_2 is roughly equal for all three THAs.

pHs. For example, the half-lives of the THAs at pH 9.0 range from about 30 min to 11 h (Table 3). This indicates that, in highly alkaline waters, little or no brominated THAs should survive in large distribution systems. Instead they will appear at the consumer's tap in the form of THMs. For THA analysis, high or neutral pHs should be avoided and samples should be acidified at the time of collection. At pH 7 the half-lives of brominated THAs in the presence of 25 μM of sulfite are on the order of minutes (Table 3). This suggests that sulfite reduction may be a feasible means of controlling brominated THAs. It will not, however, be very effective for chloral hydrate, nor will it assure complete removal of THA degradation products. Naturally, sulfite quenching should be avoided when analyzing samples for THAs.

CONCLUSIONS

1. Due to base-catalyzed hydrolysis, all THAs are unstable at high pHs. The hydrolysis rate can be expressed as $-d[THA]/dt = k_{OH}[THA][OH^-]^{0.5}$, where k_{OH} is estimated as 0.0055 $M^{-0.5}s^{-1}$ for BDCA, 0.024 $M^{-0.5}s^{-1}$ for CDBA, and 0.12 $M^{-0.5}s^{-1}$ for TBA. The hydrolysis rate increased with increasing bromine substitution. The principal halogenated by-products were the THMs.
2. The three brominated THAs were found to be unstable in the presence of sulfite. The dehalogenation rate was a function of THA concentration and pH at constant total sulfite. The dehalogenation rate can be expressed as $-d[THA]/dt = k_1[THA][HSO_3^-] + k_2[THA][SO_3^{2-}]$, where k_1 and k_2 were estimated as 2.5 ± 0.3 $M^{-1}s^{-1}$ and 41 ± 2 $M^{-1}s^{-1}$ for BDCA, 14 ± 1 $M^{-1}s^{-1}$ and 200 ± 20 $M^{-1}s^{-1}$ for CDBA. The dehalogenation rate increased with increasing the bromine substitution. The principal by-products were haloacetaldehydes containing one or two less bromine atoms.
3. The results from the present study show that sulfite dechlorination may be an effective process for controlling brominated THAs. Due to reaction with sulfite and hydroxide, sulfite quenching and/or high pH should be avoided in THA analysis. Alkaline drinking waters are likely to lose most of their brominated THAs during storage and distribution.

ACKNOWLEDGMENTS

The authors express their appreciation to the U.S. National Science Foundation and Dr. Edward H. Bryan for financial support of this research under grant #BCS-8958392.

REFERENCES

1. Uden, P. C. and Miller, J. W., *J. Am. Water Works Assoc.*, 75(10), 524, 1983.
2. Krasner, S. W., McGuire, M. J., Jacangelo, J. G., Patania, N. L., Reagan, K. M., and Aieta, E. M., *J. Am. Water Works Assoc.*, 81(8), 41, 1989.
3. Stevens, A. A., Moore, L. A., and Miltner, R. J., *J. Am. Water Works Assoc.*, 81(8), 54, 1989.
4. Daniel, F. B., Deangelo, A. B., Stober, J. A., Olson, G. R., and Page, N. P., *Fundam. Appl. Toxicol.*, 19, 159, 1992.
5. Pontius, F. W., *J. Am. Water Works Assoc.*, 85(11), 22, 1993.
6. Italia, M. P. and Uden, P. C., *J. Chromatogr.*, 438, 35, 1988.
7. Xie, Y. and Reckhow, D. A., *Analyst*, 118, 71, 1993.
8. Xie, Y. and Reckhow, D. A., *Proc. AWWA Water Qual. Techn. Conf.*, 1993, 1499.
9. Croue, J.-P. and Reckhow, D. A., *Environ. Sci. Technol.*, 23, 1412, 1989.
10. Xie, Y. and Reckhow, D. A., *Proc. AWWA Water Qual. Techn. Conf.*, 1993, 1761.
11. Zimmer, H., Amer, A., and Rahi, M., *Anal. Chem.*, 45, 735, 1990.
12. Xie, Y., Rajan, R. V., and Reckhow, D. A., *Org. Mass Spectrometry*, 27, 807, 1992.

CHAPTER 13

The Effect of Granular Activated Carbon Surface Chemistry on Bromate Reduction

Jennifer Miller, Vernon L. Snoeyink, and Shannon Harrell

INTRODUCTION

Granular activated carbon (GAC) is commonly used in drinking water treatment for the removal of organic compounds. However, it is also well known that GAC reacts with inorganic species in solution, such as free chlorine, chloramines, and chlorite.[1-4] GAC also appears to react with a related species, bromate, but this reaction has not been well studied.

Bromate is currently of much interest to the drinking water treatment industry, since it is scheduled to be regulated under the proposed disinfectant/ disinfection by-product rule.[5] It would be desirable to use GAC for bromate reduction, since GAC is already in use at many drinking water treatment plants. The reaction of GAC and bromate is influenced by modifying the carbon surface chemistry. The nature of the chemical changes on the carbon surface and the effect of these changes on bromate reduction is the focus of this paper.

METHODS AND MATERIALS

CARBON BATCH AND COLUMN TESTS

Batch tests were performed in a 4-liter Pyrex® glass container that had been leached in distilled water. The solution was stirred using a magnetic stirrer, turning at a speed sufficient to keep carbon suspended uniformly throughout the beaker. Solutions were mixed using distilled water spiked with sodium bromate. Phosphate buffers were used to maintain a constant pH during

1-56670-136-8/96/$0.00+$.50
© 1996 by CRC Press, Inc.

batch tests. The adjustment of solution pH was accomplished by adding either 1 M H_2SO_4 or 1 N NaOH. All batch tests were conducted at room temperature (20 to 25°C) unless otherwise noted.

Samples were taken during batch tests using a 10-ml glass syringe with a luer-lock connection, and filtered using a filter tip with 0.45-μm cellulose acetate filter paper. Samples were stored in high-density polyethylene screw-top bottles, and were kept in a cold room (temperature 3 to 8°C) until analysis.

At the end of each batch test, the carbon was collected for a determination of how much bromide was ion exchanged to the surface. The carbon was first rinsed with ten bed volumes of distilled water, to remove any of the bromate solution. Next, the carbon was rinsed with 20 bed volumes of 0.1 M K_2SO_4 followed by 20 bed volumes of distilled water. These 40 bed volumes of rinsate were collected and analyzed for bromide. The amount of bromide fixed to the carbon surface was used for the mass balance.

For packed bed column studies, 30 × 40 mesh carbon was placed in 1-in. diameter glass columns. Glass beads and glass wool were used to support the carbon bed. A Coleparmer Masterflex pump was used to feed solution to the columns, which were operated in an upflow configuration. Tygon tubing and polyethylene connectors were used with the Masterflex pump system. Feed solutions were buffered with 0.1 M potassium monobasic phosphate and pH adjustment was done with 1 M NaOH. Column tests were conducted at room temperature. Samples were collected into high-density polyethylene bottles and stored at 4°C until analysis.

ANION ANALYSIS

Standards for anion analysis were made according to *Standard Method* procedure 4110 B.[6] A stock solution for bromate was mixed using 1.1798 g $NaBrO_3$ in distilled water. The working standard for bromate was made as for the other anions. Working standards were mixed using 18-MΩ water, and were analyzed at the beginning of each analysis session. The following concentrations were used for working standards: 0, 5, 10, 20, 50, and 100 μg/l of each anion. Analysis was not done if the linear correlation factor for the standards was less than 0.99. The phosphate concentration was not determined for samples that contained high concentrations of phosphate (used as a buffer).

Analysis was done on a Dionex Series 300 ion chromatograph, equipped with a gradient pump, an autosampler, and an electrochemical detector. Samples were analyzed at room temperature, and the system temperature compensation factor was 1.7. The specific method used for analysis was developed by Hans van der Jagt and co-workers at KIWA, Nieuwegien, the Netherlands. The following operational parameters were used:

Eluent 1:   0.5 mM Na_2CO_3, 0.18 mM $NaHCO_3$
Eluent 2:   4 mM Na_2CO_3, 1.5 mM $NaHCO_3$

Regenerant:	25 mM H_2SO_4
Columns:	AG9-SC, AS9-SC
Flow rate:	2 ml/min
Conductivity detector range:	0.01 μs
Run time:	12 min: 4.5 min on eluent 1, 5.5 min on eluent 2, remaining time on eluent 1
Bromate retention time:	2.33 min

Eluent 1 is used to elute bromate, and eluent 2 is used to rinse longer retained ions off the column. Large amounts of chloride in a sample would interfere with bromate detection. Without interference, the bromate detection level was at least 2 μg/l. It was possible to see smaller amounts of bromate; however, the integration of very small peaks may not be accurate. Samples having less than 2 μg/l of bromate are shown in figures as zero bromate concentration.

Bromide and other anion analysis was done using the following parameters:

Eluent:	2 mM Na_2CO_3, 0.75 mM $NaHCO_3$
Regenerant:	25 mM H_2SO_4
Columns:	AG9-SC, AS9-SC
Flow rate:	2 ml/min
Conductivity detector range:	0.01 μs
Run time:	10 min
Retention times (min):	
Chloride	1.85
Nitrite	2.25
Bromide	2.97
Nitrate	3.69
Phosphate	5.73
Sulfate	9.25

The detection limits of these anions was generally a few micrograms per liter.

PRETREATMENT OF CARBONS

The carbon used in this study was Ceca GAC 40 by Atochem, batch number B70921-2B. The carbon was sieved following ASTM Standard D 2862-82.[7] Following sieving, the 30 × 40 fraction was rinsed with distilled water to remove fines, and then was dried in a 105°C oven. Dried carbon was stored in glass screw-cap bottles in a desiccator.

Carbon was modified several ways. Acid-washed carbon (designated AW) was rinsed with ten bed volumes of 1 N HCl at a flow rate of approximately 5 to 10 ml/min. The carbon was then rinsed with distilled water at the same flow rate, and dried and stored as above.

Outgassed carbon (designated OG) was prepared in a fluidized bed furnace. Approximately 10 g of previously sieved and dried 30 × 40 carbon was heated at 900°C for 3 h using a nitrogen flow rate of 30 to 40 l/min. The outgassed carbon was allowed to cool briefly under nitrogen, and was then stored in a screw-cap bottle in a desiccator.

Carbon that was acid washed and outgassed (designated AWOG) was acid washed first, according to the procedure described above. After drying, this carbon was outgassed as described above. The AWOG carbon was stored as described above.

MEASUREMENT OF SURFACE GROUPS ON CARBON

A number of different techniques have been used to characterize the carbon surface. Analytical techniques such as secondary ion mass spectrometry and X-ray photoelectron spectroscopy were conducted by technicians at the Center for Microanalysis of Materials, Materials Research Lab, University of Illinois–Urbana.

The measurement of surface acid and base groups was accomplished using a modification of the technique proposed by Boehm. The procedure used here was taken from Jagiello et al.[8] Since the number of equivalents of carboxyl groups were very small for the carbons studied here, this number was difficult to measure by titration. Therefore, the numbers for carboxyl groups reported in Table 2 were obtained by subtracting the total of carboxyl + lactonic groups, determined by reaction with sodium carbonate, from the total of carboxyl + lactonic + phenolic groups, determined by reaction with sodium hydroxide.

THE REDUCTION OF BROMATE BY GAC

Bromate can be reduced to bromide in the presence of GAC. An example of the data from a batch test is shown in Figure 1. The lines describing the data points are for illustration only, and do not represent a mathematical fit.

In general, the loss of bromate in batch test experiments follows a logarithmic type decay. The increase of bromide concentration in the water corresponds to the removal of bromate. The concentration of bromide in solution is somewhat less than the amount of bromate removed from solution because of bromide attachment to the carbon by ion exchange. Thus, at the end of a batch test, the sum of bromate and bromide in solution does not equal the initial mass of the element added to the system. A good mass balance can be achieved for the batch tests by including not only what is in solution as bromide and bromate, but also what is ion exchanged to the carbon surface. In other words:

Initial Br in the system = final bromate concentration in solution + final bromide concentration in solution + ion exchanged bromide =

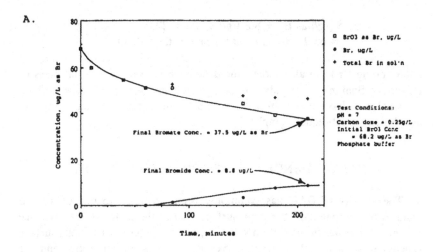

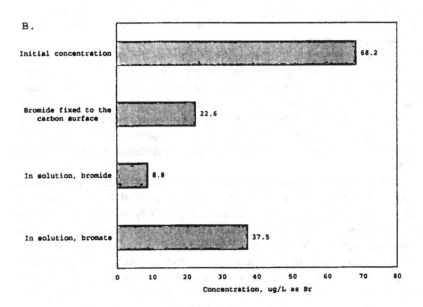

Figure 1   Mass balance for a typical batch test experiment.

$$37.5 \ \mu g/l \text{ as Br} + 8.8 \ \mu g/l \text{ as Br} + 22.6 \ \mu g/l \text{ as Br}$$
$$= 68.9 \ \mu g/l \text{ as Br at the end of the batch test}$$

After adding Br from all sources, the difference between the initial mass of Br and the final mass was only 1% for the batch test shown in Figure 1. Good mass balances have been achieved for all batch tests; however, these data are not shown on subsequent graphs.

THE INFLUENCE OF CARBON PRETREATMENT

The activity of GAC has commonly been attributed to one of several features of the material:[9] the large surface area of the material, the presence of metal impurities in the carbon structure, or the existence of various surface functional groups, such as oxygen groups. The type of impurities and the amount of impurities in a particular carbon are influenced by the source of the material used to create the carbon, as well as by the manufacturing process. The particular carbon studied here was manufactured commercially from a sub-bituminous material.

It is possible to change the amount of impurities in GAC by several simple techniques. Acid washing, or rinsing the carbon with an acid solution at room temperature, removes metals from the carbon surface. Outgassing, i.e., heating the carbon to a high temperature in an inert environment, removes surface oxygen groups. Both of these techniques, and also a combination of both acid washing and outgassing, were used to remove impurities and oxygen groups from GAC to determine whether they were responsible for the reduction of bromate.

Figure 2 shows the results of batch tests for bromate reduction by pretreated GAC. All of the tests were conducted at a solution pH of 7, with a phosphate buffer system and a carbon dose of 0.25 g/l. Only the data for bromate concentration are shown in this figure, although bromide concentrations were also measured. In Figure 2, it is readily apparent that altering the carbon surface by removing impurities has a great effect on GAC activity for bromate reduction. Using the virgin carbon as a baseline, outgassing the carbon to remove surface oxygen groups decreased the activity of the carbon greatly. Acid washing the carbon to remove metals may have increased the activity of the GAC slightly, but did not have a large effect. Doing both treatments in series, first acid washing and then outgassing the carbon, had a large effect on the activity of the carbon. The acid-washed, outgassed carbon (AWOG) is much more effective for bromate reduction than the virgin carbon.

Column tests conducted in distilled water also demonstrated the difference between the virgin and the AWOG carbon. Figure 3 is a comparison of the results for columns operated at pH 7, with an average influent bromate concentration of 470 $\mu g/l$ bromate as Br in distilled water. Both columns were

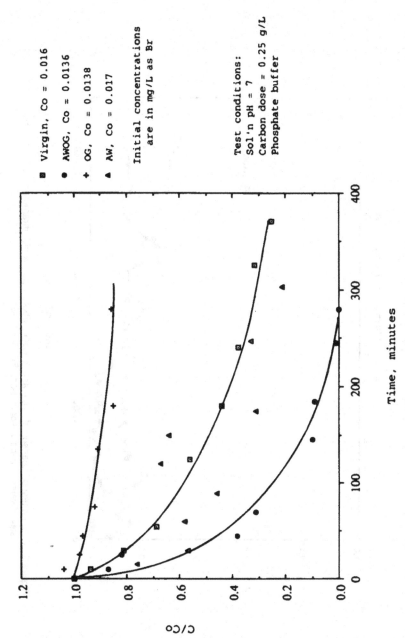

Figure 2 The effect of carbon pretreatment on bromate reduction in batch tests.

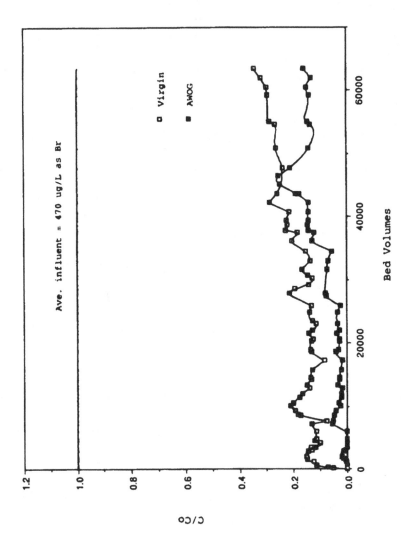

Figure 3 Column test comparison of virgin and AWOG carbon. Operational parameters: influent pH = 7; distilled water solution 1 g of carbon; EBCT = 1 min.

fed influent from the same reservoir. Only the bromate data are shown here, although the bromide data were collected. Bromide concentrations in the effluent followed the expected pattern of being somewhat less than the concentration of bromate reduced.

Although this column study was not operated under conditions that would be expected at a drinking water treatment plant, it is useful to compare the difference in performance between the virgin and the AWOG carbons. Under identical conditions, the AWOG carbon performed better than the virgin carbon for bromate reduction, as expected based on the batch test results.

PHYSICAL CHARACTERISTICS OF CARBON

One explanation for the activity of GAC is the large surface area of this material. It is possible that by pretreating the carbon in the relatively harsh manner used here the surface area and pore structure have been changed. An increase in the surface area would potentially explain the greater reactivity of the AWOG carbon. The surface area and pore size distribution have been measured with a nitrogen adsorption analyzer. This procedure is based on the BET method, which assumes monolayer adsorption of nitrogen on the carbon surface.[10] A summary of the surface areas and pore volumes is given in Table 1. A difference of more than 5% between the values is considered to be a true difference. Thus, it can be said that the acid-washed and AWOG carbons both have a larger surface area and greater pore volumes than the other two carbons. The increased surface area of carbons that have been acid treated may be due to the removal during the acid-washing step of material that blocks the carbon pores. In any case, the activity of the carbon does not correlate with the surface area, since the acid-washed carbon is not as active as the AWOG carbon. The difference in bromate reduction by the treated carbons must then be caused by differences in the surface chemistry.

CHEMICAL CHARACTERISTICS OF CARBON

The simplest type of surface chemistry analysis is to measure the surface acidity and basicity by reacting selected bases and acids with the carbon. This

Table 1   Surface Areas and Pore Volumes for Pretreated Carbons

Carbon	Surface area (m^2/g)	Mesopore vol (cc/g)	Micropore vol (cc/g)
Virgin carbon	1035	0.296	0.310
Acid washed	1120	0.326	0.340
Outgassed	990	0.295	0.300
AWOG	1135	0.329	0.347

Table 2   Boehm Acidity and Basicity Measurements for
          Virgin and AWOG Carbon

Group type	Virgin carbon (meq/g C)	AWOG carbon (meq/g C)
Carboxyl	0.045	0.017
Carboxyl + lactonic	0.311	0.222
Carboxyl + lactonic + phenolic	0.356	0.239
Basic groups	0.054	0.506

technique has been used, most notably by Boehm,[11] to infer the existence of various types of surface oxygen groups. It was used in this study in an attempt to characterize the virgin and AWOG carbons. The results for the measurement of "Boehm acidity" are given in Table 2.

The results of the acidity/basicity measurements for the virgin and AWOG carbons are quite different, indicating that the two carbons do have chemically different surfaces. In particular, the AWOG carbon has many more basic groups. It is not known at this time whether these sites play a direct role in bromate reduction.

One major drawback to the Boehm acidity measurement is that it is impossible to tell exactly what is reacting, and to quantify with certainty what is on the carbon surface. In an attempt to describe the carbon surface more accurately, more sophisticated techniques were used to characterize the carbon surface.

One surface analysis technique used was the scanning electron microscope (SEM). SEM is capable of detecting elements with atomic weights greater than nitrogen; however, it cannot detect gaseous species such as oxygen. In spite of the limitations of SEM, it is a useful technique for determining the metals on the carbon surface. Elemental scans are shown for the virgin carbon, the acid-washed carbon, and the AWOG carbon in Figures 4 through 6. It is clear from these graphs that the acid-washed and AWOG carbons contain very few metals. These were effectively removed during the acid-washing step. Because the two carbons with less metals are not equally active for bromate reduction, it can be concluded that metals on the carbon surface play neither a catalytic nor an inhibitory role in carbon reduction of bromate.

Two other surface analysis techniques, secondary ion mass spectrometry (SIMS) and X-ray photoelectron spectroscopy (XPS), were used in an attempt to determine the important differences between the virgin carbon and the more reactive AWOG carbon. SIMS is capable of detecting all elements in the parts-per-billion concentration range. The ability to detect all elements, including hydrogen, makes this technique quite powerful. Most surface analysis techniques are not capable of detecting hydrogen. SIMS uses an ion to bombard the surface of a sample. The energy from the bombardment knocks ions or groups of ions off the sample surface. These ions are then detected using a mass spectrometer. SIMS can be used for both quantitative and qualitative

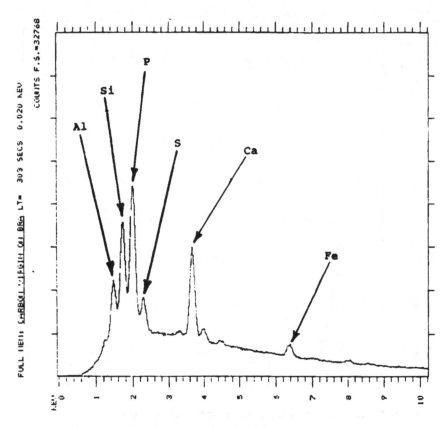

Figure 4   SEM elemental analysis of virgin Ceca carbon.

work. In this project, the primary interest was in the qualitative differences between the pretreated carbons.

The output of a SIMS analysis is a spectrograph, one for positively charged ions or groups, and one for negatively charged ions or groups. Graphs for both positively and negatively charged ions for virgin and AWOG carbon are shown in Figures 7 and 8, respectively. The horizontal dashed line on the spectrographs represents a "detection limit" of sorts; peaks below this line are too small to be considered significant. Comparing the graphs in these two figures, it can be seen that the AWOG carbon has fewer positively charged ions (i.e., metals) on the surface. This finding is consistent with the results of the SEM analysis. Comparing the graphs for the negatively charged ions shows that there is also less oxygen, nitrogen, and phosphorus on the AWOG carbon than on the virgin carbon. This finding is perhaps to be expected, since many researchers have reported that outgassing removes surface oxygen from carbon.[1] However, it is an interesting finding, since oxygen is considered to be responsible for many of the catalytic effects of carbon.

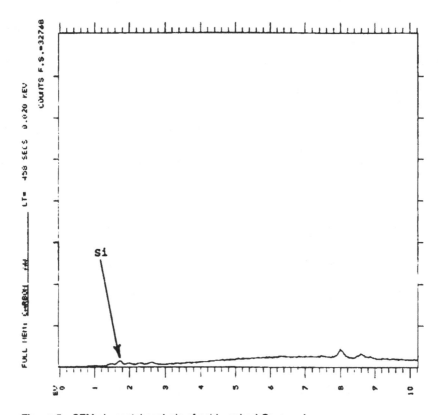

Figure 5   SEM elemental analysis of acid-washed Ceca carbon.

XPS was also used to analyze the carbon surface. The strength of XPS is that the energy of photoejected electrons identifies both the element and the chemical bonding of the atom. In this case, however, the results of XPS analysis were somewhat disappointing. Perhaps because the carbon matrix is so complex, it was difficult to detect major differences between the carbons. The elements of concern here were carbon, nitrogen, oxygen, and phosphorus. These elements were thought to be important in carbon activity, and can exist in many different chemical states. Table 3 is a summary of the atomic concentrations for carbon, oxygen, nitrogen, and phosphorus in the four pre-treated carbons.

That the AWOG carbon has less oxygen is consistent with the SIMS data. The lack of oxygen on the most reactive bromate reducing carbon suggests that perhaps the absence of oxygen is an important characteristic of the carbon. Functional groups or impurities may be blocking reactive sites on the carbon surface. The removal of oxygen during acid washing and outgassing may uncover or create sites that are reactive for bromate reduction.

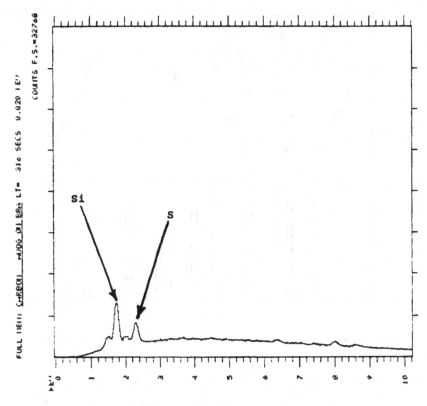

Figure 6   SEM elemental analysis of AWOG Ceca carbon.

CONCLUSIONS

1. Pretreatment of commercially available GAC can have a tremendous impact in the capacity of the carbon to reduce bromate. Acid washing followed by outgassing at 900°C increases the activity of carbon for bromate reduction.
2. The differences in the activity of pretreated carbons for bromate reduction are not caused by changes in surface area or pore volume.
3. The activity of GAC for bromate reduction does not appear to depend on metal impurities or surface oxygen groups, since the most active carbon in this study had the least amount of impurities and surface oxygen.
4. GAC activity for bromate reduction may depend on having a "clean" carbon surface, or it may be caused by species not measured here, such as free radicals.

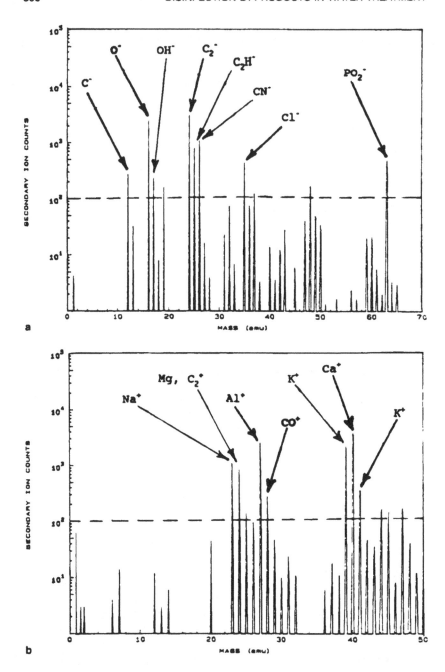

Figure 7   (a) SIMS analysis of virgin Ceca carbon, negative ions. (b) SIMS analysis of virgin Ceca carbon, positive ions.

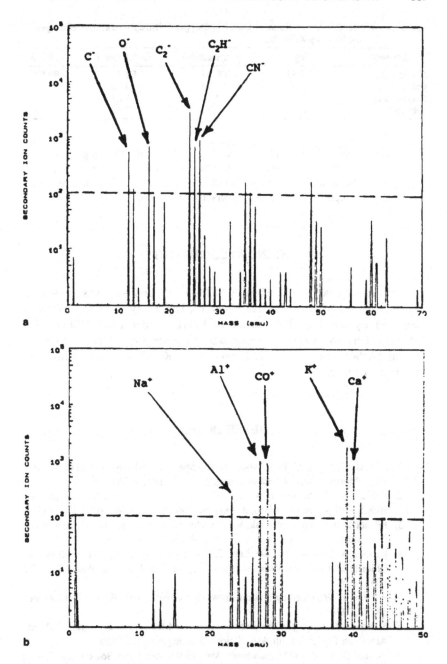

Figure 8 (a) SIMS analysis of AWOG Ceca carbon, negative ions, (b) SIMS analysis of AWOG Ceca carbon, positive ions.

Table 3 Atomic Concentrations of Carbon, Oxygen, Phosphorus, and Nitrogen Determined by XPS (%)

Element	Virgin	Acid washed	Outgassed	AWOG
Carbon	80.33	86.78	83.94	89.58
Oxygen	17.02	12.02	14.03	9.77
Phosphorus	1.90	0.48	1.34	0.13
Nitrogen	0.74	0.72	0.69	0.51

5. Surface analysis techniques such as SEM, SIMS, and XPS can provide us with a wealth of information on the chemistry of GAC. These techniques should be utilized in future studies of GAC to increase our understanding of the complex chemistry of microcrystalline carbon.

ACKNOWLEDGMENTS

The surface analysis of carbon by SEM, SIMS, and XPS was carried out in the Center for Microanalysis of Materials, University of Illinois, which is supported by the U.S. Department of Energy under grant DEFG02-91-ER45439. This work has been supported by the generosity of KIWA, Nieuwegien, the Netherlands, and of the Illinois Water Resource Center, University of Illinois–Urbana.

REFERENCES

1. Puri, B. R., Singh, D. D., Chander, J., and Sharma, L. R., Interaction of charcoal with chlorine water, *J. Indian Chem. Soc.*, 35, 181, 1958.
2. Suidan, M. T., Snoeyink, V. L., and Schmitz, R. A., Reduction of aqueous HOCl with activated carbon, *J. Environ. Eng. Div. ASCE*, 103, 677, 1977.
3. Komorita, J. D. and Snoeyink, V. L., Technical note: monochloramine removal from water by activated carbon, *J. Am. Water Works Assoc.*, 77, 62, 1985.
4. Leitner, V., Karpel, N., De Laat, J., Dore, M., Suty, H., and Pouillot, M., Chlorite and chlorine dioxide removal by activated carbon, *Water Res.*, 26, 1053, 1992.
5. Pontius, F. W., Reg-neg process draws to a close, *J. Am. Water Works Assoc.*, 85, 18, 1993.
6. *Standard Methods for the Examination of Water and Wastewater*, 17th ed., American Public Health Association, Washington, DC, 1989.
7. *Annual Book of ASTM Standards*, Vol. 15.01, American Society for Testing and Materials, Philadelphia, 1993.
8. Jacek, J., Bandosz, T. J., and Schwarz, J. A., Inverse gas chromatographic study of activated carbons: the effect of controlled oxidation on microstructure and surface chemical functionality, *J. Colloid Interface Sci.*, 151, 433, 1992.

9. Snoeyink, V. L. and Weber, W. J., The surface chemistry of active carbon, a discussion of structure and surface functional groups, *Environ. Sci. Technol.*, 1, 228, 1967.

10. Brunauer, S., Emmett, P. H., and Teller, E., Adsorption of gases in multimolecular layers, *J. Am. Chem. Soc.*, 60, 309, 1938.

11. Boehm, H. P., *Advances in Catalysis*, Vol. 16, Academic Press, New York, 1966.

CHAPTER 14

DBP Speciation and Kinetics as Affected by Ozonation and Biotreatment

Hiba M. Shukairy and R. Scott Summers

INTRODUCTION

Disinfection by-product (DBP) formation and speciation are significantly affected by the precursor compounds present in the water. Drinking water treatment processes can affect both the organic and inorganic precursor compounds. The increased use of ozonation for drinking water treatment has led to a re-evaluation of the chemistry involved in the ozonation of waters that contain natural organic matter (NOM) and bromide.

In the presence of bromide, ozonation of natural waters leads to the formation of hypobromous acid (HOBr), hypobromite ion (OBr⁻), bromate, and brominated organic by-products. Haag and Hoigné[1] have shown that ozone oxidizes bromide to form HOBr/OBr⁻ under drinking water treatment conditions. Hypobromite was found to be further oxidized to bromate or to a species that regenerates bromide, while HOBr reacted with NOM to form brominated organic by-products. The HOBr/OBr⁻ equilibrium distribution is pH dependent.

At low ozone doses, similar to what is used in microbial inactivation practice, there is little mineralization, that is, no significant change in the dissolved organic carbon (DOC) concentration occurs.[2] The formation of bromate, however, results in a decrease in the inorganic precursor bromide. The formation of bromate and brominated organic by-products increases with increasing bromide concentrations, and depends on the distribution of the HOBr/OBr⁻ species. Biological treatment results in a significant change in the concentration of the organic precursors, DOC, while the bromide concentration is conservative.

The formation of organic by-products by ozonation is dependent on the ratio of inorganic to organic precursors. In waters that contain low bromide

1-56670-136-8/96/$0.00+$.50

concentrations, ozone is believed to react with NOM before it reacts with bromide. At high bromide concentrations, some competition might exist between bromide and the organic matter. As bromide concentrations increase, at constant ozone doses and organic matter concentrations, the formation of bromo-substituted organic compounds increase, particularly at lower pH. NOM, however, differs in water sources and may react differently upon ozonation. Little information is available about this complex chemistry in natural systems.[3]

The use of chlorine disinfection further complicates the chemistry. Chlorine, similar to ozone, oxidizes bromide to form HOBr/OBr⁻. Competition will then take place between HOBr and hypochlorous acid (HOCl) for the available organic sites. The distribution of the DBP species will depend on the chlorination conditions and particularly on the ratio of bromide to DOC.[3-6]

Earlier research investigated the impact of bromide ion concentration changes on DBP formation and speciation.[7,8] These researchers found that total trihalomethane (TTHM) formation increased with increasing bromide concentrations, when all other parameters were held constant. The increase was driven by an increase in the mixed halogenated trihalomethanes (THMs) and bromoform at the expense of chloroform. Rook and Gras[9] showed that HOBr was a more potent halogenating agent than HOCl. Symons et al.,[10] in a re-examination of the effect of bromide on THM formation and speciation, have shown the importance of bromide concentration, organic matter availability, and chlorine dose and residual. THM speciation was governed by the faster reaction kinetics of the brominating species. Other researchers investigated bromine-ozone interactions and found that the formation of bromate was a function of the ozone residual and the contact time.[6,11,12] The importance of bromate formation in DBP evaluation is that, as bromate concentrations increase, bromide concentration decreases, that is, a percentage of the available bromide will be converted to bromate and will no longer be available during the chlorination reactions.

Increasing the precursor concentration, organic or inorganic, yields an increase in DBP formation. The speciation is affected by the following factors: (1) the ratio of the inorganic precursor bromide, to the organic precursors, normally measured by DOC concentration or by UV absorbance; (2) the ratio of the bromide to the free available chlorine (FAC); (3) the ratio of the chlorine dose to the DOC; and (4) the chlorination holding time. When the bromide to DOC and the bromide to FAC ratios increase and at short reaction times, a shift to the more-bromo-substituted species occurs.

To demonstrate the effect of changes in DOC and bromide concentrations, chlorine dose and holding time on DBP speciation, results from previous studies[6,13] have been reinterpreted and are reported in Figures 1 to 4. DBP speciation is summarized by bromine incorporation factors. THM speciation can be described by the bromine incorporation factor n[14] where

$$n = \text{TTHM-Br/TTHM on a molar basis}$$

and TTHM-Br is calculated from the molar concentrations of bromodi-chloromethane ($CHBrCl_2$), dibromochloromethane ($CHBr_2Cl$), and bromo-form ($CHBr_3$):

$$\text{TTHM-Br} = [1 \times CHBrCl_2 + 2 \times CHBr_2Cl + 3 \times CHBr_3]$$
in micromoles per liter.

Therefore, $n = 0$ if chloroform is the sole THM species formed and $n = 3$ when bromoform is the sole THM species formed. It is important to note that TTHM-Cl + TTHM-Br $\neq$ total THM, while TTHM-Cl + TTHM-Br = TTHM-X, where X is the halogen.

Similarly, haloacetic acid (HAA) speciation can be described by the bro-mine incorporation factor n'.[6]

$$n' = \Sigma \text{HAA-Br}/\Sigma \text{HAA on a molar basis}$$

and the sum of HAA-Br in this study is calculated from the molar concentra-tions of bromoacetic acid (BAA), bromochloroacetic acid (BCAA), and dibro-moacetic acid (DBAA). Other bromo-substituted HAAs were observed on chromatograms, but were not quantifiable by the current analytical techniques;

$$\Sigma \text{HAA-Br} = [1 \times BAA + 1 \times BCAA + 2 \times DBAA]$$
in micromoles per liter.

In this case, n' can vary between 0 if dichloroacetic acid and trichloroacetic acid are the sole species to 2 if only dibromoacetic acid is present. HAA-Cl + HAA-Br $\neq$ HAA; while HAA-Cl + HAA-Br = HAA-X.

An increase in n or n' indicates a shift in the speciation to the more bromo-substituted species, usually accompanied by a decrease in the chloro-substituted species.

The impact of ozone on THM and HAA speciation for Ohio River water (ORW) spiked with bromide to a 550-μg/l initial bromide concentration is shown in Figure 1. These results are from a study that examined the impact of ozone dose and bromide concentration on DBP formation, control, and speciation for ORW.[6] In that study, DBP formation potential was evaluated, that is, the chlorine dose was relatively high (12 mg/l) to ensure maximum formation potential. Ozonation at transferred ozone to DOC ratios ranging from 0 to 2.54 mg/mg did not result in any mineralization of the DOC.[6] Increasing ozone dose, however, resulted in increased formation of bromate and a consequent decrease in the bromide concentration, thus decreasing the bromide to DOC and the bromide to average FAC ratios. The average FAC

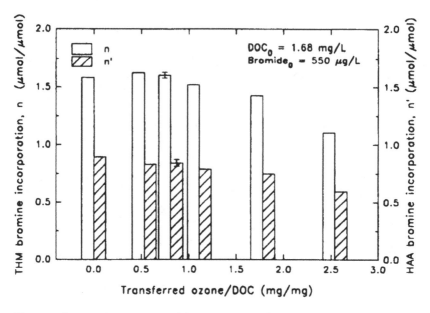

Figure 1   Impact of ozonation on DBP speciation in Ohio River water.

is defined as the sum of the chlorine dose and residual divided by 2.[10] This resulted in a shift to the chloro-substituted species and a decrease in n and n'. At low ozone doses the effect was not as significant because of small changes in the bromide concentration; only 2 to 3% of the bromide was converted to bromate. As the ozone dose increased, however, bromide concentration decreased, the chlorine demand decreased, and the bromo-substituted species decreased.

The impact of biotreatment in a batch bioreactor with 5-d contact time for the same ORW study[13] is presented in Figure 2. Increasing the bromide concentration resulted in an increase in the bromide-to-DOC ratio and a subsequent increase in n and n'. The effect of biotreatment on the bromide-to-DOC ratio can be seen in the same figure. As the DOC concentration decreased by biotreatment, at a set initial bromide concentration, the bromide-to-DOC ratio increased. This resulted in an increase in the bromo-substituted species. Therefore, for a given initial bromide concentration, biotreatment resulted in a shift to the bromo-substituted species as the bromide-to-DOC ratio increased. The impact of ozone and biotreatment on DBP speciation would have been even more significant had a lower chlorine dose been used.

The effect of chlorine dose on DBP speciation for raw and ozonated ORW (ozone-to-DOC ratio of 0.56 mg/mg) is presented in Figure 3. Increasing the chlorine-to-DOC ratio resulted in a decrease in n. As the chlorine dose increased the competition between HOCl and HOBr was in favor of the more abundant HOCl resulting in higher formation of the chloro-substituted species

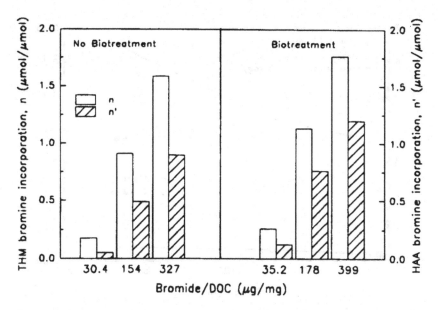

Figure 2   Impact of bromide concentration and biotreatment on DBP speciation in Ohio River water (DOC_0 = 1.68 mg/l).

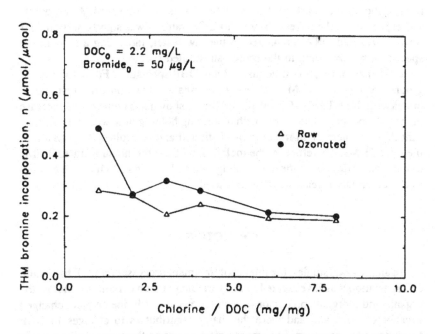

Figure 3   Impact of chlorine dose on DBP speciation in Ohio River water (ozone to DOC = 0.56 mg/mg).

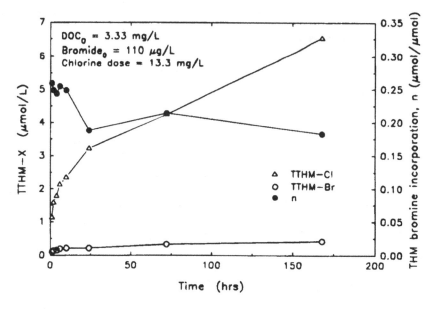

Figure 4   THM speciation reaction kinetics in groundwater natural organic matter.

and higher concentrations of the total THMs. Consequently, a decrease in the ratio of the bromo-substituted to the total THMs was observed. At all points and especially at the lowest chlorine-to-DOC ratio, n was significantly higher for the ozonated ORW, compared to the raw water, indicating a shift in the speciation by ozonation to the bromo-substituted species.

Chlorination reaction time also affects DBP speciation. For a solution of groundwater (GW) NOM, increasing chlorination reaction time resulted in increases in both TTHM-Cl and TTHM-Br, as shown in Figure 4. The increase in the chloro-substituted THMs with increasing holding time was more signifi-cant than the increase in the bromo-substituted and controlled the increase in the total THMs. Therefore, n, the molar ratio of the bromo-substituted THMs to the total THM was highest at holding times of less than 24 h and decreased as the chlorination reaction time increased.

OBJECTIVE

In all previous studies, the impact of treatment processes on DBP formation and speciation was investigated under variable reaction conditions, that is, the organic and inorganic precursor concentrations and chlorine demand changed because of treatment and were the major contributors to changes in DBP formation and speciation. The impact of treatment on the nature of the organic matter could not be delineated because of all the variability in the matrix. The objective of this study was to examine the impact of ozonation and biotreatment on the organic precursor characteristics by evaluating DBP formation, speciation, and kinetics under constant DOC, bromide, and chlorination conditions.

EXPERIMENTAL APPROACH

To chlorinate the control and treated solutions under conditions of constant DOC, bromide, and chlorine concentrations, the following approach was adopted. Solutions of GW NOM, 10 mg/l DOC, were treated by biotreatment, ozonation, and combined ozonation/biotreatment. Prior to chlorination, samples of the control and treated solutions were diluted to a constant DOC concentration of 3 mg/l, and the bromide concentrations were adjusted to 100 μg/l in all samples. Preadsorbed tap water (PAT) was used for the dilution water to provide the noncarbonaceous nutrients and the mineral matrix necessary for the growth of the microorganisms during biotreatment. For biotreatment, a batch reactor with attached biomass was used. For ozonation a transferred ozone-to-DOC ratio of 0.8 mg/mg was used, and for the combined process of ozonation/biotreatment the ozonated GW NOM solution was biotreated. Two chlorination approaches were investigated: a set dose of 4.8 mg/l (a 1.6:1 chlorine-to-DOC ratio) and a set residual of 1.0 mg/l (the chlorine dose range was 2.5 to 4.8 mg/l). For quality assurance and quality control, 4 out of 28 chlorination tests were duplicated at random. Analytical replicates were also performed according to the recommendations of the corresponding methods.

The results of the set chlorine residual chlorination conditions at three holding times (6, 24, and 72 h) are presented herein, and those of the set chlorine dose are presented elsewhere.[15] Unless otherwise stated, similar trends in the data were observed for both set dose and set residual chlorination conditions.

MATERIALS AND METHODS

ORGANIC MATTER

The organic matter used in this study was isolated by a strong basic anion resin (Lewatit MP 500 A, Bayer Chemical Co.) used in the treatment of a GW with a high humic content, 7 mg/l DOC, at Fuhrberg (Hannover), Germany. The resins were regenerated at about 50% DOC breakthrough, with a solution containing 10% sodium chloride and 2% sodium hydroxide. The regenerate has a molecular size range of 200 to 4000 with an average of 1500 as estimated by gel-permeation chromatography and a DOC concentration of 22,000 mg/l.

PREADSORBED TAP WATER

Tap water was passed through a granular activated carbon column to adsorb the organic matter. The water was then filter-sterilized by passing through a 0.22-μm pore diameter membrane filter (Millipore-HV). The DOC concentration of PAT was between 0.08 and 0.14 mg/l with an average spectral absorption coefficient (SAC) of 0.72 m^{-1} at 254 nm. Potassium bromide was added to PAT to yield bromide concentrations varying between 92 and 138 μg/l. The chlorine demand range of this dilution water was 0.4 to 2.6 mg/l resulting in the following average DBP formation at 72 h and at a chlorine dose of 4.8 mg/l: total organic halide (TOX) = 25 ± 4.6 μg Cl^-/l; TTHM = 11.0 ± 2.0 μg/l; HAA6 = 4.4 ± 3 μg/l. The DBP formation contribution of the dilution water was in the range of 5 to 10% (TOX), 7 to 10% (TTHM), and 0 to 6% (HAA6) of the control GW NOM solution at 72 h and a set dose. No corrections for the dilution water contribution were made for any of the GW NOM solutions.

OZONATION

Ozone was produced from oxygen gas by a 200-mg/h generator (Sander Model 200, Uetze-Eltze, Germany). Ozone was applied in a flow-through 1-l contactor under the following conditions: T = 22°C, pH = 7.1, liquid flow rate = 142 ml/min, and a residence time = 7.1 min. The initial DOC of the concentrated solution was 11.51 mg/l. The transferred ozone-to-DOC ratio was 0.81 mg/mg, with a transfer efficiency of 42%. The average liquid phase residual was 0.058 ± 0.015 mg/l. Applied ozone and the off-gas were measured by iodometry;[16] dissolved ozone was measured spectrophotometrically using the indigo trisulfonate method.[17] The ozone contactor was characterized as a completely mixed flow reactor.[18] Samples from the contactor were collected only after three detention times. After ozonation, the DOC concentration of the solution was measured and was found to be 10.71 mg/l, indicating that 7% mineralization by ozonation had occurred. The SAC of the solution decreased significantly, by 49%.

BIOTREATMENT

In this study, 1-l bench-scale batch bioreactors containing 150 g of ORW acclimated sand and 500 ml of the GW NOM samples were mixed on a shaker table for 5 d. No aeration was used; the available headspace in the reactor provided the necessary oxygen for the microorganisms. At the end of 5 d, samples were filtered using a 0.45-μm pore diameter membrane filter (Millipore HV). The DOC concentration and UV absorption were measured before and after biotreatment. For biotreatment of the control GW NOM solutions (DOC_0 = 10.1 mg/l), the effluents from six reactors were combined; the

DOC concentration of the mixture was measured as 8.32 ± 0.09 mg/l. For biotreatment of the ozonated GW NOM solutions (DOC_0 = 10.71 mg/l), four reactors were used and the measured DOC of the combined effluents was 7.54 ± 0.01 mg/l.

CHLORINATION

To minimize the impact of changes in chlorine demand by treatment, two chlorination approaches were selected: (1) using a set dose and (2) using a set residual. The set dose was based on a chlorine demand study for the control, no treatment, such that a residual of 0.5 mg/l was obtained after 72-h holding time. This corresponded to 4.8 mg/l, a 1.6:1 chlorine-to-DOC ratio, and was used for the control and all treated solutions. In the second approach, to obtain a set residual of 1.00 ± 0.3 mg/l, three samples were chlorinated using different doses; the chlorine residuals were then measured and the sample that fell in the required range was selected for DBP analyses. For both set dose and set residual chlorination conditions, the solutions were buffered using a phosphate buffer; the pH range was 7.0 to 7.1 before chlorination and between 7.0 and 7.2 after chlorination. Samples were held in the dark at 20°C for 6, 24, and 72 h. At the end of the holding time, and after residual measurements, the excess chlorine was quenched with sodium sulfite crystals and DBP samples were collected according to the method requirements. The water quality parameters that were evaluated were DOC, SAC, bromide concentration, chlorine demand, TOX, THMs, and HAA6 [chloroacetic acid (CAA), dichloroacetic acid (DCAA), trichloroacetic acid (TCAA), BAA, BCAA, and DBAA].

ANALYTICAL METHODS

DOC was measured according to USEPA method 415.1 with an organic carbon analyzer (Rosemount-Dohrmann 180). UV absorbance at 254 nm was measured in a 5-cm cell using a diode-array spectrophotometer (HP 8452A). UV samples were acidified to pH 2 with phosphoric acid and the results are reported as SAC which is the UV absorbance at 254 nm divided by the cell path in meters. Bromide concentration was measured on a modified high-performance liquid chromatography system (Waters 501) using a guard/separation column combination (Dionex Ionpac AG9-SC/AS9-SC). The eluant used was a 0.7-mM sodium carbonate/1.3 mM sodium bicarbonate solution. The bromide measurement technique was based on a modification of the method of Hautman and Bolyard.[19] The detection limit was 10 µg/l. Chlorine dose was measured by iodometry; and chlorine residual was measured by amperometric titration.[17] THMs were measured according to the USEPA method 502.20; HAAs, USEPA method 552; and TOX, USEPA method 450.1. The detection limit for most quantified HAAs was 0.1 µg/l (chloro- and bromoacetic acids were below detection levels); for THMs 2 μg/l (bromoform was below detection levels); and for TOX 3 μg/l.

RESULTS AND DISCUSSION

DOC AND SAC

The initial conditions and the impact of treatment on GW NOM are summarized in Table 1. Also presented in the table are the DOC, SAC, and bromide concentrations for the posttreatment, diluted solutions.

Biotreatment resulted in 20% decrease in both DOC and SAC. Ozonation resulted in a 7% decrease in DOC and 49% decrease SAC, indicating a selective oxidation of UV-absorbing functional groups. Biodegradability improved after ozonation resulting in a 30% decrease in DOC and a 23% decrease in SAC. The SAC/DOC ratio was unchanged by biotreatment, with and without ozonation, indicating equivalent removal of the DOC and UV-absorbing functional groups by biological oxidation. After treatment, these four GW NOM solutions were diluted to an average DOC concentration of 3.02 ± 0.14 mg/l and bromide was spiked to an average concentration of 96 ± 3 μg/l. While the solutions were diluted to a set DOC value, the NOM as measured by SAC was different. The SAC was 43% lower in the diluted ozonated samples as compared to the unozonated diluted samples. Although the DOC concentration was not affected by ozonation, the UV-absorbing functional groups were oxidized. Ozonation also affects the molecular size distribution of NOM. These changes in the nature of the organic precursors affect the reactivity to subsequent chlorination as some of the potential reactive sites are reacted with ozone first.

CHLORINE DEMAND

The chlorine demand kinetic results, under set residual chlorination conditions, are shown in Figure 5. The error bars shown in most figures represent relative percent differences for duplicate tests. Treatment was found to decrease

Table 1 Impact of Treatment on Organic Matter

	Concentrated GW NOM solutions			Posttreatment dilution		
Treatment	DOC (mg/l)	SAC (1/m)	SAC/DOC (l/mg.m)	DOC (mg/l)	Br⁻ (μg/l)	SAC (1/m)
Control	10.1	45.5	4.51	2.94	97	13.1
Biotreatment	8.29	35.1	4.23	3.13	97	13.0
Ozonation[a]	10.7	23.3	2.18	2.87	91	7.44
O₃/biotreatment	7.54	17.9	2.37	3.14	97	7.42

[a] DOC_0 before ozonation = 11.51 mg/l; SAC_0 = 45.5 m^{-1}.

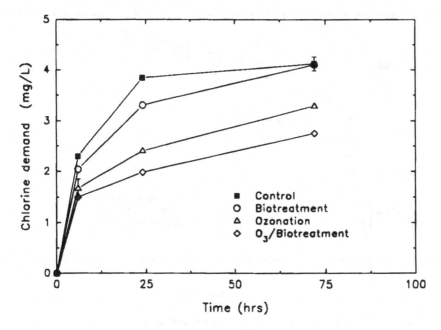

Figure 5   The effect of treatment on chlorine demand kinetics.

the chlorine demand at the constant bromide and DOC concentrations at all times investigated. The combined ozonation/biotreatment was the most effective. After 72-h holding time, the chlorine demand decreased by 20% upon ozonation and by 36% by the combined ozonation/biotreatment. Biotreatment alone resulted in a decrease in the chlorine demand except after 72 h.

A two-phase kinetic behavior was observed. There was an initial rapid demand followed by a gradual, much slower increase in demand with increasing contact time. This behavior is similar to the previously reported chlorination kinetics.[20-23] Decreases in chlorine demand by ozonation and biotreatment have been previously reported.[2,6,10,13,22,24] However, earlier reported studies on the impact of treatment on chlorine demand were done under conditions of changing organic and inorganic precursor concentrations. Under these constant reaction conditions, the change in the chlorine demand can be attributed to changes in the structure and reactivity of the organic matter.

HALOGENATED DBPs

TOX formation results are shown in Figure 6 for the set residual chlorination conditions. The most significant decrease in TOX formation was after ozonation. A decrease of 22 to 28% in TOX formation was observed at all reaction times. At any holding time, biotreatment yielded little impact on TOX formation, with or without ozonation.

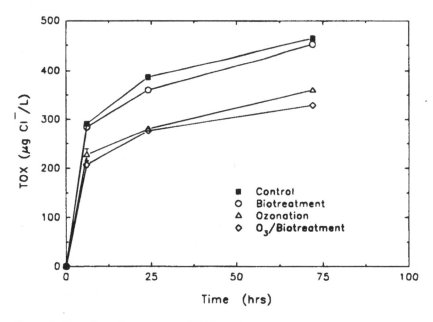

Figure 6 The effect of treatment on TOX formation kinetics.

Total THM formation results are presented in molar concentrations in Figure 7 for the set residual chlorination conditions, and in mass concentrations in Table 2 for both chlorination conditions. On a molar basis (Figure 7), total THM formation decreased only slightly with treatment. Some biological oxidation of the precursors occurred, while ozonation, especially at 24 h, resulted in better precursor removal. The change in total THM formation after combined ozonation/biotreatment was not significantly different from that for ozonation, indicating that THM precursors were selectively oxidized by ozonation. This indicates that ozone is oxidizing more of the biodegradable THM precursors than it is creating biodegradable THM precursors. Selective THM precursor oxidation by ozonation, in comparison to biotreatment, was also observed in a previous study on Ohio River water.[13]

HAA6 formation results are presented in Figure 8 and Table 3. Treatment better controlled HAA6 formation compared to that of the total THMs. HAA6 formation was equally affected by ozonation and biotreatment. The best control for HAA6 was by the combined ozonation/biotreatment. The interpretation of the data is complicated by the lack of quantitation of three other HAAs: bromodichloroacetic acid, dibromochloroacetic acid, and tribromoacetic acid. Therefore, we cannot assume selectivity of biotreatment for the HAA precursors, since a shift to other species of HAAs that are not quantifiable under the current analytical techniques could have occurred. If HAA precursors are

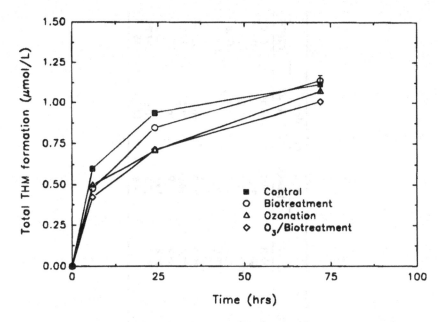

Figure 7   The effect of treatment on total THM formation kinetics.

highly polar compounds, then selectivity would be possible as such precursors are very biodegradable.

Earlier research on the impact of treatment processes on halogenated DBP formation under variable reaction conditions indicates a decrease in formation by either ozonation or biotreatment.[2,6,10,13,25] For HAA precursors, the decrease by biotreatment was shown to be very significant,[13,25] while THM precursor control was better attained by ozonation.[13] The extent of the decrease in DBP formation was dependent on the experimental conditions, particularly the nature of the organic matter, the ozone dose, and the residence time during biotreatment. These results were observed at pH 7. Ozonation at higher pH results in ozone decomposition and in a change in the reaction mechanism of ozone with the organic matter from a 1,3-dipolar cycloaddition mechanism involving molecular ozone to a hydroxyl radical mechanism that is less selective for the precursors.[2] Hydroxyl radicals will induce ozone decomposition and may even generate sites that are more precursors than the initial structures because of hydroxylation reactions.[26] pH variability during chlorination will also impact THM and HAA formation.[27] In this study, where the precursor concentrations and chlorination conditions were kept constant, DBP control was shown to be a result of selective biological or chemical oxidation of the organic precursors and not because of a change in the precursor concentration.

Table 2 The Effect of Treatment on THM Mass Concentrations in µg/l

Treatment	Time (h)	Set chlorine dose				Set chlorine residual			
		$CHCl_3$	$CHBrCl_2$	$CHBr_2Cl$	TTHM	$CHCl_3$	$CHBrCl_2$	$CHBr_2Cl$	TTHM
Control	6	65.2	17.5	3.10	85.8	58.9	15.3	2.84	77.0
	24	90.2	25.7	4.55	121	90.4	25.7	4.80	121
	72	110	29.8	4.35	144	107	31.8	5.60	144
Biotreatment	6	49.9	14.5	2.71	67.1	46.2	12.9	2.69	61.8
	24	81.0	26.3	4.36	112	78.8	26.8	4.71	110
	72	105	33.8	5.76	145	108	34.2	5.15	147
Ozonation	6	42.1	21.4	10.2	73.7	38.0	21.0	10.8	69.7
	24	60.8	28.5	12.1	101	57.4	28.0	11.7	97.1
	72	84.2	34.2	13.0	131	83.2	48.9	16.4	149
O_3/ biotreatment	6	41.4	25.4	9.64	76.4	30.2	19.4	10.5	60.1
	24	61.5	33.1	11.2	106	57.7	29.2	10.9	97.7
	72	75.9	48.3	15.2	139	87.5	34.9	12.7	135

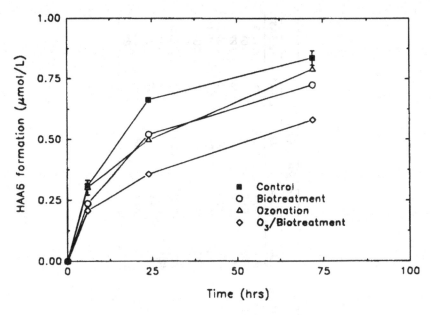

Figure 8 The effect of treatment on HAA6 formation kinetics.

SPECIATION—KINETICS

To examine the impact of treatment on THM speciation kinetics the bromine incorporation factor for THMs, n, is shown in Figure 9 for the set residual conditions. The change in n is indicative of the speciation shift that occurs after treatment. An increase in n indicates a shift to the more bromo-substituted species. The relative amount of bromo substitution did not change significantly in either the control or the biotreated GW NOM as the reaction proceeded. The increase in bromo-substituted species after ozonation was significant, with and without biotreatment, particularly under the set residual conditions where the chlorine dose was much lower than in the set dose condition, that is, the bromide to average FAC ratio was higher than at a set dose. After 6 h reaction time, n of the ozonated solutions increased by 131% for the set residual conditions. After 72 h n was still 91% higher. As the holding time was increased from 6 to 24 and 72 h, the formation of both TTHM-Cl and TTHM-Br increased and resulted in an increase in the total THMs, but the increase in the chloro products was more significant. Hence, a smaller ratio of the bromo-substituted compounds to the total, n, was observed. The combination of ozonation and biotreatment yielded results similar to those of ozonation alone. These results are similar to those reported in a study on Lake Austin water in which bromide readjustment was attempted and a shift to bromo-substituted DBPs was observed after ozonation.[28]

Page 326

DISINFECTION BY-PRODUCTS IN WATER TREATMENT

Table 3 The Effect of Treatment on HAA Mass Concentrations in µg/l

Treatment	Time (h)	Set chlorine dose					Set chlorine residual				
		DCAA	TCAA	BCAA	DBAA	HAA6	DCAA	TCAA	BCAA	DBAA	HAA6
Control	6	21.5	44.0	4.98	0.54	71.0	16.9	25.1	4.28	0.10	46.3
	24	29.2	63.3	6.04	0.43	99.0	30.6	63.3	6.32	0.36	101
	72	36.1	78.3	7.58	0.58	123	38.7	80.4	7.41	0.54	127
Biotreatment	6	14.4	29.1	3.38	<0.1	46.9	13.0	18.9	3.37	0.10	35.3
	24	27.7	50.9	5.11	0.38	84.1	24.3	48.5	5.87	0.39	79.1
	72	36.3	65.0	7.09	<0.1	108	39.5	61.7	6.84	0.41	109
Ozonation	6	18.7	27.1	6.31	0.7	52.8	16.2	21.4	7.04	0.96	45.6
	24	28.5	42.1	8.54	0.89	80.0	27.2	38.0	8.75	1.06	75.0
	72	41.9	64.7	11.0	1.22	119	42.1	64.3	11.0	1.32	119
O₃/biotreatment	6	14.8	21.4	5.85	0.66	42.7	11.5	13.3	5.56	0.77	31.1
	24	23.4	30.1	7.6	0.89	62.0	21.2	22.7	8.39	1.15	53.5
	72	37.3	42.9	10.7	1.14	92.1	35.0	36.8	10.3	1.24	83.4

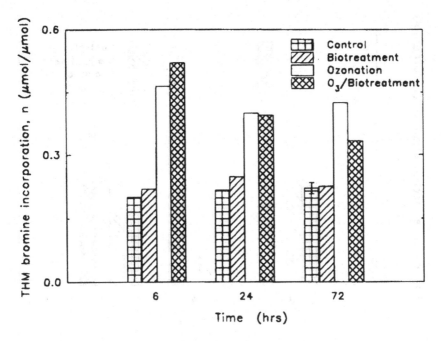

Figure 9 Impact of treatment on THM speciation kinetics.

The change in speciation behavior with treatment for the HAAs was similar to that for the THMs, as shown for the HAA bromine incorporation factor, n', in Figure 10. Similar to n, for the THMs, the most significant increase was after ozonation, with and without biotreatment. The relative amount of bromo-substituted DBPs, n and n', was highest at a 6-h reaction time compared to longer reaction times. Bromine substitution is faster than that of chlorine.[10] With increasing reaction time, the remaining reactive sites become more chloro-substituted, and the ratio of DBP-Br to the DBPs decreases as the total DBP increases. The results for the individual species, on a mass basis, are presented in Tables 2 and 3 for the THMs and the HAAs, respectively. Ozonation resulted in a shift to the bromo-substituted species. Chloroform formation decreased by 22 to 35% under both chlorination conditions, bromodichloromethane increased by 13 to 22% at a set dose and 8 to 37% at a set residual, and dibromochloromethane increased significantly, about 200% for the set dose and 193 to 280% for the set residual chlorination conditions. For the HAAs, all species decreased significantly by biotreatment. After ozonation, DCAA formation was unaffected, TCAA concentrations decreased, while an increase in BCAA and DBAA occurred.

The chlorination kinetics indicate the presence of fast- and slow-reacting fractions of NOM. These precursors can react with either chlorine or bromine

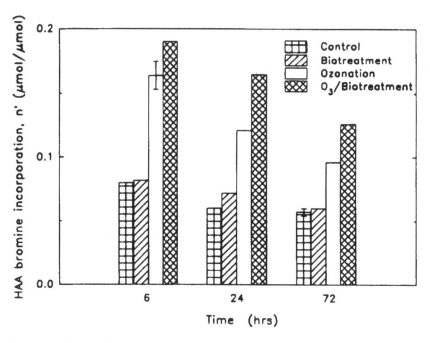

Figure 10   Impact of treatment on HAA speciation kinetics.

to form halogenated DBPs. The available precursors are most likely the same, but the extent and the rate of formation of bromo- or chloro-substituted species may be different and will be influenced by the reaction conditions such as time, temperature, pH, and bromide and chlorine concentrations. In a study by Merlet,[29] of the reaction of either chlorine or bromine with known compounds, likely THM precursors, at pH 7.2 and 15-h holding time, the following results were observed: phloroglucinol yielded the same molar concentrations of chloroform or bromoform (95%), resorcinol resulted in 93% chloroform and 55% bromoform, whereas acetyl acetone, a slower reacting compound, yielded 3% chloroform and 85% bromoform. In that study, a 100% formation was defined as 1 mol of THM formed for each mole of precursor. Methyl acetyl functions did not exhibit high reactivity at neutral pH, but increased with increasing pH. Polyhydroxybenzene structures showed maximum reactivity at neutral pH. Therefore, although the precursors may be the same, the reaction conditions and pathways, mechanisms, kinetics, steric effects, and inductive effects will dictate the speciation. This is further complicated when both chlorine and bromine are available. In this case, the speciation depends on the competition between the halogenating species for the available organic precursor sites.

The shift in speciation upon ozonation can be explained by the following: ozone reacts with organic precursors and converts fast-reacting precursors to slower ones. At the same time ozone reacts with the inorganic precursor bromide to convert it to bromine. In subsequent chlorination reactions, a competition between HOCl and HOBr occurs. With no preozonation, chlorine has to react with bromide first to oxidize it to bromine for it to substitute on the reactive sites. At the same time, chlorine will substitute on some reactive organic sites resulting in halo-substituted DBPs. The aromatic moieties, which are fast-reacting components, will tend to become chloro substituted faster. These fast-reacting precursors were found to be preferentially destroyed by ozone in comparison to the slow-reacting precursors.[30] On the other hand, ozone oxidizes the aromatic functions to yield aliphatic ketones, aldehydes, and carboxylic acids. If these aliphatic moieties contain activated sites (protons α to a carbonyl function), then these are more susceptible to bromo substitution, because of kinetics. Bromo-substituted methyl groups are bulky and would have the tendency to leave the parent compound easily because of steric effects. Merlet[28] has shown a higher yield of bromoform formation from the reaction of bromine with acetylacetone, compared to its reaction with chlorine.

With treatment and with increasing reaction time, further chlorination substitution reactions become precursor limited. The decrease in available reactive precursors by ozonation can be inferred from the SAC results which were 43% lower in the ozonated diluted samples (Table 1), even though the DOC concentration was constant.

The effect of the change in reactive precursor availability can be seen by examining both the bromo- and chloro-substituted DBPs for the THMs and HAAs, Figures 11 and 12, respectively. The results presented are for the 24-h formation set residual (1.0 ± 0.3 mg/l) scenarios. These conditions were selected for comparison because the Water Utility Database of the American Water Works Association indicates that the average mean residence time in distribution systems of utilities serving greater than 100,000 people in the U.S. is 1.3 d, while the average mean chlorine residual is 0.9 mg/l.[27] For the THM formation (Figure 11), a significant increase in TTHM-Br was observed only after ozonation, 39%. A simultaneous decrease in TTHM-Cl, 29%, occurred due to less-available organic precursor sites resulting in a 24% decrease in the total THM formation and a significant increase in n, 84%. No change in TTHM-Br occurred after biotreatment although a slight decrease in TTHM-Cl was observed.

For HAA6 formation, the behavior was similar. Treatment, ozonation, biotreatment, or the combination of both resulted in a decrease in HAA6-Cl, 22, 30 and 53%, respectively, and HAA6-Br increased significantly by ozonation, 50%. These results indicate that precursor limitation affects the chloro-substituted species more significantly than the bromo-substituted, as the formation of the latter is governed by the faster bromine substitution reactions.

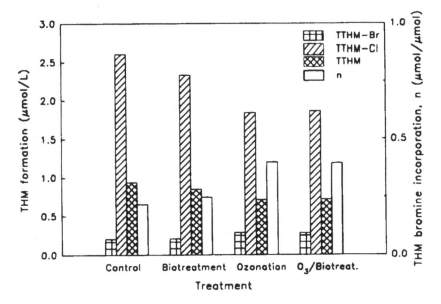

Figure 11 The effect of treatment on chloro- and bromo-substituted THMs (24-h formation).

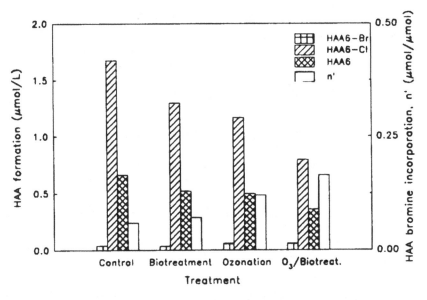

Figure 12 The effect of treatment on chloro- and bromo-substituted HAAs (24-h formation).

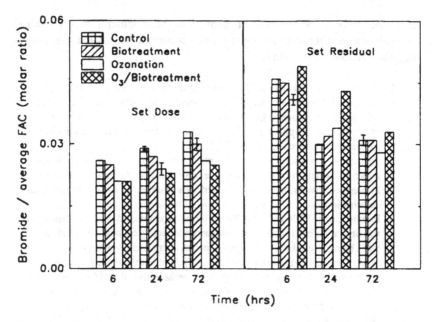

Figure 13   The change in bromide to average FAC molar ratio with treatment.

IMPACT OF FREE AVAILABLE CHLORINE

In this study, the chlorine dose and residuals were controlled by the change in the chlorine demand. The choice of chlorination at both set dose or set residual was selected to minimize the effect of chlorine demand variability and to determine its contribution to the DBP speciation. The impact of the change in the molar ratio of bromide to the average FAC with treatment, at all incubation times, for both chlorination conditions is illustrated in Figure 13. At a given reaction time, the bromide to average FAC ratio was relatively constant throughout treatment. The ratio was higher, however, at a set residual than at a set dose. The difference in the bromide to average FAC ratio between both chlorination conditions, set dose and set residual, did not seem to impact the speciation significantly, because the most significant change in speciation was caused by ozonation, Figures 9 to 12. The similarity between the set dose and set residual chlorination conditions observed in this study may be attributed to the similar chlorine doses used in both approaches.

SUMMARY AND CONCLUSIONS

DBP formation and speciation were investigated under conditions of constant organic and inorganic precursors, and chlorination conditions. Under these conditions, the impact of ozonation, biotreatment, and combined ozonation/biotreatment on the organic precursors could be isolated. The following observations were found:

1. Ozonation resulted in a significant change in the characteristics of the organic matter. SAC decreased while DOC concentration was nearly unchanged, indicating a decrease in UV-absorbing functional groups and the formation of saturated and aliphatic acetyl compounds. Biotreatment resulted in equivalent removal of both SAC and DOC.
2. Treatment, ozonation, or biotreatment decreased chlorine demand indicating selective oxidation of the organic precursors either chemical or biological.
3. In nearly all cases, treatment resulted in a decrease in DBPs formed after chlorination.
4. In most cases, TOX and total THM formation decreased more by ozonation and ozonation/biotreatment in comparison to biological treatment, indicating that chemical oxidation of organic matter decreased its reactivity.
5. HAA6 formation decreased the most by biotreatment, with and without preozonation. Biotreatment appears to be selective for HAA precursors. However, such a conclusion is only tentative as the decrease in HAA concentration could be due to a shift to the other three more-bromo-substituted HAA species that are not quantified.
6. Speciation was affected most by ozonation. A shift to bromo-substituted species occurred after ozonation and after combined ozonation/biotreatment. No such shift was observed by biotreatment alone. Both n and n', bromine incorporation factors, increased significantly with ozonation. The effect was most pronounced at 6 h. As the holding time was increased, more TTHM-Cl and HAA6-Cl were formed increasing DBP formation. The relative increase in DBP-Br with increasing holding time was much smaller.
7. SAC decreased significantly after ozonation resulting in less-available reactive aromatic unsaturated organic precursors than in the control, even when the DOC concentration was held constant. The reactive aliphatic (acetyl-) containing precursors created after ozonation seem to favor bromine substitution over chlorine substitution. For the THMs and quantifiable HAAs, DBP-Br formation increased while DBP-Cl formation was much less, relative to the control, because of decreases in the organic precursors.
8. Precursor limitation is very important is determining speciation. In the case where the reaction is precursor limited, bromine substitution is faster than chlorine substitution and will govern the DBP speciation. The formation of chloro-substituted DBPs continues as long as there are available precursors and thus control DBP distribution.

In conclusion, under conditions of constant bromide, DOC, chlorination conditions, and holding times, biotreatment did not show any selectivity for DBP precursors, that is, DOC and the precursors were removed to the same extent. Ozonation, however, had more of an impact on the organic matter characteristics, as observed by a decrease in DBP formation (reactivity) and a shift to the bromo-substituted compounds. Precursor limitations are important in assessing DBP speciation; both DOC and SAC are important parameters that should be considered in evaluating organic precursors.

ACKNOWLEDGMENTS

The authors wish to acknowledge the significant guidance of Dan Hautman, International Consultants Inc., U.S. Environmental Protection Agency (USEPA), Cincinnati, in the startup of the bromide analysis at the University of Cincinnati (UC), the assistance of Nicholas R. Dugan (UC), and the analytical support of the USEPA in HAA analyses. The authors would also like to thank and acknowledge Dr. David Reckhow, University of Massachusetts at Amherst, for his contribution and helpful discussion on data interpretation. This work was funded through a cooperative agreement CR-816700 between USEPA's Drinking Water Research Division and the University of Cincinnati. Although the research described was funded by the USEPA, it has not been subjected to agency review and therefore does not necessarily reflect the view of the agency. Mention of trade names or commercial products does not constitute endorsement or recommendation for use by the USEPA.

REFERENCES

1. Haag, W. R. and Hoigné, J., Ozonation of bromide-containing waters: kinetics of formation of hypobromous acid and bromate, *Environ. Sci. Technol.*, 17, 5, 261, 1983.
2. Langlais, B., Reckhow, D. A., and Brink, D. R., Eds., *Ozone in Water Treatment: Application and Engineering*, Lewis Publishers, Chelsea, MI, 1991.
3. Glaze, W. H., Weinberg, H. S., and Cavanagh, J. E., Evaluating the formation of brominated DBPs during ozonation, *J. AWWA*, 85, 1, 96, 1993.
4. Amy, G. L., Tan, L., and Marshall, K. D., The effect of ozonation and activated carbon adsorption on trihalomethane speciation, *Water Res.*, 25, 2, 191, 1991.
5. Summers, R. S., Benz, M. A., Shukairy, H. M., and Cummings, L., Effect of separation processes on the formation of brominated THMs, *J. AWWA*, 85, 1, 88, 1993.
6. Shukairy, H. M., Miltner, R. J., and Summers, R. S., Bromide's effect on DBP formation, speciation and control. I. Ozonation, *J. AWWA*, 86, 6, 72, 1994.
7. Minear, R. A. and Bird, J. C., Trihalomethanes: impact of bromide ion concentration on yield, species distribution, rate of formation and influence of other variables, in *Water Chlorination: Environmental Impact and Health Effects*, Vol. 3, Jolley, R. L. et al., Eds., Ann Arbor Science Publishers, Ann Arbor, MI, 1980.
8. Oliver, B. G., Effect of temperature, pH, and bromide concentration on the trihalomethane reaction of chlorine with aquatic humic material, in *Water*

Chlorination Environmental Impact and Health Effects, Vol. 3, Jolley, R. L. et al., Eds., Ann Arbor Science Publishers, Ann Arbor, MI, 1980.

9. Rook, J. J. and Gras, A. A., Bromide oxidation and organic substitution in water treatment, *J. Environ. Health,* A13, 2, 91, 1978.

10. Symons, J. M., Krasner, S. W., Simms, L. A., and Sclimenti, M., Measurement of THM and precursor concentrations revisited: the effect of bromide ion, *J. AWWA,* 85, 1, 51, 1993.

11. Siddiqui, M. S. and Amy, G. L., Factors affecting DBP formation during ozone-bromide reactions, *J. AWWA,* 85, 1, 63, 1993.

12. Krasner, S. W., Glaze, W. H., Weinberg, H. S., Daniel, P. A., and Najm, I. N., Formation and control of bromate during ozonation of waters containing bromide, *J. AWWA,* 85, 1, 73, 1993.

13. Shukairy, H. M., Miltner, R. J., and Summers, R. S., Bromide's effect on DBP formation, speciation and control: II. Biotreatment, *J. AWWA,* 87, 10, 1995.

14. Gould, J. P., Fitchorn, L. E., and Urheim, E., Formation of brominated trihalomethanes: extent and kinetics, in *Water Chlorination: Environmental Impact and Health Effects,* Vol. 4, Jolley, R. L. et al., Eds., Ann Arbor Science Publishers, Ann Arbor, MI, 1983.

15. Shukairy, H. M., The Control of Disinfection By-Product Formation by Ozonation and Biotreatment, Ph.D. dissertation, University of Cincinnati, 1994.

16. *Standard Methods for the Examination of Water and Wastewater,* 16th ed., APHA, AWWA and WPCF, Washington, D.C., 1985.

17. *Standard Methods for the Examination of Water and Wastewater,* 18th ed., APHA, AWWA and WEF, Washington, D.C., 1992.

18. Allgeier, S., Summers, R. S., and Shukairy, H. M., Formation and control of bromate during ozonation, Proc. AWWA Annu. Conf., Anaheim, Calif., 1995.

19. Hautman, D. P. and Bolyard, M., Using ion chromatography to analyze disinfection by-products, *J. AWWA,* 84, 11, 88, 1992.

20. Trussell, R. R. and Umphres, M. D., The formation of trihalomethanes, *J. AWWA,* 70, 11, 604, 1978.

21. Amy, G. L., Chadik, P. A., and Chowdhury, Z. K., Developing models for predicting trihalomethane formation potential and kinetics, *J. AWWA,* 79, 7, 89, 1987.

22. Shukairy, H. M. and Summers, R. S., The impact of preozonation and biodegradation on disinfection by-product formation, *Water Res.,* 26, 9, 1217, 1992.

23. Jadas-Hécart, A., El Morer, A., Stitou, M., Bouillot, P., and Legube, B., The chlorine demand of a treated water, *Water Res.,* 26, 8, 1073, 1992.

24. Miltner, R. J., Shukairy, H. M., and Summers, R. S., Disinfection by-product control by ozonation and biotreatment, *J. AWWA,* 84, 11, 53, 1992.

25. Speitel, G. E., Jr., Symons, J. M., Diehl, A. C., Sorenson, H. W., and Cipparone, L. A., Effect of ozone dosage and subsequent biodegradation on removal of DBP precursors, *J. AWWA,* 85, 5, 86, 1993.

26. Xiong, F. and Legube, B., Enhancement of radical chain reactions of ozone in water in the presence of an aquatic fulvic acid, *Oz. Science Eng.,* 13, 3, 349, 1991.

27. Summers, R. S., Hooper, S. M., Shukairy, H. M., and Owen, D. M., Development of uniform formation conditions for the assessment of disinfection by-product formation, Proc. AWWA Annu. Conf., New York, 1994.

28. Symons, J. M., Krasner, S. W., Simms, L. A., and Sclimenti, M., Measuring Trihalomethane Species and Trihalomethane Precursor Concentrations in the Presence of Bromide Ion, presented at the 206th American Chemical Society National Meeting, Division of Environmental Chemistry, August 1993.

29. Merlet, N., Contribution a l'Étude du Mécanisme de Formation des Trihalméthanes et des Composés Organohalogénés non Volatils Lors de la Chloration de Molécules Modèles, Ph.D. dissertation, Université de Poitiers, 1986.

30. Reckhow, D. A., Zhu, Q., Weiner, J., and MacNeil, A., Effect of Ozonation on the Kinetics of DBP Formation, presented at the IOA Conference, San Francisco, 1993.

PART IV

Chloramines and Chlorine Dioxide

Verification of a Rational Kinetic Model for Chloroform Formation from a Model Precursor during Water Chlorination

Charles N. Haas and Kirankumar Topudurti

INTRODUCTION

Formation of trihalomethanes and other disinfection by-products (DBPs) is clearly an issue of supreme importance in the design and operation of potable water treatment plants.[1-3] To date, however, the information on formation of trihalomethanes THMs and other by-products has been presented using empirical correlations of field or laboratory data[4-11] or quasi-mechanistic models with limited verification.[12,13] Currently anticipated revisions in the drinking water regulations may necessitate even closer control of DBP formation.[14] The authors have pursued the present research to develop a mechanistically based model of chloroform formation, in the hope that it may form the basis for a more rational design and optimization of water disinfection processes.

In previous work, the authors have reported on the kinetics of chloroform formation by the reaction of the model DBP precursor, phloroacetophenone ($C_8H_8O_4$—PAP) with free chlorine and monochloramine (NH_2Cl). We established the reaction stoichiometry and kinetics, and implicated a direct transfer reaction between monochloramine and PAP to form $CHCl_3$.[15,16] The prior work was conducted in systems in which only free chlorine or monochloramine was present, and in the absence of extraneous organic matter other than the model precursor compound. The objective of the present work was to verify the kinetic model in a system in which chlorine residual was dynamically changing; as would occur during water treatment processes.

MATERIALS AND METHODS

Upon addition to a water, chlorine (as $HOCl$ or OCl^-) can react with ammonia nitrogen to form inorganic chloramines (at low Cl:N ratio, principally

1-56670-136-8/96/$0.00+$.50

Table 1 Summary of Experimental Conditions Used in Verification: Batch
         Chloroform Formation Runs

Initial concentrations (μM)	Run 1	Run 2	Run 3	Run 4	Run 5
Free Cl₂	300	200	300	300	150
PAP	160	160	160	160	120
Ammonia	400	400	400	700	170
Glycine	0	0	300	0	170

monochloramine), or it can react with organic amines to form organic chloramines, or it can react with by-product precursor compounds to form halogenated and nonhalogenated DBPs. It was desired to study a system that included
each of these potential reaction pathways, yet was defined so as to permit the
determination (in separate experiments, previously reported) of reaction rate
constants and stoichiometries. Therefore, in addition to chlorine and ammonia,
glycine (as an organic amine) and 2', 4', 6'-trihydroxyacetophenone (phloroacetophenone, abbreviated as PAP) were used as reactants.

The system examined was prepared in pH 7 phosphate buffer (0.01 M)
made in demand-free water. Varying amounts of chloroform precursor (PAP),
free chlorine (FAC), ammonia nitrogen, and glycine (GLY) were added at
time zero to a well-mixed batch reactor. The time dependence (from 30 to
9000 s) of chloroform production (measured by gas chromatography) and
chlorine residual (measured amperometrically) were determined. The data
were compared to predictions from a kinetic model accounting for principal
relevant reactions, specifically including monochloramine and N-chloroglycine
(NCG). Rate constants and stoichiometries for these reactions were obtained
from the literature, or determined in separate well-controlled systems.

A glass beaker of 150-mm diameter was used as a batch reactor in experiments where chloroform formation was measured. To minimize stripping, the
reactor was operated at a low mixing intensity with a small freeboard, and
also was kept tightly covered with two layers of Parafilm™. The liquid volume
of the reactor was 4000 ml. The contents of the reactor were mixed with a
75-mm long magnetic stir bar at 80 rpm. No vortex was formed in the reactor
at this mixing condition.

Five combinations of initial concentrations of reactants were employed in
this study (Table 1). One set of conditions (labeled as run 4) was repeated
and served to provide an estimate of intrinsic experimental error. These conditions were chosen to span ranges of organic and inorganic nitrogen ratios that
might be seen under practical conditions.

In additional experiments, the predictions of kinetic models were tested
by comparing predicted combined residual chlorine concentrations (being the
sum of monochloramine and NCG) and observed combined residual chlorine
concentrations. A continuous flow completely stirred reactor (CSTR) was
chosen for this purpose. CSTRs of two different liquid volumes (250 and

Table 2 Experimental Conditions Used in CSTR Chlorine Residual
Verification Runs

Initial concentrations (μM)	Series A	Series B	Series C
Free Cl_2	200	100	300
PAP	100	150	50
Ammonia	200	200	300
Glycine	200	200	300

2500 ml) were used to facilitate the kinetic study over a wide range of residence times (30 to 1800 s). The CSTR was fed with two simultaneous streams: one contained buffered free chlorine alone and the second contained all other materials (e.g., glycine, ammonia, PAP) also in a buffer.

A step input tracer study was carried out using NaCl. It was determined that, under mixing conditions used in this study, ideal CSTR behavior was closely approximated. Furthermore, analysis of mixing inputs according to the packet diffusion model[17] showed that is was reasonable to assume complete micromixing.

Three sets of influent conditions were each duplicated in the CSTR experiments. Each set of conditions was run at mean hydraulic residence time of 4 and 20 min. Table 2 summarizes the feed concentrations of reactants for these runs. Sampling of the CSTRs for estimation of steady state composition was performed after at least three residence times had elapsed.

PREPARATION OF REAGENTS AND ANALYTICAL METHODS

All experiments were carried out at neutral pH and at $25 \pm 2°C$. Phosphate buffer of pH 7 was used as the dilution medium to prepare the reagents. The final concentration of the phosphate buffer was 0.01 M in all the solutions used in this study.

Clorox®, a commercial bleach containing 5.25% NaOCl manufactured by the Clorox Company (U.S.), was used as the source of FAC. This did not have detectable levels of bromine as determined by a method developed by Topudurti and Haas[18] at the chlorine concentrations used in this study. Stock monochloramine (NH_2Cl) was prepared from solutions of NH_4Cl and FAC, both adjusted to pH 10 with NaOH prior to mixing. A 10% stoichiometric excess NH_4Cl was employed in the preparation of NH_2Cl to suppress the hydrolysis of NH_2Cl, and the formation of dichloramine.[19] FAC and NH_2Cl were measured by forward amperometric titration. No trace of titratable FAC or dichloramine was found in the NH_2Cl prepared in this study as determined by the forward amperometric titration method.

PAP, manufactured by Aldrich Chemical Company, Inc. (U.S.) and marketed under the name 2′, 4′, 6′-trihydroxyacetophenone monohydrate (Figure 1), was measured spectrophotometrically ($\varepsilon_{max} = 14,400$ M^{-1} cm^{-1},

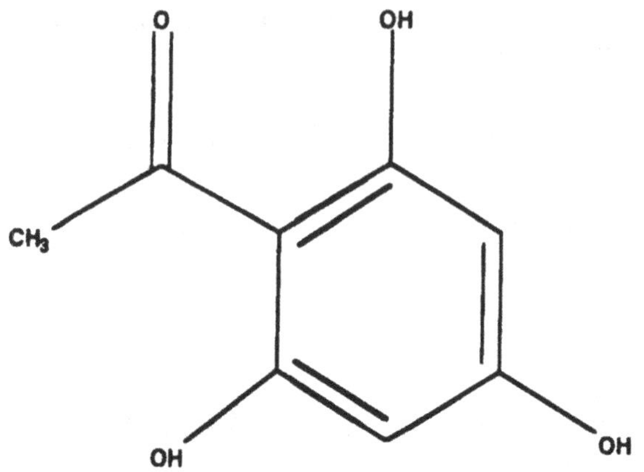

Figure 1 Structural formula of PAP.

and λ_{max} = 286 nm) with a Perkin Elmer λ3A UV/VIS spectrophotometer using a quartz cell. The absorbance due to either FAC or NH_2Cl at the concentrations present in this study was not a significant interference ($<2\%$) in the measurement of PAP.

Gas chromatographic (GC) determination of chloroform ($CHCl_3$) was performed with a Varian Model 3700GC, equipped with a ^{63}Ni electron capture detector. A 6-ft long stainless steel column with an OV-101 liquid phase was used. The column, injector, and detector temperatures were 50, 150, and 250°C, respectively. Peak area measurements were made automatically with a Varian CDS 401. The coefficient of variation (standard deviation/mean) of $CHCl_3$ measurement was about 5%. The samples (25 ml) collected for $CHCl_3$ measurement were dechlorinated with 0.2 ml of 1 M $Na_2S_2O_3$ and preserved at 4°C in tightly capped glass bottles with no headspace. The samples were brought back to room temperature before they were analyzed for $CHCl_3$. Several controls such as phosphate buffer, PAP, FAC, and sodium thiosulfate were run, and no chemical reagent used in this study was found to contain an impurity that eluted at the same time as chloroform.

RESULTS AND DISCUSSION

The experimental measurements will first be summarized, then a comparison with predictions of the model will be made. It should be reemphasized that the data set taken for model verification was independent of the data used to develop or verify individual reaction kinetics; the former data sets were generally taken on simpler systems (where only a single reaction could be isolated for study).

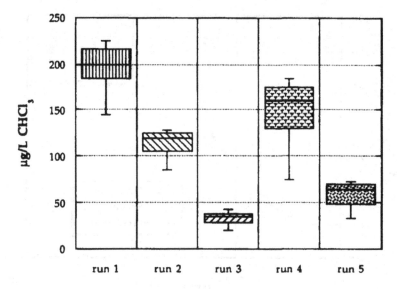

Figure 2   Summary plot of chloroform. Boxes enclose 50% of data. Bars show range of 90% of data.

EXPERIMENTAL FINDINGS

Figure 2 presents a summary plot of the measured chloroform concentrations taken in each of the runs. Figure 3 is an example (for run 4) of the time-dependent increase in chloroform concentration. Samples were taken sufficiently early so as to be able to measure the increase in chloroform, as well as sufficiently late so as to get a good estimate of the asymptotic chloroform levels.

THM FORMATION MODEL

Figure 4 presents a conceptual summary of the principal elements of the reaction model formulated on the basis of prior work.[15,16,20] Free chlorine may react with ammonia, glycine, or PAP to produce monochloramine, N-chloroglycine, or chloroform, respectively. Ammonia may react with PAP to produce chloroform and other products. PAP reacts with N-chloroglycine to produce inert products, but not active chlorine or chloroform. The rates of each of the individual reactions have either been measured by others, or in the prior work of the authors. These individual rates will now be summarized.

The following summarizes the kinetic expressions used to model the results of this study. In these expressions, concentrations are on a molar basis, and time is in seconds. Temperature is in Kelvin.

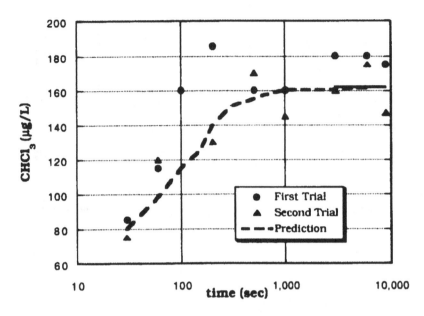

Figure 3   Chloroform formation vs. time (run 4). Results of two replicate trials compared to predictions of model are shown.

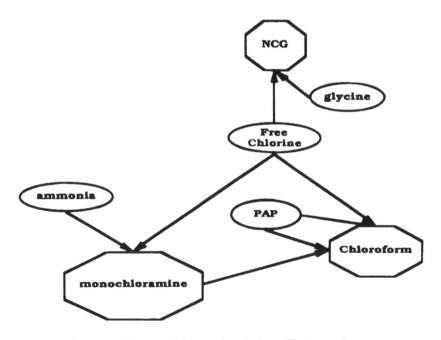

Figure 4   Conceptual diagram of the reactions in the verification system.

$$\text{I} \quad NH_3 + HOCl \Leftrightarrow NH_2Cl + H_2O$$

The forward and reverse rates were those suggested in the review of Morris and Isaac,[21] which are also consistent with the hydrolysis equilibrium constant reported by Gray et al.[22] The forward and reverse rates are given by Equations 1 and 2:

$$r_{NH_2Cl} = 6.6 \times 10^8 \exp\left(-\frac{1510}{T}\right)[HOCl][NH_3] \tag{1}$$

$$-r_{NH_2Cl} = 1.38 \times 10^8 \exp\left(-\frac{8800}{T}\right)[NH_2Cl] \tag{2}$$

These constants were verified in individual experiments run as part of this study.[20]

$$\text{II} \quad GLY + HOCl \Leftrightarrow NCG + H_2O$$

This reaction was treated as irreversible, with the following rate.[23]

$$r_{NCG} = 3.1 \times 10^{10} \exp\left(-\frac{3.5}{T}\right)[CH_2NH_2COO^-][HOCl] \tag{3}$$

$$\text{III} \quad GLY + NH_2Cl \Leftrightarrow NCG + NH_3$$

This reaction was treated as irreversible, with the rate from studies at pH 7 and 25°C determined by[23,24]

$$r_{NCG} = 1.0[glycine][free\ chlorine] \tag{4}$$

$$\text{IV} \quad NCG \Leftrightarrow unknown\ nonreactive\ products$$

The products of this reaction were found to include neither reactive chlorine nor glycine. This rate was determined by loss of residual and UV absorbancy in batch systems as follows, and was also in agreement with a prior study:[25]

$$-r_{NCG} = 1.07 \times 10^{-5}[NCG] \tag{5}$$

$$\text{V} \quad PAP + 12\ HOCl \Leftrightarrow 4CO_2 + 4CHCl_3 + 8H_2O$$

Under the pH and temperature conditions used in this work, this rate was determined to be[16]

$$-r_{PAP} = 29[HOCl][PAP] \tag{6}$$

VI   PAP + 19NH$_2$Cl $\Leftrightarrow$ 19NH$_3$ + 0.05 CHCl$_3$

+ unknown nonreactive products

The reaction rate was determined to be[16]

$$-r_{PAP} = 5.7[NH_2Cl][PAP] \tag{7}$$

VII   PAP + 5 NCG $\Leftrightarrow$ 5 GLY + unknown nonreactive products

This PAP-dependent rate of NCG decomposition was measured in prior work to be[15,20]

$$-r_{NCG} = 204[NCG][PAP] \tag{8}$$

For a batch system, the overall model can be summarized as seven simultaneous differential equations (with time in minutes) as follows:

$$\frac{d[HOCl]}{dt} = -9.42 \cdot 10^5[HOCl][NH_{3,T}] + 1.24 \cdot 10^{-3}[NH_2Cl]$$

$$- 1.13 \cdot 10^7[HOCl][Gly] + 3 \cdot 10^{-4}[NCG]$$

$$- 2.11 \cdot 10^4[HOCl][PAP]$$

$$\frac{d[NH_{3,T}]}{dt} = -9.42 \cdot 10^5[HOCl][NH_{3,T}] + 1.24 \cdot 10^{-3}[NH_2Cl]$$

$$+ 60[NH_2Cl][Gly] + 6.5 \cdot 10^3[NH_2Cl][PAP]$$

$$\frac{d[Gly]}{dt} = -1.13 \cdot 10^7[HOCl][Gly] + 3 \cdot 10^{-4}[NCG]$$

$$- 60[NH_2Cl][Gly] + 1.21 \cdot 10^4[NCG][PAP]$$

$$\frac{d[PAP]}{dt} = -1.76 \cdot 10^3[HOCl][PAP] - 342[NH_2Cl][PAP]$$

$$- 2.41 \cdot 10^3[NCG][PAP]$$

$$\frac{d[NH_2Cl]}{dt} = 9.42 \cdot 10^5[HOCl][NH_{3,T}] - 1.24 \cdot 10^{-3}[NH_2Cl]$$

$$- 60[NH_2Cl][Gly] - 6.5 \cdot 10^3[NH_2Cl][PAP]$$

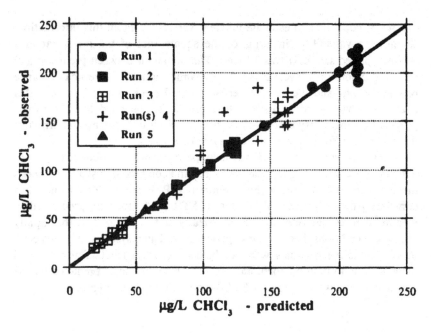

Figure 5 Comparison of predicted and observed chloroform concentrations: all experimental runs.

$$\frac{d[NCG]}{dt} = 1.13 \cdot 10^7 [HOCl][Gly] - 9.1 \cdot 10^{-4} [NCG]$$

$$+ 60[NH_2Cl][Gly] - 1.21 \cdot 10^4 [NCG][PAP]$$

$$\frac{d[CHCl_3]}{dt} = 7 \cdot 10^3 [HOCl][PAP] + 17[NH_2Cl][PAP]$$

The reaction scheme (I to VII) was solved for the conditions of each of five batch experiments using Gears method (subroutine DGEAR) of the International Mathematical and Statistical Library. Computations were conducted on a VAX 11/780 system, using tolerance and convergence criteria of 10^{-7} and 10^{-15}, respectively.

For a CSTR system (such as that used to assess chlorine residual), the above mechanisms produce a set of simultaneous nonlinear algebraic equations. These were solved for steady-state conditions and compared to observed combined chlorine levels.

COMPARISON OF MODEL PREDICTIONS AND EXPERIMENTAL RESULTS

A comparison of predicted vs. observed THM levels in the batch experiments is given in Figure 5. There were two separate runs performed under one set of conditions to estimate inherent variability. From this information, an intrinsic variability (in terms of the square root of the mean variance between paired samples) was 17 µg/l. The deviation between predicted and observed chloroform levels using the dynamic model (i.e., vertical distance between points and the line of perfect fit in Figure 1) was less than this (standard deviation of 11 µg/l). Therefore, it is concluded that the model provided a satisfactory prediction (within the limit of experimental error) of the chloroform formation in the test systems studied.

Additional confirmation of the model performance was obtained by comparison of predicted chlorine residuals (free and combined) in separate experiments on the mixed system (FAC, ammonia, PAP, and GLY) conducted in completely mixed stirred tank reactors (CSTRs). CSTRs were used to allow measurement of chlorine residual at "steady state", rather than the rapidly changing conditions that may be experienced in batch systems. As predicted, the free chlorine residuals were all below detection level. The combined chlorine (due to both monochloramine and NCG) residuals predicted by the dynamic model were within 15% of the observed values (Figure 6).

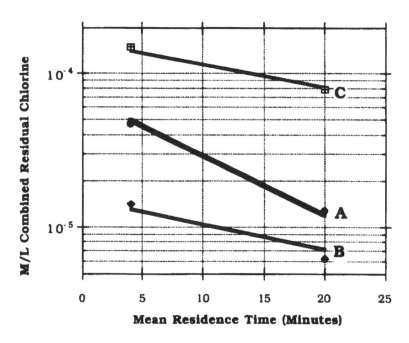

Figure 6 Comparison of predicted and observed combined chlorine residuals (monochloramine + N-chloroglycine) in CSTR series experiments. Each point is average of two separate experiments. Lines are model predictions.

CONCLUSIONS

On the basis of this work, a mechanistic kinetic model for the formation of chloroform from a model precursor has been developed. While it is believed that the precursor used in this work is significantly more active than humic materials likely to be in potable water supplies, the approach taken in this study would appear to offer promise in a development of rational kinetic models for THM and DBP formation. In particular, it is significant that, based solely on straightforward measurements of reactivity between a precursor and individual chlorinating agents (FAC, monochloramine, NCG) under controlled conditions, the behavior in a more complex system may be predicted. It may be appropriate to initiate a program to develop a spectrum of measurements for the formation of various DBPs using humics from a number of water supplies and natural waters.

ACKNOWLEDGMENTS

At the time that this work was performed Charles N. Haas and K. Topudurti were, respectively, Professor and Graduate Research Assistant in the Pritzker Department of Environmental Engineering at Illinois Institute of Technology. This work formed a portion of the doctoral dissertation of K. Topudurti at IIT.

REFERENCES

1. Krasner, S. W. *et al.*, The occurrence of disinfection by-products in U.S. drinking water, *J. Am. Water Works Assoc.*, 81(8), 41, 1989.
2. Metropolitan Water District of Southern California and James M. Montgomery Consulting Engineers, Inc., Disinfection By-Products in United States Drinking Waters, U.S. Environmental Protection Agency and Association of Metropolitan Water Agencies, 1989.
3. LARC Working Group on the Evaluation of Carcinogenic Risks to Humans, *Chlorinated Drinking-Water, Chlorination By-Products, Some Other Halogenated Compounds: Cobalt and Cobalt Compounds*, World Health Organization, Geneva, 1991.
4. Moore, G. S. et al., A statistical model for predicting chloroform levels in chlorinated surface water supplies, *J. Am. Water Works Assoc.*, 71, 37, 1979.
5. Engerholm, B. A. and Amy, G. L., A predictive model for chloroform formation from humic acid, *J. Am. Water Works Assoc.*, 75, 418, 1983.
6. Urano, K. et al., Empirical rate equation for trihalomethane formation with chlorination of humic substances in water, *Water Res.*, 17, 1797, 1983.
7. Amy, G. L. et al., Developing models for predicting trihalomethane formation potential and kinetics, *J. Am. Water Works Assoc.*, 79(7), 89, 1987a.
8. Amy, G. L. et al., Testing and validation of a multiple nonlinear regression

model for predicting trihalomethane formation potential, *Water Res.*, 21(6), 649, 1987b.

9. Morrow, C. M. and Minear, R. A., Use of regression models to link raw water characteristics to trihalomethane concentrations in drinking water, *Water Res.*, 21, 41, 1987.

10. Hutton, P. H. and Chung, F. I., Simulating THM formation potential in Sacramento delta. II, *J. Water Res. Planning Manage. ASCE*, 118(5), 530, 1992b.

11. Hutton, P. H. and Chung, F. I., Simulating THM formation potential in Sacramento delta. I, *J. Water Res. Planning Manage. ASCE*, 118(5), 513, 1992a.

12. Trussell, R. R. and Umphres, M. D., The formation of trihalomethanes, *J. Am. Water Works Assoc.*, 70(11), 604, 1978.

13. Kavanaugh, M. C. et al., An empirical model of trihalomethane formation: applications to meet the proposed THM standard, *J. Am. Water Works Assoc.*, 72(10), 578, 1980.

14. Pontius, F. W., D-DBP rule to set tight standards, *J. Am. Water Works Assoc.*, 85(11), 22, 1993.

15. Topudurti, K. V. and Haas, C. N., Chloroform formation by the transfer of active chlorine from monochloramine to phloroacetophenone, in *Water Chlorination: Environmental Impact and Health Effects*, Vol. 6, Jolley, R. L. et al., Eds., Lewis Publishers, Chelsea, MI, 1990, 649.

16. Topudurti, K. V. and Haas, C. N., THM formation by the transfer of active chlorine from monochloramine to phloroacetophenone, *J. Am. Water Works Assoc.*, 83(5), 64, 1991.

17. Nauman, E. B. and Buffham, B. A., *Mixing in Continuous Flow Systems*, John Wiley & Sons, New York, 1983.

18. Topudurti, K. and Haas, C. N., Analytical Method for Free and Combined Bromine, 9th Annu. Midwest Water Chemistry Workshop, Chicago, 1986.

19. Isaac, R. A., Transfer of Active Chlorine from Monochloramine to Organic Nitrogenous Compounds, Harvard University, 1981.

20. Topudurti, K. V., Transfer of Active Chlorine from Organic and Inorganic Monochloramines to a THM Precursor, Illinois Institute of Technology, 1988.

21. Morris, J. C. and Isaac, R. A., A critical review of kinetics and thermodynamic constants for aqueous chlorine-ammonia systems, in *Water Chlorination: Environmental Impact and Health Effects*, Vol. 4, Jolley, R. L. et al., Eds., Ann Arbor Science Publishers. Ann Arbor, MI, 1983, 49.

22. Gray, E. et al., Chloramine equilibria and the kinetics of disproportionation in aqueous solution, in *Organometals and Metalloids, Occurrence and Fate in the Environment*, Brenchman, F. E. and Bellama, J. M., Eds., Symposium #82, American Chemical Society, Washington, D.C., 1978, 264.

23. Isaac, R. A. and Morris, J. C., Transfer of active chlorine from chloramine to nitrogenous organic compounds. I. Kinetics, *Environ. Sci. Technol.*, 17(12), 738, 1983.

24. Isaac, R. A. and Morris, J. C., Transfer of active chlorine from chloramine to nitrogenous organic compounds. II. Mechanisms, *Environ. Sci. Technol.*, 19(9), 810, 1985.

25. Culver, R. H., The Reaction of Chlorine with Glycine in Dilute Aqueous Solution, Harvard University, 1955.

CHAPTER 16

Modeling the Distribution of Chloramines during Drinking Water Chloramination

Jeyong Yoon and James N. Jensen

ABSTRACT

Chloramination is likely to become an even more common form of drinking water disinfection. To date, the impact of organic chloramine formation on the chloramination process has not been assessed. The purpose of the present work was to apply an existing kinetic model for chlorine transfer to typical conditions of drinking water chloramination. In particular, the impact of treatment conditions (chlorine-to-ammonia ratio, chlorine dose, reaction time, and pH) on monochloramine residuals was assessed. Model simulations indicate that organic chloramine formation may be significant under typical conditions of drinking water chloramination. Residual inorganic monochloramine concentrations were relatively insensitive to pH, contact time, and chlorine-to-ammonia ratio, but increased with decreasing organic N and/or increasing chlorine dose.

INTRODUCTION

Chloramination is now a common method of disinfection in the water industry. Based on a recent survey,[1] 20% of the surveyed utilities used some form of the chlorine-ammonia process. It is likely that chloramination will become more popular if the disinfection by-products regulations become more stringent, as expected. For example, prior to 1990, about 19% of surveyed purveyors switched to chloramination to address trihalomethane formation problems.[1]

There has been increasing reliance in the water industry on using residual disinfectant concentrations rather than microbiological measurements for the

1-56670-136-8/96/$0.00+$.50
© 1996 by CRC Press, Inc.

assessment of water quality.[2] However, monochloramine residuals may not be good indicators of water quality. In the presence of even small quantities of organic nitrogen, it is possible for chloramination to produce organic chloramines. Organic chloramines are much weaker disinfectants than inorganic monochloramine (NH_2Cl), but are indistinguishable by the common analytical techniques employed in water treatment laboratories.[3] Interferences from organic chloramines in the measurement of monochloramine could result in a false assessment of treated water quality. New analytical methods have been developed recently that are capable of measuring monochloramine and organic chloramines without significant interferences.[4-7]

The impact of organic chloramine formation on the chloramination process has not been assessed to date. Organic chloramine formation may necessitate changing chloramination conditions (e.g., ammonia and chlorine addition order, chlorine-to-ammonia ratios, and contact time). Given the greater role that chloramination will play in the future, it is imperative to determine the impact of organic chloramines on the disinfection efficiency of chloramination.

BACKGROUND

Monochloramine is formed in drinking water treatment when aqueous chlorine combines with added ammonia. Chlorine also combines with nitrogenous organics (e.g., amino acid and proteinaceous material) to form organic chloramines. To date, little work has been done on measuring organic chloramines in treated drinking water. Gordon and colleagues[8] suggested without data that organic chloramines are formed readily in water treatment.

Most organic chloramines have little or no bactericidal activity.[3,9] Thus, to correctly assess the disinfection ability of chloraminated water, it is necessary to understand the distribution of chlorine species between inorganic chloramines and organic chloramines. The distribution of chlorine species depends on the relative chlorination rates, concentrations of nitrogenous compounds, and chlorine transfer reactions between chloramines and nitrogenous compounds.

Yoon and Jensen[10] developed a kinetic model for chlorine transfer model in a chlorine/ammonia/organic nitrogen system. The conditions employed by the authors were more representative of wastewater than drinking water. The goals of the present work are twofold. First, the mathematical model for chlorine transfer developed by Yoon and Jensen will be applied to typical conditions of drinking water chloramination. Second, modeling results will be interpreted in terms of (1) mechanisms of organic chloramine formation (direct chlorination vs. chlorine transfer from monochloramine), and (2) the impact of treatment conditions (chlorine-to-ammonia ratio, chlorine dose, reaction time, and pH) on monochloramine residuals.

APPROACH

The kinetic model for chlorine distribution presented by Yoon and Jensen[10] is summarized in Scheme I. The reactions in this system can be divided into three major categories. First, chlorine reacts with ammonia and the nitrogenous organic (forward reactions in Equations 1 and 2). Second, chloramines may hydrolyze back to liberate free chlorine (backward reactions in Equations 1 and 2). Third, active chlorine may be transferred between chloramines and nitrogenous compounds (Equation 3).

Scheme I   Kinetic Model of the Chlorine/Ammonia/Organic Nitrogen System

$$\text{HOCl} + \text{NH}_3 \underset{k_{h,a}}{\overset{k_{CL,a}}{\rightleftarrows}} \text{NH}_2\text{Cl} + \text{H}_2\text{O} \tag{1}$$

$$\text{HOCl} + \text{RNH}_2 \underset{k_{h,oc}}{\overset{k_{CL,o}}{\rightleftarrows}} \text{RNHCl} + \text{H}_2\text{O} \tag{2}$$

$$\text{NH}_2\text{Cl} + \text{RNH}_2 \underset{k_{oc,a}}{\overset{k_{m,o}}{\rightleftarrows}} \text{NH}_3 + \text{RNHCl} \tag{3}$$

In the equations above, $k_{i,j}$ is the rate constant for the reaction of species i with species j. Unless otherwise stated, the rate constants are observed constants, that is, $k_{CL,a}$ is the rate constant for the reaction of $\text{NH}_3 + \text{NH}_4^+$ and $\text{HOCl} + \text{OCl}^-$.

Initial concentrations of inorganic monochloramine and organic chloramine were found through analytical solution of the differential equations of $[\text{NH}_2\text{Cl}]$ and $[\text{RNHCl}]$ over time without considering chlorine transfer. Using the steady-state assumption for $[\text{HOCl}]$ and mass balances for total chlorine (Cl_T), total organic amine (O_T), and total ammonia (A_T), the differential equations describing the time rate of change of $[\text{HOCl}]$, $[\text{NH}_2\text{Cl}]$, and $[\text{RNHCl}]$ can be expressed as one equation in one unknown, namely, $[\text{NH}_2\text{Cl}]$. This differential equation was solved numerically by the fourth order Runge-Kutta method using experimental conditions (pH, A_T, and O_T) and literature values for $k_{CL,a}$, $k_{CL,o}$, and the acid dissociation constants for ammonium and the protonated nitrogenous organic. Further details on the kinetic model are presented elsewhere.[10]

Baseline model conditions are summarized in Table 1. Glycine was chosen as the model compound for organic nitrogen. Glycine was selected because (1) it has been found in raw waters,[11] (2) kinetic constants for glycine and N-chloroglycine are known,[6,12] and (3) Yoon and Jensen[13] reported oxidants in chlorinated wastewaters that had chlorine transfer properties similar to glycine/N-chloroglycine. The baseline pH was 6.8. Although this pH is below the median pH of 7.6 from surveyed utilities,[14] it was selected because rate constants were available at pH 6.8.[10] In addition, the chlorine transfer rate constants have been shown to be fairly independent of pH in the range of 4

Table 1   Baseline Conditions

Parameter	Baseline value
Constant Parameters	
Model nitrogenous organic	Glycine
Contact time	45 min
pH	6.8
Variable Parameters	
Chlorine dose	5 mg/l as Cl_2 (7.1×10^{-5} M)
Chlorine-to-ammonia ratio	0.8 mol/mol (4 mg Cl_2/mg NH_3-N)
Organic-N concentration	0.3 mg/l as N (2.1×10^{-5} M)
Chlorination mode	Preammoniation

Note: At chlorine dose = 5 mg/l as Cl_2 (7.1×10^{-5} M) and chlorine-to-ammonia weight ratio = 4 mg Cl_2:1 mg N, the baseline ammonia concentration was 1.25 mg/l as N (8.9×10^{-5} M).

to 8.[10,12] The baseline contact time employed was 45 min. This corresponds to about the median contact time prior to the first consumer for both chloraminating and all surveyed facilities.[1,14] In addition, preliminary runs showed little effect of contact time up to 7 d of simulated contact.

The baseline chlorine dose was 5 mg/l as Cl_2 (7.1×10^{-5} M). About 70% of chloramination plant surveyed used a dose less than 5 mg/l as Cl_2. The median dose was 3.2 mg/l as Cl_2.[14] The baseline chlorine-to-ammonia ratio was 0.8 mol/mol (4 mg Cl_2/mg NH_3-N). This is within the range of 3:1 to 7:1 mg Cl_2/mg NH_3-N employed in practice.[14]

The default organic-N concentration was 0.3 mg/l as N (2.1×10^{-5} M). This is within the range of raw water sources. Although few water utilities routinely monitor organic nitrogen in their source waters,[15] the available data show organic nitrogen in raw drinking waters ranges from 0.1 to 22 mg/l as N (the latter quantity was thought to be due to an algal bloom).[16] Free amino acid concentrations up to 0.3 mg/l as N were found in raw drinking waters in northwest France.[17]

Chloramination can be conducted in one of three modes: preammoniation (i.e., ammonia addition before chlorine), preformed monochloramine (i.e., addition of a chlorine/ammonia mixture), or postammoniation (i.e., chlorine addition before ammonia). Although postammoniation was the most common approach among the three modes for trihalomethane reduction,[1] preammoniation was selected as the default chlorination mode. Preammoniation was employed to eliminate the operational variable of chlorine contact time prior to ammonia addition in the modeling runs.

RESULTS

EFFECTS OF ORGANIC NITROGEN CONCENTRATION

The model conditions were chosen to simulate the addition of chlorine to water containing a large amount of ammonia and small organic nitrogen level

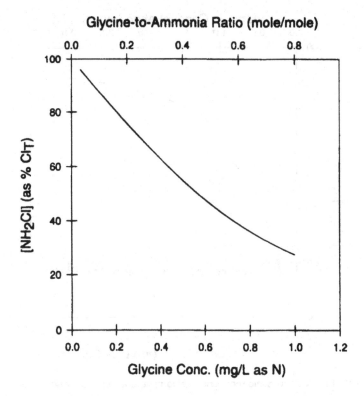

Figure 1   Residual inorganic monochloramine as a function of the glycine concentration.

(i.e., chloramination with preammoniation). The residual monochloramine (as a percentage of the chlorine dose) is shown as a function of the glycine concentration in Figure 1 (pH 6.8, 45-min contact, chlorine dose = 5 mg/l as Cl_2 [7.1×10^{-5} M], chlorine-to-ammonia ratio = 0.8 mol/mol). As shown in Figure 1, even a small glycine concentration can have a significant effect on the residual monochloramine. For example, inorganic monochloramine is 80% of the total chlorine at a glycine concentration of only 0.2 mg/l as N (1.4×10^{-5} M). At a glycine concentration of 0.8 mg/l as N (5.7×10^{-5} M), inorganic monochloramine is expected to make up less than 40% of the total chloramine because of increased formation of organic chloramine. Since common analytical methods cannot differentiate between inorganic and organic chloramines,[3,10] the bactericidal properties of this water may be overestimated by the measured monochloramine concentration.

EFFECTS OF CHLORINE DOSE

The residual monochloramine (as a percentage of the chlorine dose) is shown as a function of the chlorine dose concentration in Figure 2 (pH 6.8,

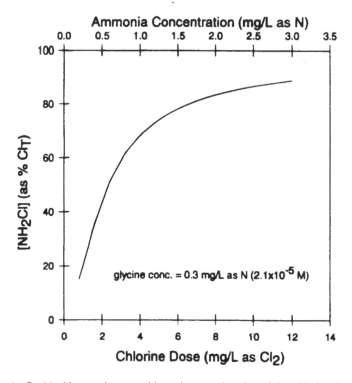

Figure 2   Residual inorganic monochloramine as a function of the chlorine dose.

45-min contact, glycine concentration = 0.3 mg/l as N (2.1 × 10^{-5} M), chlorine-to-ammonia ratio held constant at 0.8 mol/mol). At low chlorine doses (<2 mg/l as Cl$_2$ [2.84 × 10^{-5} M]), the inorganic monochloramine concentration increases with chlorine dose at a fairly constant rate (about 25% of the total chlorine per milligram per liter as Cl$_2$ chlorine added). The percentage of total chlorine that is monochloramine levels out at higher chlorine doses. For example, at a chlorine dose of 10 mg/l as Cl$_2$ (1.4 × 10^{-4} M), the inorganic monochloramine still makes up only about 85% of the total chlorine. At these higher doses, essentially all the organic nitrogen is chlorinated. Thus, the organic chloramine concentration is constant while the inorganic monochloramine increases with chlorine dose (at a fixed chlorine-to-ammonia ratio).

The combined effects of organic nitrogen concentration and chlorine dose are shown in Figure 3 (pH 6.8, 45-min contact, chlorine-to-ammonia ratio = 0.8 mol/mol). At high glycine concentrations and low chlorine doses (front left corner of Figure 3), inorganic monochloramine makes up only a few percent of the total chloramine. Organic chloramine dominates under these conditions. At low glycine concentrations and high chlorine doses (back right corner of Figure 3), inorganic monochloramine dominates.

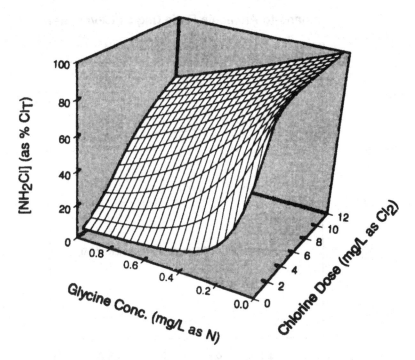

Figure 3   Residual inorganic monochloramine as a function of the glycine concentration and chlorine dose.

EFFECTS OF CHLORINE-TO-AMMONIA RATIO

The residual monochloramine (as a percentage of the chlorine dose) is shown as a function of the chlorine-to-ammonia ratio in Figure 4 (pH 6.8, 45-min contact, glycine concentration = 0.3 mg/l as N [2.1×10^{-5} M], chlorine dose = 5 mg/l as Cl_2 [7.1×10^{-5} M]). The highest molar chlorine-to-ammonia ratio simulated was 1:1. Thus, all the values in Figure 4 lie in the first leg of the breakpoint curve.

Figure 4 simulates the effect of changing the ammonia dose. An increase in the ammonia dose (corresponding to a smaller chlorine-to-ammonia ratio) would result in an increase in the percentage of the total chlorine found as monochloramine. However, the influence of the chlorine-to-ammonia ratio is small. A doubling of the ammonia dose (from chlorine-to-ammonia molar ratio of 0.8 to 0.4) would result in an increase in inorganic monochloramine from 3.6 mg/l as Cl_2 to only 3.8 mg/l as Cl_2 (5.1 to 5.4×10^{-5} M).

EFFECTS OF CHLORAMINATION MODE

The impacts of chloramination mode on inorganic monochloramine residuals are shown in Figure 5 (pH 6.8, glycine concentration = 0.3 mg/l as N

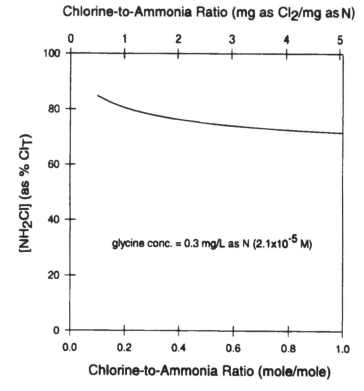

Chlorine-to-Ammonia Ratio (mg as Cl_2/mg as N)

Figure 4   Residual inorganic monochloramine as a function of the chlorine-to-ammonia ratio.

$[2.1 \times 10^{-5}\ M]$, chlorine dose = 5 mg/l as Cl_2 $[7.1 \times 10^{-5}\ M]$, chlorine-to-
ammonia ratio = 0.8 mol/mol). Residual monochloramine concentrations are
shown initially and after 45 min of chlorine contact for preammoniation,
preformed monochloramine addition, and postammoniation. The addition of
preformed inorganic monochloramine resulted in the largest residual monochloramine concentrations.

DISCUSSION

MECHANISTIC IMPLICATIONS

The results of the simulations suggest that direct chlorination (and not
chlorine transfer) is the dominant mechanism of organic chloramine formation.
Two pieces of evidence point to this conclusion. First, residual monochloramine
concentrations were calculated when chlorine transfer is ignored ($k_{m,o} = k_{oc,a}$

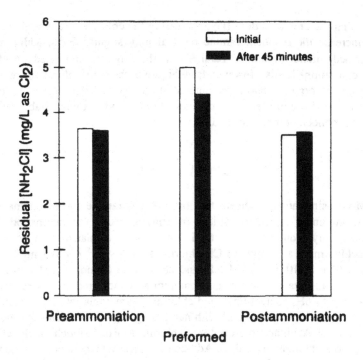

Figure 5 Effects of chlorination mode on the inorganic monochloramine concentration.

$= 0$). No significant differences were observed between these values and the concentrations reported in Figure 1. This suggests direct chlorination dominates over chlorine transfer. Second, simulated monochloramine concentrations varied little over 7 d of simulated contact time. This implies that the slower chlorine transfer reactions had little effect on the monochloramine profiles. Again, direct chlorination probably was favored over chlorine transfer.

IMPLICATIONS FOR DRINKING WATER TREATMENT

The model simulations presented here have several implications with regard to drinking water chloramination. First, organic chloramine formation may be important even at low organic nitrogen concentrations. Second, contact time is not expected to play a large role in determining the inorganic monochloramine/total chlorine ratio. It should be noted that, at longer contact time (i.e., several days), the decomposition of inorganic monochloramine through oxidation of ammonia-N may be significant.[18-20] Third, increasing chlorine dose (at constant chlorine-to-ammonia ratio) will significantly increase the percent of total chlorine that is inorganic monochloramine at low chlorine doses. However, at high chlorine doses, an increase in the chlorine dose is an inefficient way to increase the residual monochloramine. Fourth, increasing the ammonia dose (i.e., decreasing the chlorine-to-ammonia ratio) will increase the monochloramine residual only slightly. Fifth, addition of preformed monochloramine resulted in the highest simulated residual monochloramine levels. However, from Figure 5, the monochloramine residuals from preformed monochloramine addition changed the most over time. The simulated use of pre- or postammoniation provided a more stable, albeit smaller, monochloramine residual.

CONCLUSIONS

Model simulations indicate that organic chloramine formation may be significant under typical conditions of drinking water chloramination. For example, only about 70% of the total chlorine was calculated to be inorganic monochloramine at 5 mg/l as Cl_2 chlorine (7.1×10^{-5} M), 0.3 mg/l as N glycine (2.1×10^{-5} M), pH 6.8, and chlorine-to-ammonia molar ratio of 0.8:1. The residual inorganic monochloramine concentration was relatively insensitive to pH, contact time, and chlorine-to-ammonia ratio. The residual NH_2Cl concentration increased with decreasing organic N and/or increasing chlorine dose. At higher chlorine doses, chlorine dose had a much smaller effect on the monochloramine residual. Addition of preformed monochloramine was calculated to produce the highest, but most unstable, monochloramine residual.

REFERENCES

1. Water Quality Division Disinfection Committee, *J. Am. Water Works Assoc.*, 84(9), 121, 1992.
2. U.S. Environmental Protection Agency, *Fed. Reg.*, 52, 42223, 1987.
3. Wolfe, R. L. and Olson, B. H., *Environ. Sci. Technol.*, 19, 1192, 1985.
4. Hempel, C. E. and Jensen, J. N., *Proc. Am. Water Works Assoc. Water Quality Technology Conference*, American Water Works Association, Denver, 1991, 1035.
5. Yoon, J. and Jensen, J. N., *Proc. Am. Water Works Assoc. Water Quality Technology Conference*, American Water Works Association, Denver, 1993, 475.
6. Yoon, J. and Jensen, J. N., submitted.
7. Puranik, S., Harrington, L. J., and Jensen, J. N., submitted.
8. Gordon, G., Cooper, W. J., Rice, R. G., and Pacey, G. E., *Disinfection Residual Measurement Methods*, American Water Works Association, Denver, 1987.
9. Feng, T. H., *J. Water Pollut. Control Fed.*, 38, 614, 1966.
10. Yoon, J. and Jensen, J. N., *Environ. Sci. Technol.*, 27, 403, 1993.
11. LeCloirec, C. and Martin, G., in *Water Chlorination: Chemistry, Environmental Impact and Health Effects*, Vol. 5, Jolley, R. L. et al., Eds. Lewis Publishers, Chelsea, MI, 1983, 821.

12. Snyder, M. P. and Margerum, D. W., *Inorg. Chem.*, 27, 2545, 1982.
13. Yoon, J. and Jensen, J. N., *Water Environ. Res.*, 67(5), 842, 1995.
14. Haas, C. N., Presented at the American Water Works Association Preconference Workshop, Philadelphia, 1991.
15. Cooke, G. D. and Carlson, R. E., *Reservoir Management for Water Quality and THM Precursor Control*, American Water Works Association, Denver, 1989.
16. Morris, J. C., Ram, N., Baum, B., and Wajon, E., *Formation and Significance of N-Chloro Compounds in Water Supplies*, EPA-600/2-80-031, U.S. Environmental Protection Agency, Cincinnati, 1980.
17. LeCloirec, C., Thesis, University of Rennes, Rennes, France, 1984.
18. Leao, S. L. and Selleck, R. E., in *Water Chlorination: Environmental Impact and Health Effects*, Vol. 4, Jolley, R. L. et al., Eds., Ann Arbor Science Publishers, Ann Arbor, MI, 1983, 139.
19. Jafvert, C. T. and Valentine, R. L., *Water Res.*, 21, 963, 1987.
20. Jafvert, C. T. and Valentine, R. L., *Environ. Sci. Technol.*, 26, 577, 1992.

CHAPTER 17

Chloramine Loss and By-Product Formation in Chloraminated Water

Solomon W. Leung and Richard L. Valentine

INTRODUCTION

It is well known that chloramines will decompose under UV photolysis. In fact, this is one of the methods of how chlorinated chlorine-demand-free water is dechlorinated.[1] However, the questions of how chloramines decompose under photolysis and what chemical species are being produced in the process are unclear.

An understanding of the role of nitrite and nitrate in chloramine chemistry is important for several practical and fundamental reasons. Nitrite and nitrate are substances whose concentrations are regulated by the EPA and by which maximum contaminant levels (MCLs) have been designated. To safeguard our drinking water and for better control of treatment technologies, it is important to understand the sources of nitrite and nitrate in drinking water. One potential source is from the decomposition of chloramines.

Previous studies from Pressley et al.[2] and Hand and Margerum[3] postulated that very little nitrate should be produced by decomposition of chloramines in the combined region. Recent studies[4,5] showed that nitrate was produced in the decomposition of monochloramine.

In addition, nitrate and nitrite showed strong association with an unidentified product formed in slow decomposition of chloramines.[6] This unidentified product thus far could not be identified nor its basic elemental composition be determined; it was not known how much or even if chlorine and nitrogen were both constituents of this unidentified product. If nitrite and nitrate are to play an important role in the formation of the unidentified product, the study of nitrite/nitrate formation in the photolysis-driven monochloramine decomposition could provide information on the understanding of the constituent atoms of this unidentified product.

1-56670-136-8/96/$0.00+$.50

Theoretically, a thorough understanding of the reactions of the chloramine-ammonia system can be used to maximize the efficacy of the chlorination process while minimizing the generation of undesirable products. It was the objective of this study to investigate the formation of nitrate in monochloramine solutions by slow self-decomposition and photolysis. The photolysis setup resembled the methodology used in the preparation of chlorine-demand-free water. The effects of pH, buffer concentration, and chlorine-to-ammonia molar ratio (M ratio) in the slow decomposition and photodegradation processes were also studied.

EXPERIMENTATION

MONOCHLORAMINE PREPARATION

Monochloramine was prepared by dropwise addition of low-chloride, nitrate-free (<1 ppm) chlorine stock solution into a well-stirred, buffered solution containing either ammonium chloride or ammonium sulfate at room temperature. Phosphate was used as the buffer system in the chloramine solution and the pH of the solution was designed to be 7.5 or above to ensure that monochloramine decomposed slowly during preparation. Since all the reacting reagents were practically free of nitrate, the initial nitrate concentration in the monochloramine solution was less than 0.01 mM.

MEASUREMENTS

Chloramine concentrations were determined by the DPD method; each chloramine solution was analyzed in triplicate with the same procedures outlined in the *Standard Methods*.[1] Anions were analyzed by a Dionex 2000i Model ion chromatography (IC) equipped with a Spectra-Physics 4270 integrator. The anion column was an HPIC-AS4A with an HPIC-AG4A guard column and an MPIC-NG1 organic guard column. The anion suppression system was an AFSP/N 35350. For anion determination, sample solutions were compared with the average of three standard injections in duplicate. The UV spectra of the samples were also measured from 200 to 350 nm by an HP UV/VIS spectrophotometer.

SLOW MONOCHLORAMINE DECOMPOSITION

Monochloramine solutions prepared as previously mentioned were wrapped by aluminum foil and stored in the dark in a 25°C constant temperature incubator for self-decomposition. Monochloramine concentrations used were in the range of 1 to 1.5 mM, the initial pH of experiments varied from 7.5 to 8.6, and the phosphate buffer concentrations were 15 and 25 mM. The ammonia

Table 1 Formation of Nitrate in Slow Monochloramine Decomposition

Exp #	Mono-chloramine (mM)	Phosphate (mM)	Ammonia (mM)	pH	Ionic strength (M)	Nitrate/mono (mM/mM)	Reaction time (h)
1	1.42	25	3	8.17	0.080	0.0296	1776
2	1.01	25	2	8.09	0.080	0.0240	1560
3	1.04	15	2	7.82	0.100	0.0290	1395
4	1.53	15	3	8.61	0.100	0.0412	1665
5	1.52	15	3	8.58	0.052	0.0306	476
6	1.48	15	15	8.54	0.071	0.0445	240
7	1.46	15	3	7.51	0.045	0.0150	475
8	1.52	15	15	7.51	0.064	0.0150	475

Note: Nitrate generated per unit concentration of monochloramine decomposed was calculated at the end of each run.

concentration varied from 2 to 15 mM. Samples were drawn periodically for analysis of chloramines, NO_2^-, NO_3^-. Table 1 lists the experimental conditions of the eight different solutions used.

PHOTOLYSIS OF MONOCHLORAMINE

Monochloramine solutions prepared as previously stated were placed in front of an ultraviolet sunlamp at room temperature after initial measurements were taken. This experimental setup assembled the method of which chlorine-demand-free water was prepared. The UV sunlamp was equipped with a commercial 275-W GE Model RSM sunlamp bulb. The solution vessel used in the photolysis of chloramine was made by borosilicate; thus a large percentage of UV light (>85%) was expected to be masked by the vessel. Initial conditions for eight different solutions being photolyzed are listed in Table 2. Monochloramine concentrations were in the range of 1 to 1.5 mM, the initial pH of experiments varied from 7.8 to 9.0, and the phosphate buffer concentrations varied from 0 to 25 mM. The ammonia concentration varied from 2 to 3 mM, such that the molar ratio of chlorine to ammonia (M ratio) for each experiment was about 0.5 to ensure that the solution was within the combined region. Samples were drawn periodically for analysis of chloramines, NO_2^-, NO_3^-.

Table 2 Initial Conditions of Experiments for the Photolysis of Monochloramine

Exp #	Ammonia (mM)	Phosphate (mM)	Mono-chloramine (mM)	Initial pH	Final pH	Ionic strength (M)
A	3	25	1.58	8.13	7.81	0.083
B	2	15	1.04	7.81	7.59	0.045
C	3	15	1.53	8.61	7.96	0.054
D	2	15	1.55	8.52	7.91	0.067
E	2	25	1.54	8.46	8.01	0.053
F	3	15	1.54	7.86	7.59	0.049
G	2	15	1.08	8.65	8.12	0.052
H	3	0	1.53	8.95	6.26	0.007

Temperature was not controlled, but the room temperature was between 23 and 25°C which was the same temperature for the solutions initially, and the temperature of solutions rose moderately during experimentation (less than 10°C). Distance from the solutions to the light source remained unchanged throughout the experimentation. Samples were periodically drawn for analysis. Ionic strength of the solutions was not adjusted and was estimated according to the constituents of the solutions (Table 2). Irradiation time for the solutions various from 100 to 310 h. Solutions from experiments C and D were also exposed to direct sunlight for 8 h in a quartz container, Nitrite and nitrate were monitored.

RESULTS AND DISCUSSION

FORMATION OF NITRATE FROM MONOCHLORAMINE DECOMPOSITION

Nitrate was found in the slow self-decomposition of monochloramine, which concurred with the results from Dlyamandoglu and Selleck.[5] As shown in Table 1, the concentration of nitrate generated per unit concentration of monochloramine decomposed appeared to be dependent on solution conditions. Among other factors, pH seemed to be the deterministic factor in nitrate formation. Given the same experimental conditons, nitrate formation was more favorable at higher pH, higher ionic strength, and lower molar concentration ratio.

FORMATION OF NITRATE FROM PHOTOLYSIS OF MONOCHLORAMINE

Monochloramine was found to decompose under photolysis to form nitrate, but no significant nitrite was observed. Nitrite could have been generated in the solutions; it would have been oxidized into nitrate under the experimental environment. Figure 1 shows chloramine reduction as a function of photolysis time for solutions listed in Table 2 with various pH, monochloramine, and ammonia concentrations. Figure 2 shows the nitrate production with the same solutions. In each experiment, over 60% of the initial oxidant disappeared in the first 50 h. The decomposition rate then decreased drastically as concentration of monochloramine was reduced. For example, in experiment B (Figure 1), 60% of the monochloramine disappeared in the first 28 h, the next 35% took more than 110 h. Even after a total of 220 h, there were still about 3% of the original monochloramine (1.04 mM) left in the solution. This suggested that monochloramine in dilute concentration was relatively stable under the experimental conditions.

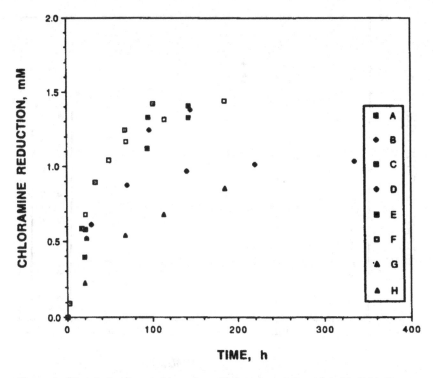

Figure 1   Photolysis of monochloramine with time for solutions listed in Table 2.

The solution pH was an important factor in photolyzed monochloramine decomposition, as is in many photolysis processes. Monochloramine at low pH (experiment C, pH 7.8) decomposed much faster than monochloramine at higher pH (experiment G, pH 8.7). From the results of experiments C and D, and experiments D and E, the molar concentration ratio and phosphate buffer used in the solutions did not seem to have significant effect on the photodecomposition process. The rate and extent of nitrate formation both appeared unaffected.

Figure 3 is a composite plot of nitrate formed by photolysis of monochloramine solutions obtained from experiments A to H. As shown, an approximately linear relationship existed between nitrate formed and monochloramine reduced. The slope of the linear curve indicated that 0.14 ± 0.012 (95% CL) mM of nitrate would be generated for each millimolar concentration of monochloramine decomposed, which constituted about 14% of nitrogen in monochloramine. The linear relationship also suggested that the controlling mechanism in the photodegradation process of monochloramine was the major pathway leading to the formation of nitrate, or the rate of nitrate formation was a linear function of monochloramine concentration, regardless of the initial condition of the solutions in the experimental range. This was not

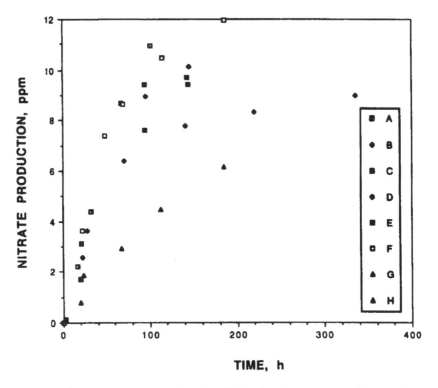

Figure 2   Simultaneous nitrate production with time for solutions listed in Table 2 during the monochloramine decomposition.

observed in the slow decomposition of monochloramine solutions; the production of nitrate is dependent on solution conditions. The nitrate formation represented a recovery of about 56% of the oxidation potential of monochloramine based on the presumed net stoichiometric reaction,

$$4NH_2Cl + 3H_2O \rightarrow 4Cl^- + 5H^+ + NO_3^- + 3NH_3$$

The rest of the nitrogen presumably went to some other oxidation product(s) such as nitrogen gas or other nitrogen and oxygen-containing compound(s).[7,8] Additionally, other products such as ClO_2^- and ClO_3^- could be produced. It should be noted that two solutions taken from experiments C and D were also exposed to direct sunlight and the amount of nitrate formed appeared to fit the linear relationship developed using artificial light photolysis.

Nitrate was formed in monochloramine solutions undergoing slow self-decomposition, and decomposition initiated by artificial light. The nitrate generation from decomposition process initiated by photolysis appeared to be

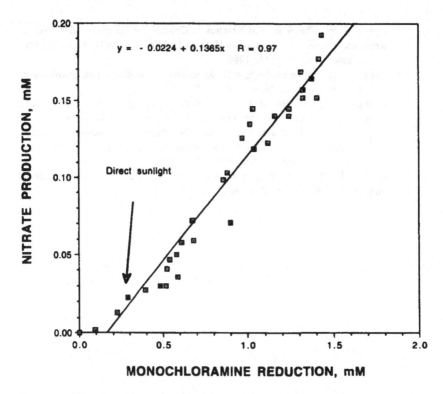

Figure 3 Nitrate production from photolysis of monochloramine. Measurements were taken from experiments A to H of Table 2, including data points from direct sunlight radiation from experiments C and D.

unaffected by pH, buffer concentration, and ionic strength in the studied conditions, and the concentration of nitrate generated per unit concentration of monochloramine decomposed was three times higher than the nitrate generated by slow self-decomposition. The results suggested that monochloramine decomposed through different pathways and more than one mechanism of nitrate formation might be involved.[6]

REFERENCES

1. *Standard Methods for the Examination of Water and Wastewater,* 17th ed. American Public Water Works Association, Washington, DC, 1989.
2. Pressley, T. A., Bishop, D. F., and Roan, S. G., Ammonia nitrogen removal by breakpoint chlorination, *Environ. Sci. Technol.,* 6, 622, 1972.
3. Hand, V. C. and Margerum, D. W., Kinetics and mechanisms of the decomposition of dichloramine in aqueous solution, *Inorg. Chem.,* 22, 1449, 1983.

4. Leung, S. W., Chemistry and Kinetics of Chloramine Decomposition: Nitrite Reactions and the Formation of an Unidentified Product, Ph.D. thesis, University of Iowa, Iowa City, IA, 1989.

5. Dlyamandoglu, V. and Selleck, R. E., Reactions and products of chloramination, *Environ. Sci. Technol.*, 26(4), 808, 1992.

6. Leung, S. W. and Valentine, R. L., An unidentified chloramine decomposition product. I. Chemistry and characteristics *Water Res.*, 28, 1475, 1994.

7. Saguinsin, J. L. S. and Morris, J. C., The chemistry of aqueous nitrogen trichlorite, in *Disinfection: Water and Wastewater*, Johnson, D. J., Ed., Ann Arbor Science Publishers, Ann Arbor, MI, 1975.

8. Saunier, B. M. and Selleck, R. E., The kinetics of breakpoint chlorination in continuous flow system, *J. AWWA*, 71, 164, 1979.

CHAPTER 18

A Study of Disinfection By-Products Formed Using Four Alternative Disinfectants as a Function of Precursor Characteristics

Huijia Teng and John N. Veenstra

INTRODUCTION

The influence of disinfection by-products (DBPs) on the operation and design of water treatment plants has been increasing since the early studies by Rook[1] and Stevens et al.[2] on trihalomethanes (THMs). Work conducted in the 1980s identified the existence of nonvolatile halogenated organics (non-THM organics), of which the majority produced by chlorination were haloacetic acids (HAAs).[3,4] Within the HAAs, dichloroacetic acid (DCAA) and trichloroacetic acid (TCAA) were the dominant members.

Concerns over potential health effects led the U.S. Environmental Protection Agency (USEPA) to set a maximum contaminant level (MCL) for total THMs. The disinfectant/disinfection by-product rule (D-DBP Rule) currently proposed by the USEPA will likely lower the MCL for TTHMs and establish MCLs for HAAs.[5] Past and pending DBP regulations have provided an impetus to seek a better understanding of how these compounds are formed as well as how their production can be controlled.

An extensive amount of research has been done on THMs. That research has indicated that several factors, such as source-related properties of aquatic organic matter, type of disinfectant, dosage of disinfectant, reaction time and conditions (i.e., pH), and treatment procedure selected, influence the amount of THM production. The studies focusing on the source-related properties of the aquatic organic matter have investigated items such as molecular weight distribution,[6,7] functional groups, carboxylic acidity,[8-11] and the humic substance content of a sample.[12] Methods developed for limiting organic halide production within water utilities have concentrated on precursor removal and disinfectant selections. Many methods have been developed to minimize THMs production. THM control methods can be divided into three main classes: (1) THM precursor removal or alteration (commonly used approaches are alum coagulation and preozonation); (2) use of alternative disinfectants, such as ozone, chloramine, chlorine dioxide, ultraviolet light, or a combination; and (3) removal of THMs produced by activated carbon adsorption.

One type of study of the source-related properties has focused on the influences of different molecular weight fractions of organics in the water source. Veenstra and Schnoor[13,14] found that an average of 87% of the THMs were formed from organics with molecular weights of 3000 or less. Collins et al.[15] determined the apparent molecular weight (AMW) distribution of samples obtained from four sources using ultrafiltration. The authors fractionated the samples into six AMW groups ranging from <500 to >30,000 and then tested each fraction for THM yield. They discovered that the THM yield generally increased as a function of molecular weight, although there were several departures from this trend. Sinsabaugh et al.[16] fractionated two waters into six apparent molecular weight groups. Their work showed that the specific yields from dissolved organic carbon (DOC) under 1 kilodalton (kD) and over 30 kD were 40 to 60% lower than yields from organics between 1 and 30 kD. Besides the research on molecular weight (MW) fractions the effects on THM production of another source-related property, the hydrophobic and hydrophilic organic content of water, have been investigated. Both types of organics do not have the same potential THM yield. Kuo and Amy's[17] work showed that the hydrophobic fraction accounted for more of the THM formation potential than that of the hydrophilic portion.

Less work has been conducted on HAAs than on THMs. Recent investigations have indicated higher formation potentials for nonvolatile compounds than those of volatile organics connected with the drinking water disinfection process. Uden and Miller[18] found that the individual concentration of TCAA and DCAA (30 to 160 μg/l) in tap water samples were comparable to and sometimes exceeded the concentration of chloroform. The research of Reckhow and Singer[4] showed that the formation amount of nonvolatile organic compounds was about three times that of the volatile ones using the free chlorination process. Additional work by Reckhow et al.[19] on aquatic humic and fulvic acids extracted from five sources showed the humic acids produced higher concentrations of chloroform, TCAA, and DCAA than the corresponding fulvic acids. The average specific yields of individual compounds from both the humic and fulvic acids (chlorinated at pH = 7) were greatest for TCAA followed by chloroform and DCAA. Johnson and Jensen[20] noted that the chlorination of diverse naturally occurring organics produced from 1.5 to 11 times as much nonvolatile organohalide compounds as chloroform.

This study focused on the influence of precursor characteristics on the formation potential of DBPs using different disinfectant alternatives. The specific objectives of this research were to:

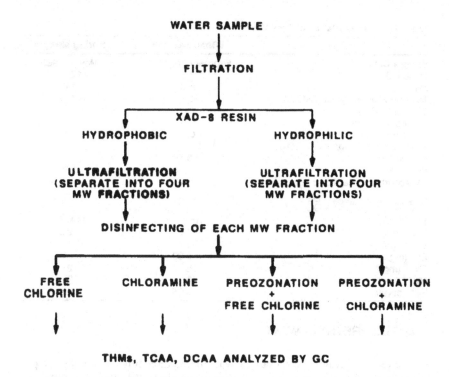

Figure 1   Research design flow diagram.

- Investigate the production of THMs and HAAs (DCAA and TCAA) from different apparent molecular weight fractions of hydrophilic and hydrophobic organics isolated from Kaw Reservoir, Oklahoma.
- Study the formation potential of DBPs using four different disinfection alternatives: free chlorine, chloramines, preozonation plus free chlorine, and preozonation plus chloramines.

EXPERIMENTAL METHODS

To illustrate the whole experimental procedure clearly, a process diagram is shown in Figure 1. The sample water originated from Kaw Reservoir, Oklahoma, and the two samples, one for winter and one for spring, were taken from the raw water storage tank at the City of Stillwater, Oklahoma, Water Treatment Plant. Table 1 presents information concerning the average chemical characteristics of the two samples.

Table 1   Characteristics of Raw Water

	Winter sample	Spring sample
Sample Date	February 10, 1989	May 1, 1989
Temp (°C)	8	17
pH	8.3	8.4
Total alkalinity (mg/l as $CaCO_3$)	163	179
Turbidity (NTU):		
Before filtration	5.8	6.4
After filtration	0.1	0.1
Total hardness (mg/l as $CaCO_3$)	257	280
Fluoride (mg/l)	0.59	0.62
DOC (mg/l)	3.75	8.0
UV absorbance (at 254 nm)cm^{-1}	0.071	0.075
Specific absorbance $\frac{A_{254}}{DOC}$	.019	.009
SO_4^{-2} (mg/l)	109	114
NO_3^- (mg/l)	2.2	2.3
Cl^- (mg/l)	155	170
Br^- (mg/l)	0.33	—
Ca^{+2} (mg/l)	68	72
Mg^{+2} (mg/l)	19	18
Fe^{+3} (mg/l)	0.05	—

SAMPLE PRETREATMENT

To remove turbidity, each sample was filtered three times. First, the sample was run through a sand filter, at the rate of 1 gal/min/ft^2. This was followed by filtering the sample through a pre-rinsed glass microfiber filter paper with a 1.5-μm pore size filter paper (Whatman, 934AH), then concluding with the sample being filtered through a pre-rinsed 0.45-μm filter (Millipore Type HA).

SEPARATION OF HYDROPHILIC AND HYDROPHOBIC MATERIALS

A resin extraction method[21] was utilized to fractionate the dissolved organic matter present in the water samples into hydrophobic (HB) and hydrophilic (HI) groups. A column packed with XAD-8 resin (40 to 60 mesh, from Rohm-Haas) was used. The column was 2.54 cm in diameter with a resin depth of 28 cm.

The sample was acidified to pH 2.0 with concentrated HCl. Three liters of sample (filtered raw water) were run through the XAD-8 resin at a 4 ml/min (0.2 gpm/ft^2) flow rate. The hydrophilic substances were contained in the effluent from the column. To obtain the hydrophobic component sorbed on the resin, a 0.1 N NaOH eluent was used at a flow rate of 1 to 2 ml/min (0.05 to 0.1 gpm/ft^2) to back elute the column. Both the HI and HB fractions were neutralized to pH 7 prior to storage.

Measurement of the total organic carbon (TOC) of the raw water, HI and HB fractions was done using a Beckman Model 915 total organic carbon analyzer. The TOC of HB fraction was concentrated during the XAD-8 resin sorption process, so it was necessary to dilute the TOC of the HB fraction to the same value as the TOC of the HI fraction to create solutions that would be used during the preozonation and chlorination experiments.

ULTRAFILTRATION

The method set out by Anderson et al.[22] was used to separate the HI and HB samples into the MW fractions. A 150-ml sample of the solution was placed in an Amicon stirred ultrafiltration cell. Less than 90% of this solution was filtered through the ultrafiltration membrane under a pressure of 276 to 415 kPa. The membranes used in this work were YM2 (cutoff MW <1000), YM10 (cutoff MW <10,000), and YM30 (cutoff MW <30,000) produced by Amicon Corporation. The stepwise ultrafiltration procedure is shown in Figure 2.

OZONATION

Ozonation of the samples was performed in a semi-batch operation by bubbling ozone through a 500-ml glass reactor. Fractionated samples were ozonated at pH 7 while whole water samples were conducted at ambient pHs. All ozonation was conducted at room temperature. Ozone was generated from air with a Griffin ozone generator (Griffin Technics Corporation, Lodi, NJ). Gas-phase ozone concentration in the influent and effluent gas streams was measured by bubbling the gas stream through two 500-ml gas washing bottles each containing a 10-g/l KI solution. The absorbed ozone doses were determined from the KI traps using the iodometric titration method[23] by taking the difference between the influent and effluent gas phase concentrations. The ozone generation conditions were maintained throughout the experiment using a 1-l/min flow rate for a contact time ranging from 1.5 to 2.5 min to obtain the desired absorbed ozone dosages of 0.7 to 0.8 mg O_3 per milligram of TOC. The ozonated water was allowed to stand at room temperature for 2 to 3 h to dissipate residual ozone prior to being subjected to chlorination.

DISINFECTION

All samples were buffered with a phosphate buffer (0.1 M) and adjusted to pH 7 prior to chlorination/chloramination. The chlorination/chloramination experiments were conducted in 40-ml glass vials with PTFE-faced silicone septa and screw caps.

Four different disinfection processes were studied in this research:

1. Using free chlorine as the sole disinfectant.
2. Using chloramines as the sole disinfectant.

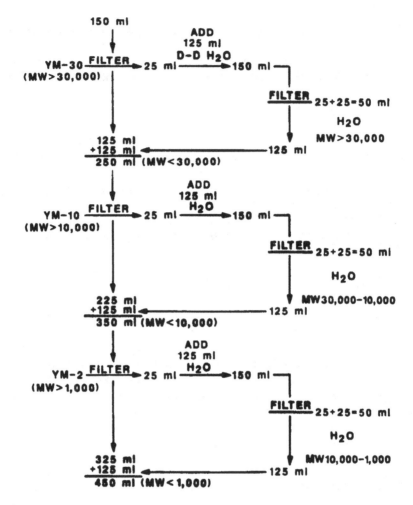

Figure 2 Ultrafiltration operation procedure.

3. Using a combination of preozonation and free chlorine to disinfect the water.
4. Using a combination of preozonation and chloramines to disinfect the water.

THMs and TCAA, DCAA production was monitored at 24 h for all MW fractions of both the HB and HI compounds. All experiments were conducted at room temperature. The chlorine and chloramine doses were selected to provide a residual after 24 h.

The disinfectant residuals were checked at the end of the incubation period to make sure residual disinfectant was left in each fraction at the end of the reaction and to estimate the amount of disinfectant consumed. The disinfectant residuals were measured using a spectrophotometer at 530-nm wavelength, by the DPD colorimetric method.[24] A 4 to 6% NaOCl solution (Fisher Scientific) was diluted with Milli-Q water to create the stock solution that served as the source of free chlorine. The free chlorine disinfectant dosage was set at a 5 to 1 mass ratio of free chlorine to total organic carbon (TOC).

The same dosage (mass ratio) of chloramine (5:1) was used in the study. The monochloramine solutions were prepared by reacting ammonium chloride (NH_4Cl) with a previously prepared aqueous chlorine solution at a three to one molar ratio (NH_3 to OCl^-), at pH 10.[25] The chloramine solution pH was then adjusted to pH 7 to 7.5 by use of a 5 N NaOH solution. The combined chlorine residuals were measured on a spectrophotometer using the DPD colorimetric method.[24] Greater than a 98% combined chlorine solution was obtained by this method. A fresh monochloramine solution was prepared on the day of each chloramination experiment.

DISINFECTION BY-PRODUCT ANALYSIS

At the end of the desired reaction time the samples to be analyzed for DBPs were quenched with ascorbic acid in excess of the stoichiometric amount. The samples were acidified and stored headspace-free at 4°C for no more than 2 d before analysis for THMs and HAAs.

TRIHALOMETHANES

Samples were analyzed for THMs on a Perkin-Elmer Sigma 2000 gas chromatograph (GC) equipped with an electron-capture detector using a glass column packed with 3% SP-1000 on 100/120 mesh Supelcoport. Prior to the GC analysis samples containing the THMs were subjected to a liquid/liquid extraction using pentane as the solvent.[23] The GC operating conditions were oven temperature = 160°C; detector temperature = 350°C, and carrier gas (95% argon and 5% methane) flow rate = 60 ml/min.

HALOACETIC ACIDS

Because TCAA and DCAA are not readily detected by the GC directly, a methylating derivation process was necessary to convert the acidic form to the more volatile methyl esters that can then be detected by the GC. The analytical methodology used was a borontrifluoride (BF_3)/methanol derivatization that was modified from the work of Calabrese et al.[26]

The HAAs GC methodology utilized a glass column packed with GP 10% SP-2330 on 100/120 mesh Chromosorb WAW. The operating conditions of the instrument were oven temperature = 105°C; injector temperature = 160°C; and detector temperature = 350°C with a carrier gas (95% argon and 5% methane) flow rate = 60 ml/min.

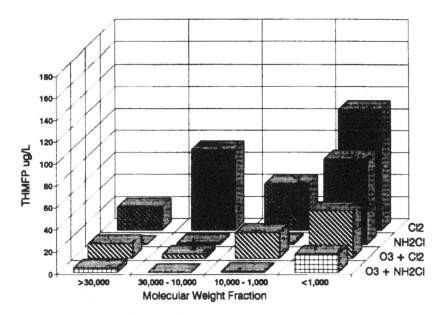

Figure 3   THM formation from winter HI sample.

PHYSICAL CHARACTERISTIC MEASUREMENT

Ultraviolet (UV) absorbance of the samples was measured using a Perkin-Elmer Lambda 3 UV/VIS Spectrophotometer at 254-nm wavelength. The pH of these samples was adjusted to 7 prior to the UV measurement.

The anionic ion concentrations of the samples were detected by ion chromatography (Dionex Corporation, Model 2000i/SP).

RESULTS AND DISCUSSION

Data generated after 24 h of contact between the HB and HI fractions and the four disinfectants were used to examine the impact of the process variables. The trihalomethane formation potential (THMFP) and haloacetic acid formation potential (HAAFP) for the fractionated winter and spring samples are shown in Figures 3 to 6, and 7 to 10, respectively.

The THMFP of the HI winter is shown in Figure 3. Under the free chlorination treatment the precursors in the <1000-MW range produced the largest amount of THMs. A significant production of THMs was also seen in the MW range of 30,000 to 10,000. The free chlorine disinfectant produced the largest total THMFP of all disinfectants for the hydrophilic winter sample.

Coupling preozonation and chlorination for the same sample showed a large decrease in the THM production from the 30,000 to 10,000-MW range

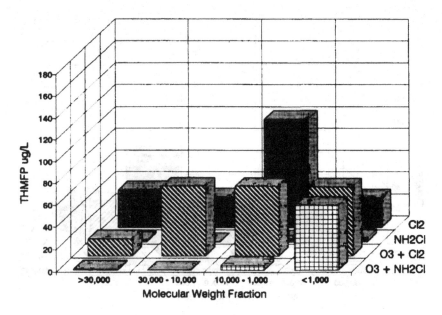

Figure 4   THM formation from winter HB sample.

and a corresponding decrease in the <1000-MW range. This disinfection scenario was an acceptable THMFP control methodology for the winter sample.

Chloramination produced only minimal levels of THMFP except in the <1000-MW range. The preozonation plus chloramination produced the smallest amount of THMFP, with the <1000-MW range yielding the largest amount.

The chlorinated hydrophobic winter THMFP maxima occurs in 10,000 to 1000-MW range (Figure 4). The preozonation plus chlorination option increased the THMFP of the 30,000 to 10,000-MW range, as opposed to chlorination alone, and caused each fraction from 30,000 to <1000 to produce approximately equal THM levels. The preozonation plus chlorination scenario produced the largest total THMFP for the hydrophobic winter sample.

Chloramination effectively controlled the THMFP production in all molecular weight ranges. The preozonation plus chloramination showed an increase in THMFP from the <1000-MW range compared to chloramination alone. The chloramination alone produced the lowest THMs of all disinfectants used on the hydrophobic winter sample.

The molecular weight range of 10,000 to 1000 produced the largest amount of THMs using free chlorine as the disinfectant for the spring HI sample (Figure 5). This disinfectant also produced the greatest quantity of THMs of all four disinfection options used on the spring hydrophilic sample. Coupling preozonation and chlorination reduced the THM production in 10,000 to 1000-

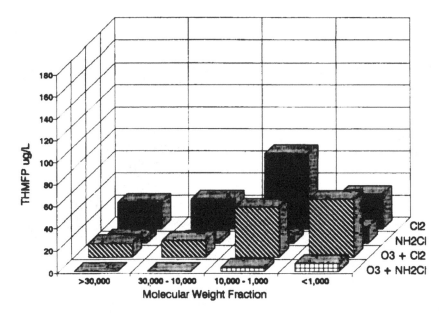

Figure 5   THM formation from spring HI sample.

MW range, but increased the THMFP of the <1000-MW range. The increase
in the THMFP production in <1000-MW range was to the point where the
total produced using this disinfectant method was nearly equal that of the
sample subjected to chlorination alone. The smallest MW range was normally
the top-producing THM range under this disinfecting regime for all samples
evaluated in this work.

Chloramination effectively limited the THM production by the various
MW fractions of the spring HI sample. The <1000-MW range did produce
the maximum amount of THMs under this disinfection option. Ozonation plus
chloramination produced the lowest THM total of any disinfectant tried. THM
production of all MW fractions was reduced from that obtained using chloramines alone.

THMFP production for the spring hydrophobic (Figure 6) sample was
maximum in the 10,000 to 1000-MW range. Chlorination alone of this fraction
produced the largest concentration of THMs seen in any MW range of the
spring or winter samples. The <1000-MW range also produced a significant
quantity of THMs. The addition of ozone, via the preozonation plus chlorination disinfection scheme, reduced the THM production in both the 10,000 to
1000- and <1000-MW ranges.

Chloramination produced low levels of THMs except for the <1000-MW
range. A similar raise in THMFP of this molecular weight fraction under
chloramination was also seen in the hydrophilic winter sample. The preozonation plus chloramination produced very low levels of THMs.

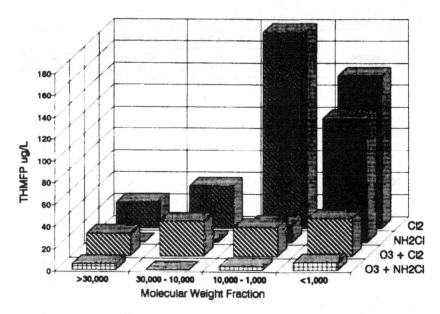

Figure 6 THM formation from spring HB sample.

Overall, this portion of the study showed that the 10,000 to 1000-MW range generally contained the highest THMFP using free chlorine. After ozonation, the highest THM formation range tended to shift to the lightest molecular weight fraction. An increase in TOC in the lightest molecular weight fraction was also seen following ozonation. A similar increase in the TOC of the lowest molecular weight fractions following ozonation of Kaw Reservoir water was also seen in a previous study.[27] The increase in TOC in the lightest molecular weight fraction resulted from a shift in TOC from higher MW fractions as a result of oxidation by ozone. Under the chloramination scheme, the MW <1000 range yielded the highest THM formation. Using chloramines effectively controlled THMs. However, the spring HB and winter HI samples showed significant production in the <1000-MW range that could be the result of some free chlorine residual during the chloramination step. The quantity of THMs formed in this MW fraction (<1000) using chloramines was less than that seen using free chlorine. Generally, lowest total THM formation resulted under preozonation plus chloramination conditions. The general trend was for hydrophobic fractions to produce more total THMs under various disinfection modes. There were exceptions to this trend. The data plotted in Figures 3 to 6 are given in Table 2.

The hydrophilic winter sample (Figure 7) under the free chlorination condition showed the largest general production of HAAs with the exception of the preozonated and chlorinated sample. Using chlorine only the greatest production of HAAs occurred in the 10,000- to 1000-MW range, whereas

Table 2   THM Formation Potential as a Function of MW Fraction

MW fraction (Da)	THMs formation potential (μg/l)			
	Chlorine	Chloramine	Ozone and chlorine	Ozone and chloramine
Winter HI				
>30,000	21	1	13	4
30,000–10,000	73	1	3	1
10,000–1000	42	2	23	1
<1000	110	77	43	16
Total	246	81	82	22
Winter HB				
>30,000	34	1	15	1
30,000–10,000	28	1	63	1
10,000–1000	99	3	63	4
<1000	27	3	62	58
Total	188	8	203	64
Spring HI				
>30,000	24	4	12	1
30,000–10,000	27	3	15	1
10,000–1000	68	6	45	4
<1000	31	14	52	7
Total	150	27	124	13
Spring HB				
>30,000	23	1	20	5
30,000–10,000	38	2	31	1
10,000–1000	179	16	26	3
<1000	139	113	34	7
Total	379	91	111	16

when preozonation was added the largest production was seen in the <1000-MW range. The increase in HAAFP displayed by the <1000-MW range using ozone and chlorine was due mostly to DCAA. As a result of the large increase in HAAs formed in the <1000-MW fraction the preozonation plus chlorination disinfection option produced the greatest quantity of DBPs of interest for the HI winter sample.

Chloramination of this water sample produced the lowest total HAA concentration of all disinfection options. Coupling preozonation chloramination also exhibited good control over the production of HAAs. With this disinfection scheme an increase in HAA production was observed in <1000-MW range. The DCAA production in this MW range was larger than that for the TCAA.

Experiments conducted on the hydrophobic winter sample (Figure 8) show the largest production of HAAs in 10,000- to 1000-MW range. The total HAAFP (for the winter HB sample) was 561 μg/l using free chlorine as the disinfectant. A potential contributing factor to this large value could be the amount of residual free chlorine. Under the preozonation and chlorination disinfection routine the precursors in the <1000-MW range produced the largest concentration of HAAs. This represents a shift to a lower MW fraction when compared to the chlorination experiments. The total HAAFP was slightly larger than that seen using free chlorine alone. For the HB fraction of the

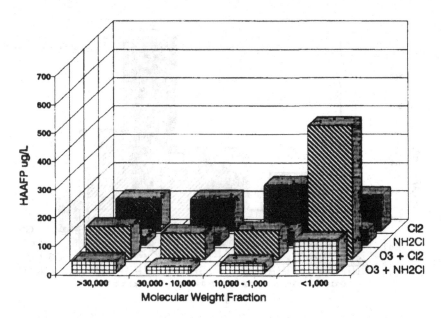

Figure 7   HAA formation from winter HI sample.

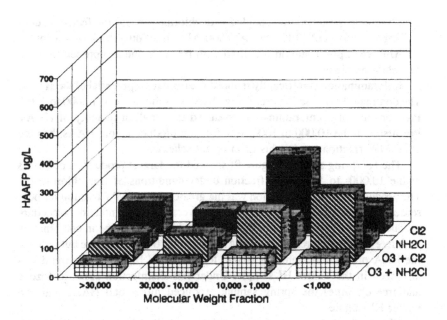

Figure 8   HAA formation from winter HB sample.

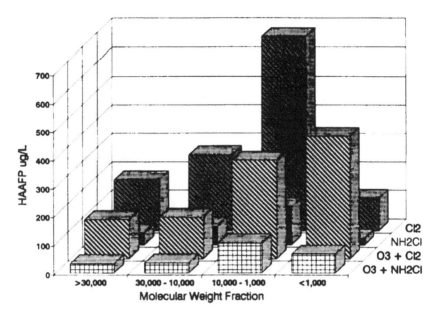

Figure 9   HAA formation from spring HI sample.

winter sample, preozonation coupled with chlorination was ineffective at controlling the total HAAFP. The >30,000 MW fraction did show a drop in HAAFP when preozonation was used with free chlorine as opposed to using free chlorine alone.

Chloramination significantly reduced the formation potential of the HAAs. The dominant MW fraction producing HAAs was the range of 10,000 to 1000. Preozonation plus chloramination produced the smallest quantity of HAAs. Precursors in the 10,000 to 1000-MW fraction produced more HAAs than the other MW fractions under this disinfectant scheme.

The HI spring sample (Figure 9) showed the largest production of HAAs in the 10,000- to 1000-MW fraction under conditions of free chlorination. The production from this fraction was the greatest of any MW fraction for both seasonal samples using any disinfectant. Overall, this disinfection procedure produced the largest amount of HAAs. The addition of ozone in combination with free chlorine increased the HAAFP of the <1000-MW range but reduced the HAA production from the 10,000 to 1000-MW fraction. This trend was also seen in the winter HI sample. Under this disinfection scenario (ozone and free chlorine) the spring sample produced more total HAAs than the winter HI sample.

Chloramination as a disinfection regime controlled the formation of HAAs. The 10,000- to 1000-MW range had the largest production of HAAs. As before the spring fraction produced more total HAAs than the corresponding winter fraction. Preozonation plus chloramination showed maximum HAAFP

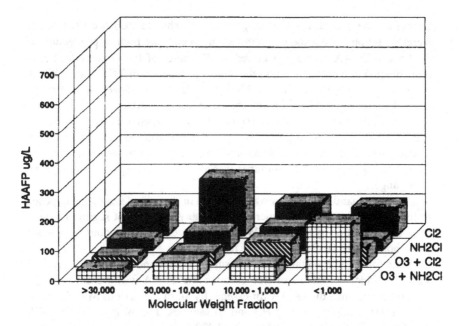

Figure 10 HAA formation from spring HB sample.

in the 10,000- to 1000-MW range. This disinfection option produced the lowest total HAAs for this sample. Overall, under all disinfectant conditions the production of HAAs for the spring sample was larger than that for the winter sample.

Chlorination of hydrophobic spring sample (Figure 10) showed the highest production occurred in the 30,000 to 10,000-MW range. The total HAAFP of this water fraction under free chlorination was slightly larger than that of the corresponding HB winter sample. Preozonation plus chlorination shifted the production maximum to the 10,000 to 1000-MW range. The HAA production was reduced using preozonation and chlorination as opposed to using chlorination alone. This is in contrast to the winter HB sample, when using preozonation plus chlorination showed a raise in total HAA production with the largest amount coming from the <1000-MW group.

Chloramination exhibited good control of HAAFP. The largest production occurred in the 10,000 to 1000-MW range. Greater than 50% of HAA in 10,000- to 1000-MW range was due to DCAA. Preozonation plus chloramination showed a sharp increase in HAAFP in the <1000-MW range. This rise was due mainly to increased production of DCAA. The corresponding winter sample did not exhibit this sharp increase in the <1000-MW range using ozone and combined chlorine.

The precursor study showed that MW 10,000 to 1000 generally contained the highest DCAA and TCAA formation potential using free chlorine. After ozonation, the highest HAAFP range tended to shift to the lowest molecular weight fraction. Under the chloramination scheme the highest production of TCAA and DCAA normally occurred in MW range of 10,000 to 1000. Using preozonation and chloramination the largest production of TCAA and DCAA occurred in either the 10,000 to 1000 or <1000-MW range. No definitive trend was firmly established using this disinfection regime.

In both the spring and winter HI samples a substantial increase in DCAA was observed in the <1000-MW range under preozonation plus chlorination as compared to chlorination alone. In the spring HB sample an increase in DCAA using ozone plus chloramines was also observed, again in the <1000-MW range.

A larger quantity of HAAs was produced in the spring sample as opposed to the winter sample. Also generally under any disinfectant regime the sum of HAA production by the four MW groups showed the HB fraction to produce more nonvolatile DBPs than the HI fraction.

Data for the individual HAAs measured, TCAA and DCAA, are given in Table 3. Under conditions of free chlorine the TCAA was present in larger concentrations than the DCAA for essentially all of the various MW fractions for both the winter and spring HB and HI samples. The most productive MW fractions were the 10,000 to 1000 and 30,000 to 10,000. The use of ozonation ahead of chlorination increased the formation of DCAA in both the HB and HI winter fractions. TCAA production under these conditions increased for the winter HI fraction but decreased for the winter HB fraction. For both compounds ozone caused a shift in the MW fraction producing the largest quantity of the contaminants to the <1000-MW fraction. The decrease in DCAA and TCAA formation as the result of preozonation was more pronounced for the spring hydrophobic fraction. The impact of preozonation on DCAA and TCAA formation in this study was mixed. Zhu and Reckhow[28] found, under short chlorine contact times, that preozonation decreased the concentrations of DCAA and TCAA, while under long chlorine contact times the formation of DCAA increased and TCAA decreased as a result of preozonation.

The use of chloramines produced less DCAA and TCAA for both the winter and spring HI and HB fractions compared to the free chlorine disinfectant scenario. The use of ozonation ahead of chloramine addition increased the production of TCAA and DCAA for the <1000-MW fraction as opposed to chloramines alone. In addition the production of DCAA and TCAA in the 1000 to 30,000-MW range was generally lower when chloramine addition was preceded by ozonation; however, there was some variation in this trend. Generally, under the ozone and chloramine condition the DCAA concentration increased and the TCAA concentration decreased compared to that found using chloramines alone.

Table 4 presents the specific formation potential (24 h, pH = 7) for TTHM as a function of molecular weight fractions for both the winter and spring HB

Table 3 DCAA and TCAA Formation Potential as a Function of MW Fraction

MW fraction (Da)	HAA formation potential (µg/l)							
	Chlorine		Chloramine		Ozone and chlorine		Ozone and chloramine	
	DCAA	TCAA	DCAA	TCAA	DCAA	TCAA	DCAA	TCAA
Winter HI								
>30,000	33	81	15	14	37	76	23	18
30,000–10,000	33	80	16	19	13	77	10	13
10,000–1,000	67	96	26	20	28	72	17	17
<1,000	48	77	26	25	303	168	72	73
Total	181	334	83	178	381	393	122	121
Winter HB								
>30,000	33	79	21	28	34	27	20	19
30,000–10,000	19	63	16	23	21	71	29	14
10,000–1,000	139	135	45	62	64	112	39	31
<1,000	11	82	14	24	111	128	32	28
Total	202	359	96	139	230	338	120	92
Spring HI								
>30,000	67	112	20	22	40	97	15	17
30,000–10,000	130	136	35	30	48	98	17	19
10,000–1,000	373	303	75	58	177	168	61	42
<1,000	29	85	14	8	236	193	40	26
Total	599	636	144	118	501	556	133	104
Spring HB								
>30,000	15	76	12	23	10	17	13	19
30,000–10,000	74	118	24	37	26	15	28	27
10,000–1,000	90	118	36	42	47	31	28	22
<1,000	21	76	13	23	39	20	126	62
Total	200	388	85	125	122	83	195	130

and HI samples using the four disinfection scenarios evaluated. The lowest MW fraction (<1000) possessed the largest formation potential for all samples utilizing any of the disinfectants tested. The only exception was >30,000-MW fraction for the spring hydrophilic sample using free chlorine.

The specific formation potential of TCAA and DCAA in terms of micrograms per milligram TOC for the HI and HB fractions for each individual molecular fraction, for the winter and spring samples, are shown in Table 5. The results show the yield, under free chlorination conditions, for TCAA to be greater than that for DCAA for both the HI and HB fractions. This trend is in agreement with data presented (average of five values, pH = 7.0, 24 h; free chlorine; fulvic acids = 30.1 µg TCAA per milligram TOC and 12.8 µg DCAA per milligram TOC; humic acids = 67.6 µg TCAA per milligram TOC and 20.3 µg DCAA per milligram TOC) by Reckhow et al;[19] under conditions of free chlorination the TCAA yield was significantly greater from the HB fraction compared to the HI fraction. This also agrees with data presented by Reckhow et al.[19] Under conditions of free chlorination the DCAA yields were similar for both fractions in this work as opposed to those of Reckhow et al., whose DCAA yields were greater for the humic acids than the fulvic acids. The data of Reckhow et al. did show the difference between

Table 4 Specific Formation Potential (μg/mg TOC) as a Function of MW Fraction (normalized to 1 mg TOC/l)

MW fraction (Da)	Chlorine TTHM/TOC	Chloramine TTHM/TOC	Ozone and chlorine TTHM/TOC	Ozone and chloramine TTHM/TOC
Winter HI				
>30,000	7.75	0.37	4.81	1.48
30,000–10,000	26.07	0.36	1.07	0.36
10,000–1000	9.13	0.43	5.00	0.22
<1000	28.21	19.74	11.03	4.10
Average	15.85	5.23	5.48	1.54
Winter HB				
>30,000	13.08	0.38	5.77	0.38
30,000–10,000	14.74	0.53	33.16	0.53
10,000–1000	11.79	0.36	7.50	0.48
<1000	27.00	3.00	62.00	58.00
Average	16.65	0.84	27.10	14.85
Spring HI				
>30,000	30.00	5.00	15.00	1.25
30,000–10,000	2.9	0.33	1.65	0.11
10,000–1000	3.22	0.28	2.13	0.19
<1000	14.09	6.36	23.64	3.18
Average	12.55	2.99	10.60	1.18
Spring HB				
>30,000	16.43	0.71	14.29	3.57
30,000–10,000	8.26	0.43	6.74	0.22
10,000–1000	31.40	2.81	4.56	0.53
<1000	115.83	94.17	28.33	5.83
Average	42.96	24.53	13.48	2.54

the yields of the two fractions for DCAA to be less than that for TCAA. Data from the current study showed the preozonation plus free chlorination disinfectant scenario generally increased the DCAA yield from both the HB and HI fractions while it slightly increased the yield for TCAA from the HI fractions and winter HB fraction but reduced the TCAA yield from the spring HB fraction. TCAA precursor material in this spring fraction appeared to preferentially destroyed by the initial ozonation process.

The chloramine disinfectant scenario produced reduced yields for both TCAA and DCAA compared to those observed using free chlorine. The HB fractions produced larger yields than the HI fractions. The ozone and chloramine system produced increased average yields of TCAA and DCAA from both the winter and spring HI and HB fractions compared to the use of chloramines alone. This increase in the average yields was influenced greatly by the substantial increase in the yields seen for the compounds in the <1000-MW fraction of both the HB and HI fractions.

The technique, mentioned by Reckhow and Singer,[29] of employing THMs as a benchmark for the production of other DBPs utilizing ratios of the specific by-products to THMs was used and the results are presented in Table 6. Under conditions of free chlorination more TCAA than DCAA was produced with

Table 5   Specific Formation Potential (μg/mg TOC) as a Function of MW Fraction
(normalized to 1 mg TOC/l)

MW fraction (Da)	Chlorine		Chloramine		Ozone and chlorine		Ozone and chloramine	
	DCAA/ TOC	TCAA/ TOC	DCAA/ TOC	TCAA/ TOC	DCAA/ TOC	TCAA/ TOC	DCAA/ TOC	TCAA/ TOC
Winter HI								
>30,000	12.22	30.00	5.56	5.19	13.70	28.15	8.52	6.67
30,000–10,000	11.79	28.57	5.71	6.79	4.64	27.50	3.57	4.64
10,000–1000	14.57	20.87	5.65	4.35	6.09	15.65	3.70	3.70
<1000	12.31	19.74	6.67	6.41	77.69	43.08	18.46	18.72
Average	12.72	24.79	5.89	5.68	25.53	28.60	8.56	8.43
Winter HB								
>30,000	12.69	30.38	8.08	10.77	13.08	10.38	7.69	7.31
30,000–10,000	10.00	33.16	8.42	12.11	11.05	37.37	15.26	7.37
10,000–1000	16.55	16.07	5.36	7.38	7.62	13.33	4.64	3.69
<1000	11.00	82.00	14.00	24.00	111.00	128.00	32.00	28.00
Average	12.56	40.4	8.96	13.56	35.69	47.27	14.9	11.59
Spring HI								
>30,000	17.63	29.47	5.26	5.79	10.53	25.53	3.95	4.47
30,000–10,000	14.29	14.95	3.85	3.30	5.27	10.77	1.87	2.09
10,000–1000	17.68	14.36	3.55	2.75	8.39	7.96	2.89	1.99
<1000	13.18	38.64	6.36	3.64	107.27	87.73	18.18	11.82
Average	15.69	24.35	4.76	3.87	32.86	33.0	6.72	5.09
Spring HB								
>30,000	10.71	54.79	8.57	16.43	7.14	12.14	9.29	13.57
30,000–10,000	16.09	25.65	5.22	8.04	5.65	3.26	6.09	5.87
10,000–1000	15.79	20.70	6.32	7.37	8.25	5.44	4.91	3.86
<1000	17.50	63.33	10.83	19.17	32.5	16.67	105.00	51.67
Average	15.0	41.1	7.73	12.75	13.38	9.38	31.32	18.74

DCAA showing less production that the THMs for the two HB samples. The use of ozone coupled with free chlorine generally showed larger ratios for TCAA than DCAA especially in the HI samples. In the two HI fractions it appears the preozonation had a greater impact on the THM precursors than the HAA precursors. The HB fractions show a decrease in the TCAA ratios with a small increase in the DCAA ratio.

The chloramine disinfection system produced a large increase in both the TCAA and DCAA ratios for all the samples tested, compared to free chlorine. The increase was more a function of the small amount of THMs formed using

Table 6   By-Product Formation Ratios

	Chlorine		Chloramine		Ozone and chlorine		Ozone and chloramine	
	DCAA/ TTHM	TCAA/ TTHM	DCAA/ TTHM	TCAA/ TTHM	DCAA/ TTHM	TCAA/ TTHM	DCAA/ TTHM	TCAA/ TTHM
Winter HI	1.02	1.99	11.09	10.83	3.86	9.64	9.31	9.77
Winter HB	0.87	2.24	14.17	19.92	1.35	1.69	14.83	10.31
Spring HI	3.51	4.23	7.54	6.44	3.75	5.51	13.24	12.55
Spring HB	0.81	1.90	6.59	11.08	1.08	0.78	14.48	11.75

chloramines than a significant increase in the production of TCAA and DCAA. Ozone coupled with chloramination showed larger increases in the DCAA ratio than TCAA compared chloramines alone.

CONCLUSIONS

The following conclusions can be drawn from the data collected for this project:

1. In fractionated Kaw Reservoir water, under conditions of free chlorination, the MW fraction of 10,000 to 1000 generally produced the largest THMFP and HAAFP.
2. Ozonation tended to shift the highest THMFP and HAAFP to the smallest MW range (<1000).
3. Both chloramination and preozonation plus chloramines effectively controlled THM production. The preozonation plus chloramines generally produced the lowest observed THMFP levels.
4. Under the chloramination scheme the highest production of TCAA and DCAA normally occurred in the MW range of 10,000 to 1000.
5. In many cases, preozonation enhanced the formation of DCAA in the <1000-MW range resulting from chlorination or chloramination.
6. Both chloramination and preozonation plus chloramines controlled the production of HAAs.
7. The yield (micrograms per milligram of TOC) for TCAA was greater than that for DCAA using either chlorination alone or preozonation plus chlorination.
8. The use of preozonation coupled with free chlorination increased the yield of DCAA for both the HI and HB fractions compared to those using free chlorination only.

REFERENCES

1. Rook, J. J., Formation of haloforms during chlorination of natural waters, *Water Treat. Exam.*, 23, 234, 1974.
2. Stevens, A. A., Slocum, C. J., Seeger, D. R., and Robeck, G. G., Chlorination of organics in drinking water, *J. AWWA*, 68, 615, 1976.
3. Miller, J. W. and Uden, P. C., Characterization of nonvolatile aqueous chlorination products of humic substances, *J. Environ. Sci. Technol.*, 17, 150, 1983.
4. Reckhow, D. A. and Singer, P. C., The removal of organic halide precursors by preozonation and alum coagulation, *J. AWWA*, 76, 151, 1984.

5. Pontius, F. W., Complying with the new drinking water quality regulations, *J. AWWA,* 82, 232, 1990.
6. Oliver, B. G. and Visser, S. A., Chloroform production from the chlorination of aquatic humic material: the effect of molecular weight, environment and season, *Water Res.,* 14, 1137, 1980.
7. Davis, J. A. and Gloor, R. Adsorption of dissolved organics in lakewater by aluminum oxide. Effect of molecular weight, *Environ. Sci. Technol.,* 15, 1223, 1981.
8. Rest, C. R., Hoehn, R. C., Knocke, W. R., and Novak, J. T., The removal of specific molecular weight fractions of THMs precursors by alum coagulation, in *Proceedings of the Annual AWWA National Conference,* Las Vegas, NV, June 5–9, 1983.
9. Vik, E. A., Carson, O. A., Eikum, A. S., and Gjessing, E. T., Removing aquatic humus from Norwegian lakes, *Am. Water Works Assoc.,* 773, 58, 1985.
10. Collins, M. R., Amy, G. L., and King, P. H., Removal of organic matter in water treatment, *ASCE J. Environ. Eng.,* 111, 850, 1985.
11. Schnitzer, M. and Khan, S. U., *Humic Substance in the Environment,* Marcel Dekker, New York, 1972.
12. Collins, M. R., Amy, G. L., and Steelink, C., Molecular weight distribution, carboxylic acidity, and humic substances content of aquatic organic matter: implications for removal during water treatment, *Environ. Sci. Technol.,* 20, 1028, 1986.
13. Schnoor, J. L., Nitzschke, J. L., Lucas, R. D., and Veenstra, J. N., Trihalomethane yields as a function of precursor molecular weight, *Environ. Sci. Technol.,* 13, 1134, 1979.
14. Veenstra, J. N. and Schnoor, J. L., Seasonal variations in trihalomethane levels in an Iowa river water supply, *J. AWWA,* 72, 583, 1980.
15. Collins, M. R., Amy, G. L., and King, P. H., Removal of organic matter in water treatment, *ASCE J. Environ. Eng.,* 111, 850, 1985.
16. Sinsabaugh, R. L., III, Hoehn, R. C., Knocke, W. R., and Linkins, A. E., Precursor size and organic halide formation rates in raw and coagulated surface waters, *ASCE J. Environ. Eng.,* 112, 139, 1986.
17. Kuo, C. J. and Amy, G. L., Factors affecting coagulation with aluminous sulfate. II. Dissolved organic matter removal, *Water Res.,* 22, 863, 1988.
18. Uden, P. C. and Miller, J. W., Chlorinated acids and chloral in drinking water, *J. AWWA,* 75, 524, 1983.
19. Reckhow, D. A., Singer, P. C., and Malcolm, R. C., Chlorination of humic materials: byproduct formation and chemical interpretations, *Environ. Sci. Technol.,* 24, 1655, 1990.
20. Johnson, J. D. and Jensen, J. N., THM and TOX formation: routes, rates, and precursors, *J. AWWA,* 78(4), 156, 1986.
21. Thurman, E. M. and Malcolm, R. L., Preparative isolation of aquatic humic substances, *Environ. Sci. Technol.,* 15, 463, 1981.
22. Anderson, L. J., Johnson, J. D., and Christman, R. F., Extent of ozone's reaction with isolated aquatic fulvic acid, *Environ. Sci. Technol.,* 20, 739, 1986.
23. *Standard Methods for the Examination of Water and Wastewater,* 15th ed., American Public Health Association, Washington, D.C., 1980.
24. *Water Analysis Handbook,* Hach Company, Loveland, CO, 1982.

25. Jacangelo, J. G. and Olivieri, V. P., Reactivity of monochloramine with nucleic acids and proteins, *Proceedings of Annual AWWA National Conference*, Dallas, June 10–14, 1984.

26. Calabrese, E. J., Chamberlain, C. C., and Young, M., The effects of trichlorocetic acid, a widespread product of chlorine disinfection, on the dragonfly nymph respiration, *J. Environ. Sci. Health*, A22(4), 343, 1987.

27. Veenstra, J. N., Barber, J. B., and Khan, P. A., Ozonation: its effect on the apparent molecular weight of naturally occurring organics and trihalomethane production, *Ozone: Sci Eng.*, 5, 225, 1983.

28. Zhu, Q. and Reckhow, D. A., Investigation of Pre-Ozonation and Chloramination on Haloacetic Acid formation, Presented at the Annual AWWA National Conference, San Antonio, TX, June 1993.

29. Reckhow, D. A. and Singer, P. C., Chlorination by-products in drinking waters: from formation potentials to finished water concentrations, *J. AWWA*, 82, 173, 1990.

CHAPTER 19

The Use of ClO$_2$ in Drinking Water Treatment: Formation and Control of Inorganic By-Products (ClO$_2$$^-$, ClO$_3$$^-$)

N. Karpel Vel Leitner, J. De Laat, M. Doré, and H. Suty

ABSTRACT

The effect of drinking water processes on the concentration of chlorite, chlorate, and chlorine dioxide has been examined. Laboratory studies indicate that under drinking water conditions (neutral pH and low concentrations of chlorite), chlorite and chlorine dioxide are oxidized rapidly by ozone or are converted into chloride by activated carbon. Chlorine (HOCl, ClO$^-$) can oxidize ClO$_2^-$ and ClO$_2$ into chlorate. However, taking into account the concentrations involved and the residence time, the kinetic rate constants are too low to allow chlorate formation in the distribution system. UV irradiation of ClO$_2^-$ and ClO$_2$ at 254 nm also produces chlorate. On the basis of kinetic results, the production of chlorate under UV disinfection conditions (irradiation doses $\approx$ 250 J/m^2) is unlikely.

Analyses of samples collected from two drinking water utilities that use ClO$_2$ for preoxidation and for disinfection indicate that, as expected from laboratory data, chlorite ion was reduced into chloride by activated carbon or oxidized into chlorate by ozonation. The analyses also revealed the presence of ClO$_2^-$ and ClO$_3^-$ at the outlet of ClO$_2$ generators.

INTRODUCTION

In Europe, many drinking water treatment plants use chlorine dioxide for the pretreatment and/or for the postdisinfection of water. In preoxidation, ClO$_2$ is used to control taste and odors, to prevent algal growth, and to remove iron and manganese.[1,2] Moreover, ClO$_2$ does not form trihalomethanes (THMs) and also reduces the concentration of the precursors of organohalogenated compounds.[3,4] However, the application of ClO_2 produces inorganic by-products, chlorite (0.5 to 0.7 mg ClO_2^-/mg ClO_2 consumed), and chlorate.[3,5] There are also other possible sources of chlorate ions in finished drinking water such as raw water contamination, presence of chlorate in feedstock solutions of hypochlorite or photodecomposition of residual hypochlorous acid during the disinfection process.[6,7] Chlorite and chlorate may pose potential health effect hazards.[8-11] The World Health Organization has proposed a provisional guideline value of 200 $\mu g/l$ for the concentration of chlorite in drinking water.[12] No guideline has been established for chlorate because of insufficient toxicological studies.

The objective of this study was to evaluate the evolution of oxychlorine species through the train of drinking water processes. In the first part of the study, laboratory experiments were carried out to investigate the reactions of ClO_2 and ClO_2^- with ozone, activated carbon, chlorine, and UV light. In the second part, the concentrations of ClO_2, ClO_2^-, and ClO_3^- were determined at different stages of drinking water treatment in two plants that pretreated their water with chlorine dioxide.

MATERIALS AND METHODS (LABORATORY EXPERIMENTS)

Aqueous solutions of chlorite ($NaClO_2$, purity >98%, Prolabo, France) or chlorine dioxide were prepared in phosphate buffered ultrapure water (6.2 < pH < 8.5) or in tap water (pH = 7.8; dissolved organic carbon = 1.2 mg C/l).

Stock solution of chlorine dioxide was prepared from the reaction between H_2SO_4 and $NaClO_2$. ClO_2 gas was purified by bubbling through a saturated solution of $NaClO_2$ prior to absorption in ultrapure water. Concentration of ClO_2 in stock solution was measured spectrophotometrically by taking the extinction coefficient at 360 nm equal to 1200 l mol^{-1} cm^{-1}.

Ozonation was carried out in a 1-l ozone contactor. Ozonated air produced by an ozone generator was introduced continuously through a porous sparger as microbubbles at the bottom of the column (gas flow rate: 0 to 10 l/h). The reactor was fed by water (liquid flow rate was 10 l/h).

UV experiments were carried out in a cylindrical reactor equipped with a low-pressure mercury vapor lamp in axial position (HANAU TNN 15/32 lamp, emission: 2 W at 253.7 nm; reactor: volume: 4 l, annular path length 6.25 cm; 20.0 ± 0.5°C).

Experiments with activated carbon were carried out under batch and continuous flow conditions with different granulometric fractions of CECA 40 (CECA, France).

Chloride, chlorate, and chlorite were analyzed by high-pressure liquid chromatography by using a UV detector (ClO_2^-) or a conductimetric detector (Cl^-, ClO_3^-). The detection limits were 20 ppb.

Residual ClO_2 and ozone in water were analyzed by the ACVK and Carmin Indigo spectrophotometric method respectively.[13,14] When both species were present in solution, adequate amounts of sodium metaarsenite were added to remove dissolved ozone before analysis with ACVK.

Chlorine was analyzed by a colorimetric method after reaction with a mixture of pyridine and barbituric acid in presence of KCN.[15]

RESULTS AND DISCUSSION

PILOT-SCALE EXPERIMENTS

Reactions with Ozone

As shown by the data in Figure 1, ozonation of chlorite in dilute aqueous solution (1.9 mg/l in tap water) primarily gives chlorine dioxide and then chlorate. For ozone doses higher than 2.0 to 2.5 mg O_3/l, chlorite and chlorine dioxide were not detected in solution and the concentration of chlorate (2.3 mg/l) indicates a production of 1 mol of chlorate per mole of chlorite.

According to the literature[16-18] the oxidation pathway of chlorite by ozone in water involves molecular ozone and the hydroxyl radical (Table 1).[18] Hydroxyl radicals can be produced from the decomposition of ozone initiated

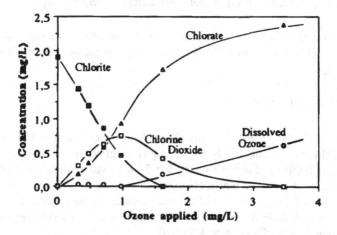

Figure 1 Ozonation of tap water spiked with chlorite (ClO_2^- = 1.9 mg/l, pH = 7.8, TOC = 1.2 mg/l; continuous flow reactor: volume 1 l; liquid flow rate 10 l/h; contact time: 6 min). Concentration of oxychlorine species and of dissolved ozone vs. ozone dose.

Table 1   Reactions of Ozone with Chlorite or Chlorine Dioxide in Water

Reactions	Kinetic constants
$ClO_2^- + O_3 \rightleftarrows ClO_2 + O_3^{\cdot-}$	$k_f = (4 \pm 1)\ 10^6\ M^{-1}\ s^{-1}$
	$k_r = (1.8 \pm 0.2)\ 10^5\ M^{-1}\ s^{-1}$
$ClO_2 + O_3 \rightarrow ClO_3 + O_2$	$k = (1.05 \pm 0.1)\ 10^3\ M^{-1}\ s^{-1}$
$ClO_2 + ClO_3 + H_2O \rightarrow 2ClO_3^- + 2H^+$	
$O_3^{\cdot-} \rightleftarrows O_2 + O^{\cdot-}$	$k_f = 3.3\ 10^3\ s^{-1}$
	$k_r = 3.0\ 10^9\ M^{-1}\ s^{-1}$
$O^{\cdot-} + H^+ \rightleftarrows OH^{\cdot}$	$pK_a = 11.9$
$O_3 + OH^- \rightarrow O_2^{\cdot-}/HO_2^{\cdot} + O_3 \rightarrow OH^{\cdot}$	
$O_3 + OH^{\cdot} \rightarrow O_2 + O_2^{\cdot-} + H^+$	$k = 1.1\ 10^8\ M^{-1}\ s^{-1}$
$ClO_2^- + OH^{\cdot} \rightarrow ClO_2 + OH^-$	$k = 4.2\ 10^9\ M^{-1}\ s^{-1}$
$ClO_2 + OH^{\cdot} \rightarrow ClO_3^- + H^+$	$k = (4.0 \pm 0.4)\ 10^9\ M^{-1}\ s^{-1}$

by hydroxide ion or by chlorite. Kinetic constants in Table 1 show the great reactivity of molecular ozone toward chlorite [k = $(4 \pm 1)\ 10^6\ M^{-1}\ s^{-1}$] and chlorine dioxide [k = $(1.05 \pm 0.1)\ 10^3\ M^{-1}\ s^{-1})$] and predict that chlorite and chlorine dioxide do not coexist with aqueous ozone more than a few seconds in dilute aqueous solution.

During ozonation of neutral and dilute aqueous solutions of chlorite, aqueous ozone reacts rapidly with chlorite to yield chlorine dioxide (initial step). Further reactions of molecular ozone and hydroxyl radicals with chlorite and chlorine dioxide take place to produce chlorate. The stoichiometry of the overall reactions of ozone with chlorite and chlorine dioxide is

$$ClO_2^- + O_3 \rightarrow ClO_3^- + O_2 \qquad (1)$$

$$2ClO_2 + O_3 + 2OH^- \rightarrow 2ClO_3^- + H_2O + O_2 \qquad (2)$$

which corresponds to the consumption of 1 mol of O_3 per mole of ClO_2^- (0.71 mg O_3/mg ClO_2^-) or of 0.5 mol of O_3 per mole of ClO_2 (0.35 mg O_3/ mg ClO_2).

Reactions with Chlorine

Figures 2 and 3 present results obtained for the chlorination ($Cl_2 \sim 5$ mg/ l) of chlorine dioxide ($ClO_2 \sim 3.5$ mg/l) in ultrapure water (pH = 7.6, phosphate buffer) and in tap water (pH = 7.8). To satisfy the chlorine demand of the tap water (~ 0.9 mg/l), chlorine was introduced in water (Cl_2 dosage ~ 6 mg/l) before the addition of chlorine dioxide. In the absence of chlorine, the chlorine dioxide demand of the two waters was below 0.5 mg/l after 72-h reaction time (ClO_2 dose ~ 4.5 mg/l).

In the presence of chlorine, the data show that reactions between chlorine and chlorine dioxide leads to chlorate (as an end by-product) and chlorite (as an intermediate). This decomposition of chlorine dioxide which is catalyzed

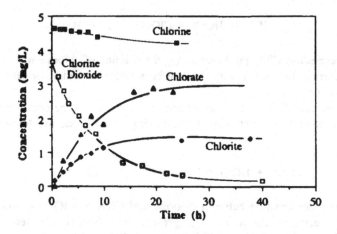

Figure 2 Chlorination of CIO$_2$ in ultrapure water. Evolution of residual chlorine and oxychlorine species (CIO$_2$, CIO$_2^-$, CIO$_3^-$) with reaction time (pH = 7.6, phosphate buffer; 20°C).

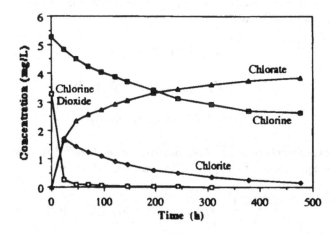

Figure 3 Chlorination of CIO$_2$ in tap water. Evolution of residual chlorine and oxychlorine species (CIO$_2$, CIO$_2^-$, CIO$_3^-$) with reaction time (pH = 7.8, 20°C).

by chlorine is accompanied by a small consumption of chlorine according to the following reaction (Figure 2):[19]

$$2CIO_2 + 1HOCl + H_2O \rightarrow (2 - x)CIO_3^- + x\ CIO_2^- + x\ HOCl$$
$$+ (1 - x)Cl^- + (3 - x)H^+ \quad (3)$$

Further reactions between residual chlorine and chlorite lead to the production of chlorate (Figure 3):[20-22]

$$ClO_2^- + HOCl \rightarrow ClO_3^- + Cl^- + H^+ \qquad (4)$$

The intermediate ClO_2, produced during the reaction of chlorine with chlorite in more concentrated solutions, has not been observed under our experimental conditions.

At the end of the run, the production of chlorate and the consumption of chlorine were consistent with the following overall reaction (chlorine in excess):

$$2ClO_2 + HOCl + H_2O \rightarrow 2ClO_3^- + Cl^- + 3H^+ \qquad (5)$$

Therefore, the reaction between chlorine and ClO_2 or ClO_2^- in dilute and neutral aqueous solutions and in the presence of excess chlorine leads to the formation of chlorate (1 mol ClO_3^-/mole of ClO_2 or ClO_2^- decomposed; 1 mole of chlorine consumed/mole of ClO_2^- and 0.5 mole of chlorine consumed/ mole of ClO_2). The reactions between chlorine and chlorite or chlorine dioxide could result from parallel routes leading to the intermediate Cl_2O_2 according to the following pathway.[19,23]

$$2ClO_2 + Cl_2 \rightleftarrows 2Cl_2O_2 \qquad (6)$$

$$Cl_2O_2 + H_2O \rightleftarrows Cl^- + ClO_3^- + 2H^+ \qquad (7)$$

$$Cl_2O_2 + Cl^- \rightleftarrows ClO_2^- + Cl_2 \qquad (8)$$

Kinetic experiments that have been described elsewhere[19,22] showed that the rate of disappearance of ClO_2 and ClO_2^- are second order:

$$-\frac{d[ClO_2^-]}{dt} = k_{(Chlorine, ClO_2^-)}[ClO_2^-][Chlorine]$$

$$-\frac{d[ClO_2]}{dt} = k_{(Chlorine, ClO_2)}[ClO_2][Chlorine]$$

The kinetic constants are very low and vary with pH:

$$k_{Chlorine/ClO_2} = 0.12 \; M^{-1} \, s^{-1} \text{ at pH 6.5 and } 1.0 \; M^{-1} \, s^{-1} \text{ at pH 8.5}$$

$$k_{Chlorine/ClO_2^-} = 9 \; 10^{-2} \; M^{-1} \, s^{-1} \text{ at pH 6.2 and } 8 \; 10^{-3} \; M^{-1} \, s^{-1} \text{ at pH 8.5}$$

These kinetic values indicate that a postdisinfection with chlorine or sodium hypochlorite (oxidant dose $\leq$ 5 mg/l) of drinking waters pretreated by ClO_2

will have little effect on the concentrations of residual chlorine dioxide and of chlorite in the distribution system.

UV Disinfection

Several workers have studied the photochemistry of oxychlorine species in aqueous solutions.[24-28] The photodecomposition of chlorite and chlorine dioxide by UV irradiation leads to the production of chloride and chlorate, and oxygen as stable end products, and to other analyzable by-products (chlorite, chlorine dioxide, chlorine) as intermediates.

In a recent study,[29,30] we showed that the productions of chloride and chlorate from the photodecomposition of chlorite and chlorine dioxide solutions in a UV reactor equipped with a germicidal low-pressure mercury lamp (emission at 253.7 nm) could be described by the following overall reactions (5.6 < pH < 8.3).

$$10\ CIO_2 + 5H_2O \xrightarrow{h\nu} 4Cl^- + 6ClO_3^- + \frac{7}{2} O_2 + 10\ H^+ \qquad (9)$$

$$10\ CIO_2^- \xrightarrow{h\nu} 6Cl^- + 4ClO_3^- + 4O_2 \qquad (10)$$

These global reactions are the result of a series of complex reactions initiated by UV irradiation. As far as the rate of photodecomposition is concerned,[30] the kinetic data showed that irradiation doses at 253.7 nm of about 4500 ± 600 and 8000 ± 600 J m^{-2} are needed to have a 50% reduction in CIO_2 and CIO_2^- concentrations, respectively (Figure 4). These data suggest that no significant reduction in CIO_2 and CIO_2^- concentrations can be expected at an irradiation dose of 250 J m^{-2} which is usually applied for the UV disinfection of drinking waters.

Reactions with Activated Carbon

Reactions between chlorite and activated carbon yield oxygen, chloride, and chlorate as end products. The productions of chloride and chlorate depend on the initial concentrations of CIO_2^- (Figure 5), the GAC concentration, pH, and on the type and the granulometry of GAC.[31,32] When low concentrations of CIO_2^- are used (<5 mg/l), chloride is the predominant end product of the decomposition of chlorite by GAC (>0.9 mol Cl$^-$/mol CIO_2^- removed).

According to the literature on the interactions between oxychlorine species and activated carbon,[33,34] the decomposition of chlorite by activated carbon may lead initially to the formation of radical entities (Cl·, ClO·) by reactions of chlorite with free radicals present at the surface of the activated carbon. Other reaction sites such as oxygen-containing functional groups and metal

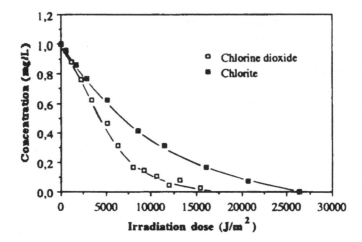

Figure 4 Photodecomposition of chlorite and chlorine dioxide vs. irradiation dose.
Initial concentrations: ClO_2^- or ClO_2: 1 mg/l; ultrapure water: pH = 6.8 to 7.5
(phosphate buffer); batch reactor: volume: 4.0 l; lamp: HANAU TNN 15/32;
Photonic flux at 253.7 nm: 2 W.

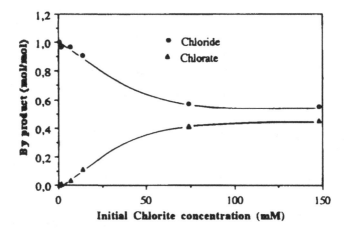

Figure 5 Effect of the initial concentration of chlorite on the production of chloride and
chlorate (mol of Cl^- or ClO_3^-/mol of ClO_2^- decomposed). Batch reactor: GAC
CECA 40 (0.8 to 1.0 mm), 10 g/l; pH = 7.0 ± 0.2; 20°C, contact time 24 h;
chlorite removal >99%.

ions present as impurities in activated carbon may also initiate the decomposition of chlorite. The primary by-products (Cl^-, ClO^-, ClO_2) react with activated carbon to give chloride and through secondary reactions, lead to the formation of a number of intermediates (Cl_2O_2, Cl_2O_3, $HOCl$, . . .) which then decompose to yield chloride, chlorate, and oxygen as final products (Figure 6). Increasing chlorite concentrations will increase the yield of chlorate because secondary reactions are favored.

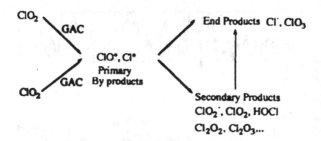

Figure 6 Proposed reaction scheme for the interactions between chlorite, chlorine dioxide, and activated carbon in the absence of organic compounds.

In the presence of organic compounds in solution or adsorbed on activated carbon, organic by-products may also be formed.[35,37] In a recent study on the interactions between phenol (20 mg/l), chlorite (50 mg/l), and granular activated carbon (GAC columns), a variety of organic by-products were identified by GC/MS analyses (Table 2). The formation of these products demonstrates the occurrence of hydroxylation, carboxylation, oxidation, halogenation, dehydroxylation, and dimerization reactions.

Under drinking water conditions ($CIO_2^- < 1$ to 2 mg/l), chlorite is mostly converted into chloride by powdered or granular activated carbon. Figure 7 shows the evolution of chlorite at the outlet of a granular activated carbon column fed with tap water spiked with 1 mg/l chlorite (GAC: CECA 40; granulometry >0.5 mm; hydraulic loading: 9 m h⁻¹; empty bed contact time: 4 min). At the end of the run (60,000 bed volumes of water), the cumulative amount of CIO_2^- removed reached 128 mg CIO_2^-/g GAC. From these results, a lifetime of GAC for CIO_2^- removal of about 8 to 12 months can be expected for practical applications (EBCT ~ 10 to 15 min; maximum concentration of CIO_2^- in the effluent: 0.2 mg/l).

Since GAC filters can be used for the reduction of chlorite concentrations of water pretreated with chlorine dioxide, further investigations are needed to

Table 2 Organic By-Products of the Interactions among Phenol, Chlorite, and Activated Carbon (pH = 7.2)[39,40]

1,4-Benzoquinone	Hydroxylated compounds
1,4 Naphthoquinone	4-Hydroxy 3-chloro benzoic acid
Methyl 2-hydroxybenzoate	Hydroxybenzaldehyde
Aromatic acids	Hydroxyphenol
Benzoic acid	Biphenyls
2-Hydroxybenzoic acid	2-Phenoxyphenol
3-Hydroxybenzoic acid	2,2'-Dihydroxybiphenyl
Halogenated compounds	2,5-Cyclohexadiene 1,4-dione
2-Chlorophenol	2,2(hydroxyphenyl)
4-Chlorophenol	4-Phenoxyphenol
Chloroparabenzoquinone	2-Hydroxyphenyl 4-Hydroxyphenyl
2,4 Dichlorophenol	methanone
Dichlorohydroxy benzoic acid	Bis (2-hydroxy phenyl) methanone
	2,2'-Hydroxyphenoxy benzoic acid

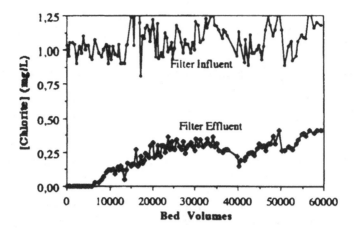

Figure 7 Chlorite breakthrough curve. Influent: tap water spiked with 1 mg ClO_2^-/l; granular activated carbon: CECA40; GAC column: bed height: 0.56 m; internal diameter: 3.76 cm, liquid flow rate: 10 l/h; EBCT: 4 min.

confirm the performances of GAC and to evaluate the risk of formation of undesirable organic by-products under such conditions (total organic carbon ≤ 5 mg/l; $ClO_2^- \leq 2$ mg/l). Moreover, care must be taken since several authors showed the formation of chlorate in significantly high concentrations when water containing chlorite and $HOCl/ClO^-$ were introduced together on a GAC filter.[38] It appears that chlorite could be oxidized to chlorate on the GAC by free chlorine present in the applied water.

FULL-SCALE RESULTS

To confirm the results obtained from laboratory experiments, oxychlorine analyses were performed on samples collected from two water utilities that use chlorine dioxide, ozone, and GAC (Figure 8). Samples were taken at different treatment steps; before ClO_2 preoxidation: RW (raw water) and after preoxidation with ClO_2 (PW); coagulation-flocculation-decantation (CFDW); sand filtration (SFW); ozonation (OW); granular activated carbon filtration (GACFW); remineralization; and final ClO_2 disinfection (DW) (Figure 8).

Residual ClO_2 was determined colorimetrically just after sampling. Samples for ClO_2^- and ClO_3^- analyses were purged with an inert gas (helium) to remove ClO_2 and dissolved ozone and to prevent ClO_2^- and ClO_3^- formation during storage. ClO_2^- and ClO_3^- concentrations were determined by ion chromatography.

Plant A

Analyses carried out on preoxidized water samples showed the presence of ClO_2^- (0.5 to 0.7 mg ClO_2^-/mg ClO_2 consumed) and of chlorate (0.2 to

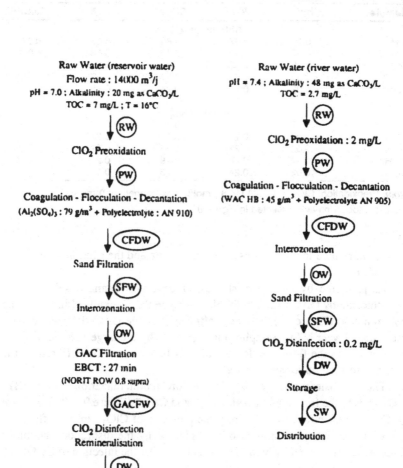

Figure 8 Water treatment scheme and sample locations (plants A and B).

0.35 mg ClO_3^-/mg ClO_2 introduced). A control of the concentration of the oxychlorine species at the outlet of chlorine dioxide generator (chlorine/chlorite system) revealed high concentrations of chlorate ($\approx$ 0.3 mg ClO_3^-/mg ClO_2).

After coagulation-flocculation-decantation, an increase in the concentration of ClO_2^- was observed. This can result from the reactions of residual ClO_2 with organic and inorganic reducing compounds. A small decrease in ClO_2^- concentration without ClO_3^- formation was often observed during sand filtration ($\leq$ 5 to 10%). This decrease which has not been clearly explained

Table 3 Results Obtained from Plant A (June 1992)

Samples[a]	RW	PW	CFDW	SFW	GACFW
			Chlorite (mg/l)		
8 H	<0.02	0.67	0.80	0.66	<0.02
11 H 30	<0.02	0.52	0.78	0.68	<0.02
14 H	<0.02	0.73	0.82	0.68	<0.02
17 H	<0.02	0.52	0.58	0.57	<0.02
Mean	<0.02	0.61	0.74	0.65	<0.02
			Chlorate (mg/l)		
8 H	<0.02	—	0.29	0.27	0.24
11 H 30	<0.02	0.24	0.26	0.27	0.25
14 H	<0.02	0.29	0.26	0.27	0.26
17 H	<0.02	0.20	0.19	0.21	0.27
Mean	<0.02	0.24	0.25	0.25	0.25

Note: Preoxidation: ClO_2 = 1 mg/l (14.8 μmol/l); interozonation: 0 mg O_3/l.

[a] Sample locations are presented in Figure 8.

may be attributed to interactions between chlorite and the filter medium (sand or biofilm).

As predicted by laboratory studies, chlorite was quantitatively reduced into chloride by GAC filtration (Table 3, Figure 9a) or oxidized into chlorate by ozonation (Figure 9b). The high efficiency of the GAC (in operation for more than 1 year before sampling) is explained by the large volume of GAC with regard to the flow rate. If ozonation is followed by GAC filtration, the chlorate formed no longer evolves.

The postdisinfection step by ClO_2 before remineralization by $Ca(OH)_2$/ CO_2 leads to the formation of both ClO_2^- and ClO_3^- (Figure 9a). The increase in ClO_3^- concentration results from the presence of ClO_3^- in the effluent of the ClO_2 generator and may also be explained by the disproportionation of ClO_2 which is favored by high pH values [just after the injection of $Ca(OH)_2$].

Plant B

On this site, the evolution of chlorite and chlorate after preoxidation and sand filtration was comparable to plant A. From the data reported in Table 4, it should be noted that ozonation leads to a decrease of 62% of the concentration of ClO_2^- with an increase in the ClO_3^- concentration (0.5 to 0.6 mol of ClO_3^-/mol of ClO_2^- eliminated) which is lower than the stoichiometry of 1 mol of ClO_3^-/mol of ClO_2^- eliminated. The presence of ClO_2^- in ozonated water indicates that the dose of ozone applied was too low to oxidize completely ClO_2^- into ClO_3^- (residual ozone = 0). The ClO_2 formed by ClO_2^-/O_3 reaction could have reacted with reducing species thus explaining the lower ClO_3^- production than expected.

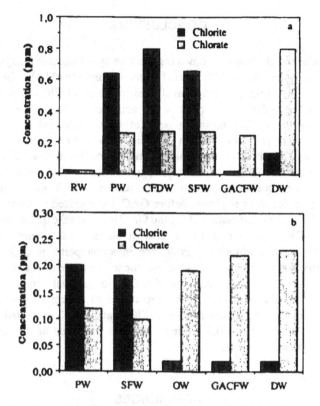

Figure 9 Results obtained from plant A (August 1992) (a): Preoxidation ∼ 1 mg ClO_2/l, interozonation = 0 mg O_3/l; final disinfection ∼ 0.6 mg ClO_2/l injected before remineralization. (b): Preoxidation ∼ 0.3 mg ClO_2/l; interozonation ∼ 2 mg O_3/l. Key to abscissa is presented in Figure 8.

Table 4   Results Obtained from Plant B (December 1992)

Samples[a]	RW	PW	CFDW	OW	SFW	DW	SW
			Chlorite (mg/l)				
8 H	<0.02	1.15	1.17	0.51	0.39	0.37	
11 H	<0.02	0.97	1.14	0.44	0.48	0.42	
14 H	<0.02	1.07	1.19	0.40	0.38	0.48	0.48
17 H	<0.02	1.12	1.23	0.43	0.40	0.50	
Mean	<0.02	1.08	1.18	0.45	0.41	0.44	0.48
			Chlorate (mg/l)				
8 H	<0.02	0.21	0.19	0.64	0.67	0.71	
11 H	<0.02		0.22	0.73	0.68	0.76	
14 H	<0.02	0.14	0.14	0.66	0.65	0.63	0.60
17 H	<0.02	0.14	0.13	0.70	0.65	0.65	
Mean	<0.02	0.16	0.17	0.68	0.66	0.69	0.60

Note: Preoxidation: ClO_2 ∼ 2 mg/l (29.6 μmol/l); final disinfection: ClO_2 ∼ 0.2 mg/l.

[a] Sample locations are presented in Figure 8.

CONCLUSIONS

Chlorite and chlorate are known disinfection by-products of ClO_2 treatment in drinking water. When ClO_2 is used as preoxidant, the chlorite formed (0.6 to 0.7 mg ClO_2^-/mg ClO_2 consumed) and the residual ClO_2 in preoxidized waters can be reduced into chloride by activated carbon or oxidized into chlorate by ozone. The optimum point at which activated carbon should be used with respect to ozonation must be chosen in relation with meeting requirements. If filtration on granular activated carbon follows clarification immediately, filtration allows a complete elimination of the residual ClO_2 and of ClO_2^-. Subsequent ozonation no longer results in the formation of ClO_3^-. If the ozone treatment is placed before GAC, as practiced in many drinking water utilities, the combination of O_3 and GAC improves the quality of finished water but ClO_3^- is formed. In the case of utilities that disinfect water by ozone followed by chlorine dioxide, the injection point of ClO_2 must be sufficiently distant from the ozone contactor to allow the ozone to disappear with time and to avoid any reaction of ClO_2 with residual ozone.

This work also emphasized the importance of optimized generation of ClO_2 to minimize the concentrations of ClO_2^- and ClO_3^- in treated water. A chlorate level as high as 0.3 mg/mg ClO_2 was measured at the outlet of a ClO_2 generator.

REFERENCES

1. White, G. C., *The Handbook of Chlorination*, 2nd ed., Van Nostrand Reinhold, New York, 1986.
2. Masschelein, W. J., *Chlorine Dioxide, Chemistry and Environmental Impact of Oxychlorine Compounds*, Rice, R. G., *Ann Arbor Science*, Publications Ann Arbor, MI, 1979.
3. Werdehoff, K. S. and Singer, P. C., *J. Am. Water Works Assoc.*, 9, 107, 1987.
4. Ben Amor, H., De Laat, J., and Doré., M., *Environ. Technol. Lett.*, 9, 1105, 1988.
5. Ben Amor, H., De Laat, J., and Doré, M. *Rev. Sci. Eau*, 3, 83, 1990.
6. Nowell, L. H. and Hoigné, J., *Water Res.*, 26, 593, 1992.
7. Bolyard, M., Snyder Fair, P., and Hautman, D. P., *J. Am. Water Works Assoc.* 9, 81, 1993.
8. Lubbers, J. R. and Bianchine, J. R., *J. Environ. Pathol. Toxicol. Oncol.*, 5, 215, 1984.
9. Lubbers, J. R., Chauhan, S., Miller, J. K., and Bianchine, J. R., *J. Environ. Pathol. Toxicol. Oncol.*, 5, 229, 1984.
10. Lubbers, J. R., Chauhan, S., Miller, J. K., and Bianchine, J. R., *J. Environ. Pathol. Toxicol. Oncol.*, 5, 239, 1984.
11. Causi, D., *Environ. Health Perspect.*, 46, 13, 1982.
12. *Revision of the WHO Guidelines for Drinking-Water Quality*, Report of the final task group meeting, Geneva, 1992.

13. Masschelein, W. J., Fransolet, G., Laforge, P., and Savoir, R., *Ozone Sci. Eng.*, 11, 209, 1989.
14. Bader, H. and Hoigné, J., *Ozone Sci. Eng.*, 4, 169, 1982.
15. Karge, H., Z *Analy. Chem.*, 200, 57, 1964.
16. Richard, Y. and Brener, L., *T.S.M. Eau*, 12, 627, 1981.
17. Richard, Y. and Brener, L., in *Handbook of Ozone Technology and Applications*, Rice, R. G. and Netzer, A., Eds., Ann Arbor Science Publishers, Ann Arbor, MI, 1982, 277.
18. Klaning, U. K., Sehested, K., and Holcman, J., *J. Phys. Chem.*, 89, 760, 1985.
19. Karpel Vel Leitner, N., De Laat, J., Dore, M., Suty, H., and Pouillot, M., *Environ. Technol.*, 12(9), 803, 1991.
20. Tang, T. F. and Gordon, G., *Environ. Sci. Technol.*, 18(3), 212, 1984.
21. Aieta, E. M. and Roberts, P. V., *Environ. Sci. Technol.*, 20(1), 50, 1986.
22. Karpel Vel Leitner, N., De Laat, J., Dore, M., and Suty, H., *Environ. Technol.*, 12(6), 477, 1991.
23. Taube, H. and Dodgen, H., *J. Am. Chem. Soc.*, 71, 3330, 1949.
24. Bowen, E. J. and Cheung, W. M., *J. Chem. Soc.*, 159, 1200, 1932.
25. Basco, N. and Dogra, S. K., *Proc. R. Soc. Ser. A*, 323, 29, 1971.
26. Buxton, G. V. and Subhani, M. S., *J. Chem. Soc. Faraday Trans.*, 68, 947, 1972.
27. Buxton, G. V. and Subhani, M. S., *J. Chem. Soc. Faraday Trans.*, 68, 958, 1972.
28. Mialocq, J. C., Barat, F., Gilles, L., Hickel, B., and Lesigne, B., *J. Phys. Chem.*, 77(6), 742, 1973.
29. Karpel Vel Leitner, N., De Laat, J., and Dore, M., *Water Res.*, 26(12), 1655, 1992.
30. Karpel Vel Leitner, N., De Laat, J., and Dore, M., *Water Res.*, 26(12), 1665, 1992.
31. Karpel Vel Leitner, N., De Laat, J., Dore, M., Suty, H., and Pouillot, M., *Water Res.*, 26(8), 1053, 1992.
32. Karpel Vel Leitner, N., De Laat, J., Dore, M., Suty, H., and Pouillot, M., *Environ. Technol.*, 13(7), 621, 1992.
33. Chen, A. S. C., Larson, R. A., and Snœyink, V. L., *Carbon*, 22(1), 63, 1984.
34. Voudrias, E. A., Larson, R. A., and Snœyink, V. L., *Carbon*, 25, 503, 1987.
35. Chen, A. S. C., Larson, R. A., and Snœyink, V. L., *Environ. Sci. Technol.*, 16(5), 268, 1982.
36. Voudrias, E. A., Dielman, L. M. J., Snœyink, V. L., Larson, R. A., McCreary, J. J., and Chen, A. S. C., *Water Res.*, 17(9), 1107, 1983.
37. Jackson, D. E., Larson, R. A., and Snœyink, V. L., *Water Res.*, 21(7), 849, 1987.
38. Dixon, K. L. and Lee, R. G., *J. Am. Water Works Assoc.* 83(5), 48, 1991.
39. Karpel Vel Leitner, N., Ph.D. thesis, Ecole Superieure d'Ingenieurs de Poitiers, Poitiers University, France, 1991.
40. Karpel Vel Leitner, N., De Laat, J., Dore, M., Suty, H., and Pouillot, M., *Environ. Sci. Technol.*, 28(2), 222, 1994.

PART V

Influence of Natural Organic Matter on By-Products

CHAPTER 20

Using Fractionated Natural Organic Matter to Study Ozonation By-Product Formation

S. A. Andrews and P. M. Huck

ABSTRACT

Natural organic matter was isolated and fractionated from surface water for subsequent ozonation and disinfection by-product (DBP) identification under conditions typical of drinking water treatment. Aquatic natural organic matter (NOM) was isolated from two source waters by adsorption on Amberlite®* XAD-8 and XAD-4 macroreticular resins, and on Bio Rad®** AG MP-50 ion exchange resin. These fractions accounted for approximately 50 to 60% of the dissolved organic material and 60 to 75% of the color. They were characterized and then ozonated under a range of pH, alkalinity, and ozone dosages encountered during drinking water treatment. By-products of either health concern or which may contribute to biological instability of treated drinking water were investigated, including aldehydes, oxoacids, and low molecular weight carboxylic acids, and the results compared with those obtained by ozonating a standard material obtained from the International Humic Substances Society and by ozonating the natural waters. Comparison and correlation of by-product yields with reductions in ultraviolet absorbance are also reported. On the basis of ultraviolet absorbance measurements, the fulvic acid fractions studied accurately represented the natural water and may be the primary sources of precursor material for aldehydes and oxoacids formed during ozonation. However, the NOM fractions that were isolated in this research contributed less significantly to carboxylic acid formation.

* Rohm and Haas, Philadelphia.
** Bio Rad, Mississauga, ON.

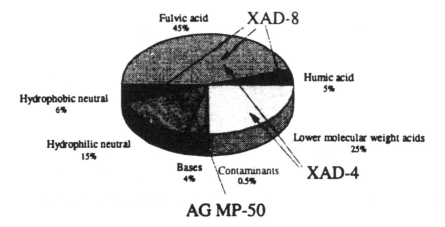

Figure 1 Typical composition of dissolved organic carbon from surface water in the U.S. and likely resins on which the fractions could be isolated.[1]

INTRODUCTION

To study DBP formation from the oxidation of NOM under controlled conditions, NOM must first be isolated from natural sources. The use of NOM fractions allows results to be related more directly to the nature of the matter found in different types of raw waters. It also provides a link to treatment, because processes such as coagulation and adsorption remove different fractions of NOM. The NOM from surface waters has been classed into the general constituents shown in Figure 1.[1] Approximately half of this material is made up of humic and fulvic acids, present in a 1:9 mass ratio. Humic and fulvic acids are operationally defined by their solubility at low pH. Both of these classes of NOM are of relatively high molecular weight (180 to 10,000 Da for fulvic acids, 10^2 to 10^6 Da for humic acids),[2] and fulvic acids generally contain greater oxygen content, primarily as carboxyl groups, than humic acids. Humic and fulvic acids are also highly colored, and several hypothetical structures which include various chromophoric groups have been proposed.[2-4] Because of their high color content and abundance, and because methods for their isolation are available, much NOM research has utilized the humic and fulvic acid fractions.

The choice of NOM isolation method adds another modifier to the operational definition of humic and fulvic acids in that isolation of different portions of the NOM will be either enhanced or reduced depending on the method used. The most common means to isolate aquatic humic and fulvic acids has been by adsorption onto nonpolar polymeric resins such as Amberlite® XAD-8 (a methyl methacrylate polymer) and is the method of choice for the International Humic Substances Society.[5,6] Data are available detailing these procedures and describing the materials isolated.[1] Acidic functional groups in these molecules are protonated by acidification to pH 2, and the neutral molecule is then adsorbed. There are limitations to this particular method of isolation. For example, some nonhumic material can be co-adsorbed, particularly that which is hydrophobic or aromatic, and XAD-resins exhibit size exclusion characteristics such that the largest of the humic acid molecules may not be concentrated on these resins. Also, some humic material may irreversibly form aggregates at low pH and become trapped within the porous structure of the resin.[7] Also, while NOM adsorbed to XAD-8 resin typically accounts for 40 to 60% of the organic carbon content of surface waters,[8] variability in source water characteristics can result in NOM recoveries of between 17 and 78% of the TOC or DOC,[9-11] and, while these materials account for most of the color, they likely underestimate DBP formation. Despite these limitations, the XAD isolation method provides a high-capacity, economical, and reproducible means of isolating aquatic humus, and remains the current method of choice for the isolation of aquatic NOM. Methods to improve recovery of this material or to further fractionate it for various purposes, including DBP studies and evaluation of treatment options, continue to be developed. For example, inclusion of resins such as XAD-4 and ion exchange in the isolation procedure can improve the recovery of certain NOM classes as indicated in Figure 1.

The use of ozone in drinking water treatment has been shown to be effective in many applications, including taste and odor control, color, iron and manganese removal, THM precursor reduction, and as a coagulant aid. Ozone has been used successfully as a drinking water disinfectant in many European countries since the early 1900s, and has recently shown much promise in reducing the formation of DBPs in studies in Europe and North America.[12,13] While DBP production from the application of chlorinated disinfectants has been the subject of much research since Rook described THM formation in 1976, research into DBPs related to ozonation practice has only relatively recently been initiated because ozone was considered a more "natural" oxidant that produced primarily biodegradable by-products. Application of early ozonation literature to the problem of identifying ozonation DBPs of possible health concern has been met with limited success because most of this information has been obtained using nonaqueous solutions, and it has only been recent advancements in analytical chemistry that have made aqueous ozonation research more feasible. Therefore, the subject of by-product identification from the aqueous ozonation of organic compounds is a relatively recent one.

During drinking water disinfection, NOM reacts both with molecular ozone and with hydroxyl radicals that are produced by the reaction of ozone in water, to form oxygenated DBPs having reduced color and molecular weight[14-16] and can lead to the production of by-products that are either toxicologically or otherwise biologically significant. Aldehydes, ketones, epoxides, phenols, quinones, furans, alcohols, and carboxylic acids are all formed during ozonation of aquatic NOM.[17-20] Ozone reacts with carbon-carbon double bonds forming peroxidic intermediates which release hydrogen peroxide and leave carbonyl compounds. Research indicates that ozone probably also reacts quickly with chromophores, e.g., hydroxyl and amino chromophores.[14,21] While the classes of by-products can sometimes be predicted, only a few of the actual compounds have been identified, the most commonly reported ones being the low molecular weight aldehydes (e.g., References 22 to 24). However, the variability of by-products expected from ozonation of natural waters suggests that generalizations are difficult to make. Water quality parameters and ozonation conditions both affect DBP production. For example, ozone dose has a major impact, and the low dosages used for drinking water treatment do not generally produce the two-carbon by-products observed for high doses, but rather partially oxidized products are observed. Unsaturated fatty acids commonly yield aldehydes and carboxylic acids upon ozonation (formaldehyde, aldehydes detected as ranging in size from six to ten carbons in length), with formaldehyde yields about ten times those of other aldehydes.[17] Also formaldehyde, and perhaps other species, may be formed hours after ozonation has been terminated, complicating interpretation of results.[18]

Toxicological data on organic ozonation by-products are limited, but there has been some concern over possible adverse health effects of these DBPs for humans. For example, formaldehyde, acetaldehyde, glyoxal, and methylglyoxal are known or suspected mutagens and carcinogens.[25-28] Carboxylic acids are not generally considered a direct toxicological hazard,[25] and most of the oxoacids identified to date are either not of concern regarding human health or have unknown toxicological properties, but both classes of these DBPs may be significant components of the substrate (termed assimilable or easily biodegradable organic carbon) for growth of microbes in the distribution system which can deteriorate the quality of the distributed water.[29] Therefore, interest in by-products such as carboxylic acids and oxoacids could be considered minimal from a health perspective but may be greater with regard to biotreatment technologies and distribution system stability. While ambient concentrations of most of these by-products would be of the same order of magnitude as those encountered during natural oxidation of organic material and could be considered harmless,[30] it has been recognized that potentially harmful by-products simply may not have been detected by available analytical methods.

The relative importance of the free radical vs. ozonolysis (molecular ozone) mechanisms for typical drinking water treatment conditions is dependent on various raw water quality parameters. Bicarbonate ion scavenges free radical species and thereby promotes the molecular pathway. This has important implications with respect to drinking water treatment DBP production as the concentration of total carbonate species (which is often referred to in the drinking water treatment industry as alkalinity) varies from source to source. Reckhow et al.[31] found that small amounts of bicarbonate significantly improve trihalomethane precursor destruction by shifting the oxidation pathway away from the free radical toward the direct oxidation, or molecular, pathway. Similar shifts in reaction mechanism from radical toward molecular were observed by Legube et al.,[32] Doré et al.,[33] and Xiong et al.[21] These results are attributable to the selectivity of molecular ozone reactions with aromatic and unsaturated components of natural humic materials, those which contribute to their color and UV absorbance characteristics. Glaze[17] stated that, under drinking water treatment conditions, it is likely that the free radical pathway dominates; however, consideration must be given to the individual source water bicarbonate concentration. In addition, even when molecular reactions are initially predominant, organic and inorganic radicals are generated that participate in subsequent oxidations.[30,34-36] Humic matter can also act as free radical reaction initiators, generating radical intermediates to further react in nonselective radical reactions to, for example, hydroxylate benzene rings otherwise thought to be unreactive. Further reaction yields the aldehydes and acids mentioned previously.

In this research, an improved procedure was applied to isolate and fractionate 65 to 80% of the NOM from two drinking water sources. Isolation using Amberlite® XAD-8 resin followed by XAD-4 allowed approximately 100% more material to be isolated than when using XAD-8 alone. Bio Rad® AG MP-50 cation exchange resin placed after XAD-4 isolated 5 to 15% more material than XAD-8 plus XAD-4. Both natural waters and solutions of fractionated material were ozonated under conditions typical of current drinking water treatment practice to examine DBP formation and color removal characteristics. Ozonation experiments were conducted in semi-batch mode to study the effects of pH, alkalinity, and ozone dosage on the formation of the following organic DBPs: aldehydes, oxoacids, and carboxylic acids. Ozonated material was also chlorinated or chloraminated to determine halogenated DBP formation potentials (trihalomethanes, haloacetic acids, chloral hydrate, cyanogen chloride). The results of these latter experiments are reported elsewhere.[37-40]

METHODS AND MATERIALS

All laboratory bench test procedures were performed using natural waters or solutions of fulvic acids. The water used to prepare fulvic acid solutions and aqueous reagent solutions was distilled, deionized with a Milli-Q®* system (Millipore Corp.), and buffered to specified pH values using phosphate buffers prepared in-house unless otherwise stated. DBP standards and laboratory reagents (acids, bases, phosphate for buffers, etc.) were obtained at greater than 98% purity and were used as received. Indigo trisulfonate was obtained and prepared for ozone residual determinations as per *Standard Methods.*[6] Amberlite® XAD-8 and XAD-4 and Bio Rad® AG MP-50 resins were Soxhlet

* Millipore Corp., Mississauga, ON.

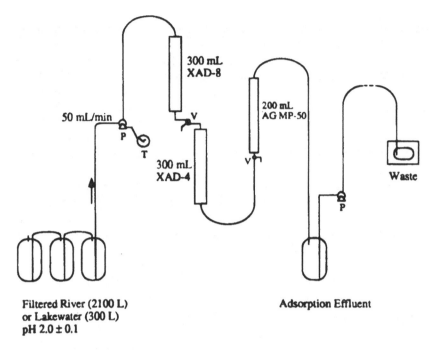

Figure 2   Apparatus for isolating natural organic matter (P = pump, T = timer, V = sample valve).[39]

cleaned by the procedures described by Malcolm.[1] Humic and fulvic acid reference materials isolated from the Suwannee River were purchased from the International Humic Substances Society (IHSS, Denver) and were used as received.

NOM ISOLATION

Approximately 2100 l river water and 300 l lake water NOM were isolated using the XAD-8 resin adsorption procedures described by Malcolm[1] which are also used by the IHSS, except that XAD-4 resin was used sequentially with XAD-8.[41] During isolation of river water NOM, AG MP-50 cation exchange resin was added following XAD-4 to isolate hydrophilic bases as described by Leenheer.[42] This resin and associated equipment were not available for lake water NOM isolation efforts. The apparatus used to isolate the NOM is shown in Figure 2. Water quality data for the two water sources were determined in our laboratory and from monthly utility reports and are summarized in Table 1. Upon return to the laboratory, samples were immediately acidified to pH 2.0 ± 0.1 with concentrated sulfuric acid (maximum 1 h delay) to preserve them prior to and during the lengthy filtration and adsorption procedures. Sample stability during this process was ascertained by measuring

Table 1   Water Quality Data for Driedmeat Lake and the North Saskatchewan River at Edmonton

Parameter	Driedmeat Lake	North Saskatchewan River
During NOM isolation[a]		
Alkalinity (mg/l as CaCO₃)	260	122
Calcium (mg/l)	84	117
Color (TCU)	55	2.2
Conductivity (μmho/cm)	316	261
Total hardness (mg/l)	156	167
pH	8.54	7.94
Temperature (°C)	13.6	2.18
Turbidity (NTU)	7.4	3.1
Phosphorus (μg/l)	116.6	NA[b]
Total dissolved solids (mg/l)	261	NA
During raw water ozonation[c]		
pH	7.9	6.9
NPOC (mgC/l)	16.4	1.67
Total alkalinity (mg/l as CaCO₃)	304	144
Conductivity (μmho/cm)	280	230

Note: Source: City of Camrose and City of Edmonton Utility Summary Monthly Reports (sampling dates were September 26, 1991, for Driedmeat Lake and December 13, 1991 to February 21, 1992, for the North Saskatchewan River).

[a] Mean of twelve samplings for river water samples, single sampling for lake water.
[b] NA = not available.
[c] Determined in the University of Alberta Environmental Laboratories.

NPOC, color, and UV absorbance on each portion of sample at collection and during filtration and adsorption. Sample was vacuum filtered through a 1.5-μm Whatman 934-HA prefilter followed by a 0.45-μm Durapore® membrane filter. NPOC measured before and after prefiltration showed a 1 to 2% loss in NPOC across the filter. This loss was within the accepted experimental error of the analyzer (2%) and was, therefore, not considered significant. Filtered sample at pH 2.0 ± 0.1 was pumped through two adsorption trains at nominal flow rates of 50 ml/min each. Influent and effluent levels of NPOC and UV absorbance were monitored for each resin cartridge throughout the isolation procedure, and the isolated NOM was desorbed as necessary prior to breakthrough as described by Andrews.[43]

The isolated NOM was recovered from the resins and further processed as described by Andrews et al.[39] Each of the XAD-4 and XAD-8 resin cartridges was individually eluted with 0.1 M sodium hydroxide as per Malcolm[1] except that non-"center cut" fractions were not collected separately but rather for each desorption the total elution volume was collected and reconcentrated on smaller (200-ml) resin columns prior to humic and fulvic acids separation. Ion-exchange resins were eluted by forward passage of 1 M ammonium hydroxide as described by Leenheer.[42] The humic acids from each of the XAD-8 and XAD-4 combined reconcentrated eluates were separated from the fulvic acids according to the procedure outlined by Malcolm.[1] Each was acidified

Table 2   Parameters and their Values Employed in the Factorial Design Experiments

Parameter	Level		
	_−[a]_	0[b]	+[c]
pH	6	6.3[d]	8
Alkalinity (mg/l as CaCO₃)	0	100	200
Ozone dose (mg:mg ozone:NPOC)	1:1	2:1	3:1

[a] Lower level.
[b] Midpoint.
[c] Upper level.
[d] pH 6.3 is the midpoint hydrogen ion concentration between pH 6 and 8.

to pH 1.0 ± 0.1 using concentrated sulfuric acid and cooled to 2°C in a refrigerator. Following centrifugation, humic acid precipitate and dissolved fulvic acids were separated and then desalted and hydrogen saturated as described by Malcolm.[1] Losses of isolated material during these steps were determined by NPOC measurements to be less than approximately 2%. The XAD-4 fraction likely contained significant quantities of low molecular weight acids in addition to fulvic acids (Figure 1); however, for simplicity this fraction is referred to in some places herein as XAD-4 fulvic acids. All fractions were subsequently freeze-dried for long-term storage. In general, the freeze-dried fulvic acid fractions were easily redissolved to prepare concentrated stock solutions for the ozonation experiments. Occasionally, it was necessary to predissolve them in 2 to 5 ml of Milli-Q water at pH 8 to 10, and then adjust the solution pH to near neutral with phosphate buffer.

The materials isolated locally were characterized by ultraviolet (UV) absorption (254 and 270 nm) and elemental analyses. FT-IR spectra were obtained from cast films of the isolated material. Average molecular weights were determined by vapor pressure osmometry. ¹³C NMR analyses were performed on selected samples. Similar data characterizing the Suwannee River fulvic acids were obtained from the IHSS where available, or from Aiken et al.[44]

FACTORIAL EXPERIMENTAL DESIGN

Full 2^3 factorial experiments with three midpoints were performed as described by Box et al.[45] and Davies[46] to investigate the effects of pH, alkalinity, and applied ozone dosage on DBP formation and color removal (as measured by UV absorbance at 254 nm). The factorial design enables an evaluation of the effects of several variables in a system with a greatly reduced number of experiments by examining the variables in carefully selected combinations. In the simplest design, each of the parameters is examined at two levels, an upper level (+) and a lower level (−). In the present work, the three parameters (pH, alkalinity, and ozone dose) were examined at the upper and lower levels shown in Table 2, constituting a full 2^3 factorial design (2 levels, 3 variables). These parameters were considered to impart the greatest effect on the formation

Table 3 The 2^3 Factorial Design

Experiment	Set parameters			Determinants for each experiment
	pH	Alkalinity	Ozone dose	
1	−[a]	−	−	UV absorbance[c]
2	+[b]	−	−	
3	−	+	−	Aldehydes
4	+	+	−	Formaldehyde acetaldehyde,
5	−	−	+	glyoxal, methylglyoxal
6	+	−	+	Oxoacids
7	−	+	+	Pyruvic, ketomalonic,
8	+	+	+	glyoxylic, oxalacetic
9	0[d]	0	0	Carboxylic acids
10	0	0	0	Formic, acetic propanoic,
11	0	0	0	benzoic

[a] − = Lower parameter level.
[b] + = Upper parameter level.
[c] Normally measured at 254 nm, also collected at 270 nm for comparison with Reference 21.
[d] 0 = Midpoint.

of DBPs from ozonation. Their upper and lower levels were based on current drinking water treatment practice and estimated water quality parameters, but included a relatively high applied ozone dose for comparative purposes.

The full 2^3 factorial design utilizes eight primary experiments defined as shown in Table 3. For each condition, there is another one that utilizes the same parameter levels for all parameters except one. For example, in experiments 1 and 2, only the pH level is varied. By comparing the results (DBP formation, UV absorbance reduction) obtained utilizing this pair of conditions, one can determine the effect of the parameter that was different (pH in the above example). Similarly, the pH level is the only difference between experiments 3 and 4, between experiments 5 and 6, and between experiments 7 and 8. Theoretically, then, the effect of changing pH might be similar regardless of which of the above pairs of experiments is used to estimate it, and in standard factorial analysis an average of them is used as the best estimation. In this research, the average of the halves of these pairs of experiments determined at one parameter level (e.g., those at pH 6) is called the "mean effect" of the parameter at that level. The mean difference between the results obtained from the two parameter levels evaluated over the entire factorial design is called the "main effect" of that parameter.

In addition to the experiments just described, it is advantageous to replicate experiments using parameter values that are intermediate between the upper and lower levels (midpoints). The information resulting from these experiments provides an estimate of the precision associated with the factorial experiment values, and allows quantitative estimation of the significance of differences in observed mean effects. While ANOVA calculations also provide estimates of error associated with factorial experiments, the significance of the differences between the pairs of responses was clearly discernible by comparison with the precision determined from the factorial midpoint experiments. When possible, midpoint parameter values are selected exactly midway between the upper and lower levels used, as shown in Table 2. In this research, midpoint pH was 6.3, representing the midpoint of the hydrogen ion concentration values; however, a choice of the midpoint hydroxide ion concentration (pH 7.7) would have been equally valid.

BENCH-SCALE OZONATION EXPERIMENTS

Experiments were performed in semi-batch mode, which may approximate plug flow cocurrent ozonation, and all experiments were performed at room temperature. A PCI ozone generator provided ozone at an applied concentration of 21.9 mg/l to the test solutions. The concentration of ozone in the gas phase was determined by bubbling it through 2×500 ml solutions of 5% potassium iodide in series, and titration of the iodine formed with sodium thiosulfate as described in *Standard Methods*.[6] Residual ozone in the aqueous samples was measured using potassium indigo trisulfonate.[6]

The factorial experiments were conducted on natural waters and on solutions of isolated fulvic acids and hydrophilic bases in Milli-Q® water. Isolated humic acids were not used in these experiments because their supply was severely limited. Similarly, since yields of hydrophilic base were also low, only the effects of pH were investigated for that fraction. Tests with fulvic acid solutions at 20 mg/l nonpurgeable organic carbon (NPOC) employed a solution volume of 125 ml, and an ozone flow rate of 100 ± 1 ml/min. The high NPOC concentration was employed to facilitate DBP detection. A PCI ozone generator provided ozone at 21.9 mg/l to the test solutions. Tests using either fulvic acid solutions at 5 mg/l NPOC or natural waters employed 500-ml sample volumes and an ozone flow rate of 50 ± 1 ml/min. For the natural waters, alkalinity setpoints shown in Table 3 were relative to ambient concentrations. Ozonation times were adjusted to achieve either a 1:1 or 3:1 ratio of applied ozone to measured NPOC (30 to 90 s for experiments at 20 mgC/l NPOC and 1 to 3 min for experiments at 5 mgC/l). A 10-min postozonation time was employed in the experiments with natural waters and fulvic acids at 5 mgC/l prior to sampling for ozonation DBPs to approximate typical water treatment practice.

To determine the effect of NPOC concentration on DBP formation, lake water fulvic acids isolated on XAD-8 resin were ozonated at a constant 3:1 applied ozone to NPOC ratio at pH 6 and 8 and without additional alkalinity. NPOC concentrations of 2.5, 5, 10, and 20 mgC/l were employed, and ozonation times were adjusted to achieve a constant 3:1 applied ozone:NPOC ratio. Ozonation DBPs and UV reduction were measured. NPOC was not routinely included as a measured parameter because initial experiments produced a minimal (less than 2 to 3%) reduction in NPOC, indicating partial oxidation of the NOM rather than mineralization to carbon dioxide at the ozone dosages employed.

UV ABSORBANCE AND BY-PRODUCTS ANALYSIS

Measured parameters for the factorial experiments included organic ozonation DBPs and UV absorbance. UV absorbance was determined using a Spectronic 601®* at 254 and 270 nm and 1-cm quartz cells. Ozonation by-products determined are listed in Table 3 and included several aldehydes, oxoacids, and carboxylic acids. Unfortunately, dicarboxylic acids such as oxalic acid could not be included because the necessary equipment only became available in our laboratory after this work had been nearly completed. All ozonation DBPs were determined by gas chromatography following derivatization. Aldehydes were determined using O-2,3,4,5,6-pentafluorobenzyl-hydroxylamine hydrochloride (PFBHA) derivatization as described by Sclimenti et al.[47] Oxoacids derivatization employed PFBHA as an initial step, followed by extraction with methyl-t-butyl ether and diazomethane esterification. Carboxylic acids were derivatized to their pentafluorophenol esters using an aqueous derivatization method developed in this research.[43] Based on a method described by Wong et al.,[48] carboxylic acids in aqueous samples (20 ml) were derivatized with pentafluorophenol using dicyclohexylcarbodiimide as a coupling reagent, and the derivatives were extracted with 4 ml 9:1 toluene:acetonitrile. All DBP derivatives were analyzed using an HP-5790A gas chromatograph equipped with a 30 m × 0.25 mm I.D. DB-5.625 capillary column (1-μm film thickness, column head pressure 12 psi) and electron capture detector.

RESULTS AND DISCUSSION

NOM ISOLATION AND CHARACTERIZATION

The yields of humic and fulvic acids and hydrophilic bases from the isolation procedure are shown in Table 4. In all, 118 mg humic acid was isolated, and 2.1 g of fulvic acid was recovered from the river water samples (1.1 g from XAD-8 resin and 1.0 g from XAD-4 resin) while 171 mg humic acid, 1.3 g XAD-8 fulvic acid, and 1.1 g XAD-4 fulvic acid were isolated from the lake water. The mass of hydrophilic bases isolated from the river water was 265 mg, which was approximately 11% of the total mass of NOM isolated. The freeze-dried river and lake water humic and fulvic acids were similar in appearance to the IHSS reference humic and fulvic materials obtained from the Suwannee River; the relative amounts of humic and fulvic acid were as expected (approximately 9:1 or 10:1 fulvic to humic acid).[1] These

---

* Milton Roy Co., Don Mills, Ontario.

Table 4　Yields, Elemental Analysis, Average Molecular Weight, and UV Absorbance of Local and IHSS NOM Fractions[43]

Fraction	Humic acid yield (mg)	Fulvic acid yield (mg)	C	H	N	O	S	UV254 abs.[a]	Average molecular weight
XAD-8 fractions									
IHSS Suwannee River	NA[b]	NA	53.60	4.25	0.70	40.90	0.55	0.031	1110[d]
North Saskatchewan River	118	1093	47.05	4.66	0.72	41.97	3.91	0.032	900
Driedmeat Lake	171	1265	42.79	4.55	3.72	40.26	4.38	0.030	1045
XAD-4 Fractions									
North Saskatchewan River	0	1004	43.37	4.67	1.07	44.55	3.04	0.027	467
Driedmeat Lake	0	1138	45.34	4.88	4.24	41.64	1.67	0.023	627
AG MP-50 Fractions									
North Saskatchewan River	NA	265[c]	16.66	6.44	17.65	25.42	23.35	0.007	NA

[a] UV absorbance units/cm/mg/l NPOC, pH 6, 254 nm.
[b] NA = not available or not applicable.
[c] Hydrophilic bases, not fulvic acids.

observations indicated that the objective of reproducing the IHSS portion of the isolation procedure had been met. In addition, mass balance calculations on the yields indicated that the lengthy isolation procedure and associated sample manipulations involved in the river water NOM isolations had little adverse effect on the recovery of this material.

Molecular weights, elemental analysis, and UV absorbance characteristics for these fractions are also indicated in Table 4. The XAD-4 isolated fulvic acids were less colored than those isolated on XAD-8 resin, which correlated with differences in UV absorbance. Average molecular weights determined for each fraction indicate that the XAD-8 fulvic acids are larger molecules than the XAD-4 fractions, as expected. XAD-4 fractions were determined to have average molecular weights of 467 Da for river water fulvic acids and 627 Da for lake water fulvic acids, whereas average values of 900 and 1045 Da were obtained for the XAD-8 river and lake water fulvic acids, respectively. No humic acids were isolated on the XAD-4 resin indicating that the larger humic acid molecules were efficiently adsorbed to the XAD-8 resin. Average molecular weights of the NOM isolated with XAD-8 resin compare well with that determined for the IHSS material (1110 Da).[49] Molecular weight data were not available for the hydrophilic base fraction.

Elemental analyses performed on the river water NOM fractions showed them to be generally similar to the IHSS material (Table 4) except that the local fractions contained approximately six times the sulfur content and only

85 to 90% of the carbon content. This relatively lower carbon content is a result of higher contributions from other elements. The nitrogen content of these samples is similar for the XAD-8 fractions but that of the XAD-4 fractions is approximately 1.5 times that of the material isolated on XAD-8. For both the lake water and river water isolates, the XAD-4 fractions contain approximately 1.5 times the amount of nitrogen as do the XAD-8 fractions. The nitrogen content is also a major distinguishing feature between the river and lake water NOM, the lake water containing four to five times the levels as the river water fractions. As for the sulfur content, all four of the local fractions contain three to eight times the sulfur of the IHSS material. XAD-4 fractions contain less than XAD-8 fractions (22% less for river NOM and 62% less for lake water NOM).

The hydrophilic bases isolated on the AG MP-50 ion-exchange resin displayed different characteristics from those of the hydrophobic and acidic components isolated on the XAD resins. Lower carbon and oxygen contents were compensated by higher nitrogen and sulfur levels. The higher sulfur content might have been from mercapto compounds, but was more likely due to sulfate not removed from the isolated NOM during the clean-up procedures associated with the hydrophilic base isolation procedure. The higher nitrogen content (18.65%) was not surprising since amino compounds were expected to be isolated by cation exchange under the conditions employed.

UV_{254} absorbance data are also shown in Table 4. They indicate the degree of unsaturation of each of the fractions and may predict the extent of reaction with ozone by molecular mechanisms. XAD-8 fractions had higher UV_{254} absorbances than did the XAD-4 fractions which, in turn, absorbed more UV light than did the ion-exchange material. This data corroborated the IR data and also agreed with the general trend of color intensity for each of the fractions. The more highly colored fractions were also associated with higher UV molar extinction coefficients. Measurements were also made at 270 nm, the results of which compare favorably with those reported by Xiong et al.[21] in which XAD-8 and XAD-4 fractions were compared.

IR spectra of the river water fulvic acids were very similar to each other and are consistent with published data for similar substances[44] and have been discussed in detail by Andrews et al.[50] Spectra of the XAD-4 fractions showed that they were more aliphatic and indicated a greater degree of carboxyl content than did those of their XAD-8 counterparts. Spectra of the AG MP-50 fractions were quite different from those of the XAD fractions. They contained very little aromatic character by comparison, but did have some carboxyl content. A very large absorbance at 1100 cm^{-1} indicated possible sulfate contamination; however, this contamination would not be expected to affect ozonation reaction mechanisms with the isolated NOM.

^{13}C nuclear magnetic resonance (NMR) spectra were obtained for selected fulvic acid fractions. This technique requires long periods of signal averaging, 2 to 10 h or more, to obtain well-resolved spectra.[51] In the present research,

2-h analyses could be accommodated. The resulting spectra illustrated fulvic acid characteristics which corroborate those obtained by FT-IR and UV analyses and are qualitatively similar to those provided by the IHSS for Ohio River fulvic acid in their relative peak intensities.[49] The large resonance observed between chemical shifts of 20 and 45 ppm (relative to trimethylsilane) originates from unsubstituted aliphatic and methylene carbons, and that from 65 to 85 ppm represents carbon in hydroxylated compounds, ring carbons of polysaccharides, or aliphatic ether carbons.[52] The peak at 162 to 190 ppm represents carbonyl carbons including esters and amides, and that at 125 to 130 ppm is from aromatic carbons (120 to 140 ppm being unsubstituted and alkyl substituted aromatic carbon, 118 to 120 ppm being aromatic carbon *ortho* to oxygen-substituted aromatic carbon). Because of their expense and because these data corroborated those obtained from other analyses, these tests were not continued.

OZONATION OF NOM FRACTIONS AND NATURAL WATERS

Ozonation experiments were conducted on two local surface waters and on 5 mg/l NPOC solutions of all five fulvic acid fractions obtained from those sources. The concentrations of the 5 mg/l NPOC solutions were confirmed by NPOC analysis to be within 0.1 mg/l of the target value. The results discussed in the following sections were obtained from a full 2^3 factorial experimental design, except for the ion-exchange fractions for which only the effect of pH was examined because of limited supply. In accordance with this type of design, experiments were performed using the combinations of set points for pH, alkalinity, and applied ozone dosage as shown in Tables 2 and 3. The average or mean effect of each parameter was calculated from the resulting data, as discussed previously. The effects thus calculated (e.g., for pH) included equal numbers of measurements made at both the upper and lower levels of the other parameters (alkalinity and ozone dose), effectively canceling out the effects of these other parameters.

While initial experiments were performed at 20 mg/l NPOC to facilitate DBP detection and results indicate what could happen in high NPOC waters, the following discussion concentrates on results of experiments performed on fulvic acid solutions at 5 mg/l NPOC which is more representative of actual conditions. Similar results were obtained at either NPOC concentration except where indicated. Additional details of experiments performed at 20 mg/l NPOC can be found in Reference 50. Data obtained from the ozonation of the relevant natural waters are also discussed to address the possible effects of nonhumic material, which is present in natural water but absent from the NOM fraction experiments.

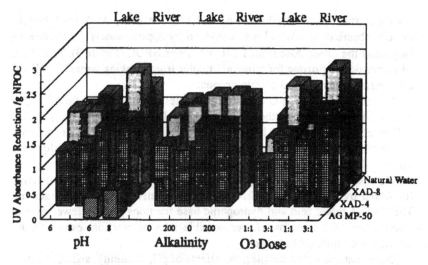

Figure 3 Mean effects of pH, alkalinity, and ozone dosage on the reduction of UV absorbance during ozonation of natural waters and respective NOM fractions; (NOM fraction concentration 5 mg/l NPOC).

Ozonation and UV Absorbance

Ozonation Parameter Effects—For the following discussion, the significance of the effect of each parameter was calculated based on the standard deviation of the results of the corresponding replicated factorial midpoint samples. The effects were statistically significant at the 5% level when the difference in UV absorbance at the upper and lower parameter set points was greater than 0.193 absorbance units (normalized to their initial values) for data collected using NOM fractions, and 0.087 for ozonated natural waters.

The mean effects of pH, alkalinity, and ozone dosage on UV absorbance reduction of ozonated NOM fractions are illustrated in Figure 3. Because of the factorial design of the experiment, the data allow the effects of each of the three factors to be determined separately. Effects of pH and alkalinity were generally small in that differences between UV absorbance reductions obtained at the upper and lower set points were less than even 10% of the average responses. Small to moderate increases in UV absorbance reduction were achieved on changing from pH 6 to 8 for all fractions except the lake water NOM isolated on XAD-8. The largest effect was observed for river water NOM isolated on XAD-8.

Ozone dose was the parameter that most affected UV absorbance reduction in experiments performed at either NPOC concentration, although the effects were more pronounced at the higher NPOC concentration.[50] At 20 mg/l NPOC, the amount of color removed (measured as the change in UV absorbance per gram of fulvic acid) on ozonation at a 3:1 ozone:NPOC dose was approximately ten times that removed when using a 1:1 ozone dose. At 5 mg/l NPOC, the UV absorbance removal was enhanced by approximately 30 to 50% by increasing the ozone dose from 1:1 to 3:1 ozone:NPOC (Figure 3). Therefore, when considering ozone for removal of color from drinking water, the NPOC concentration must be taken into account.

Ozone removed more UV absorbing species from NOM fractions isolated on XAD-8 resin than from those isolated on XAD-4 or AG MP-50 as indicated in Figure 3 by the size of the bars for this fraction relative to the other fractions. Similarly, the UV absorbance values of XAD-8 fractions were generally more susceptible to the effects of pH and ozone dose than were other fractions. This can be attributed to the greater degree of unsaturation of this material (indicated by IR, UV, and ^{13}C NMR results) relative to the other fractions. The XAD-4 fulvic acid and hydrophilic base fractions contain fewer sites of unsaturation and, therefore, fewer possible sites of reaction, especially for reaction with molecular ozone.

In the natural water samples, the effects of pH, alkalinity, and ozone dose on the reduction of UV absorbance (254 nm) were similar regardless of the water source or its NPOC concentration (Figure 3).[53] For both the lake water and river water samples, there was a 20 to 25% greater reduction in UV absorbance at pH 8 than at pH 6. A similarly enhanced removal was observed when the ozone dose was increased from 1:1 to 3:1 ozone:NPOC. These results were similar to those observed from the ozonation of corresponding fulvic acid fractions in reagent water in that the UV absorbance of the fractions was reduced to a greater extent when ozone was applied at higher pH and ozone dosages, and alkalinity did not appear to be a factor. UV absorbance removal for the isolated fulvic acids typically increased in the range of 10 to 60% when pH was increased from 6 to 8, whereas corresponding color removal increase for the ozonated natural waters was in the range of 20 to 25%. Ozone dose effects were also much more marked for the isolated fulvic acids (45 to 100% reductions in UV absorbance) than for the natural waters (20 to 25%). Alkalinity effects were negligible for both natural water samples, although it must be remembered that the alkalinity in both of these samples was relatively high prior to commencing the experiment (144 and 304 mg/l as $CaCO_3$).

There are several possible reasons why the pH and ozone dose effects were greater for the NOM fractions than for the natural waters. For example, pH-alkalinity interactive effects[50] might be partially responsible since both waters exhibited high alkalinity. The presence of natural alkalinity could not easily be reduced for the ozonation experiments of natural waters, only increased; therefore, an examination of DBP production at low alkalinity was not possible. Other sample matrix effects not tested (organic and inorganic effects, presence of natural phosphates) or the presence of more intensely colored material not represented by the fulvic acids (e.g., humic acids) could also cause changes in the magnitudes of the effects observed. While the isolated humic acid contributed to only approximately 10% of the isolated

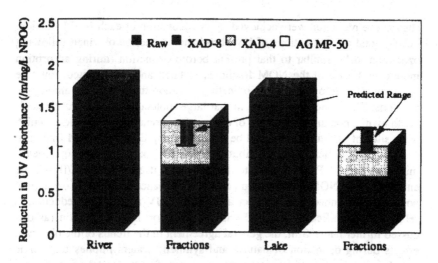

Figure 4 Contribution of NOM fractions to UV absorbance reduction from ozonation of raw waters (AG MP-50 fraction not collected for lakewater). (From Andrews, S. A. and Huck, P. M., *Ozone Sci. Eng.*, 16, 1, 1994a. With permission.[53])

NPOC, this fraction was much more highly colored than the fulvic acid fraction exerting approximately three times the UV absorbance per unit mass as did the fulvic acids.

Quantitative Aspects of NOM Ozonation — Calculations were performed to determine the extent to which results of ozonation of the NOM fractions could be directly applied to natural water samples. From adsorption monitoring data, XAD-8 resin isolated approximately 25% of the NPOC, XAD-4 another 25%, and AG MP-50 another 5%. Considering that UV absorbance is an additive quantity, and assuming that interfraction reactions to alter the spectrophotometric properties of the individual fractions do not occur, an average reduction in UV absorbance was calculated for each NOM fraction experiment and multiplied by these proportions to estimate the respective contributions of each NOM fraction to the overall UV absorbance reduction. The sum was then compared to corresponding data from the natural waters.

Typical data (Figure 4) indicate that these summed contributions correspond well to the results for the natural waters. The sum of the NPOC-adjusted UV absorbance reductions obtained for the XAD-8 and XAD-4 fractions of lake water NOM was 62% of that for the ozonated lake water. For river water, for which an additional AG MP-50 fraction was also obtained, the isolated fractions produced 76% of the UV absorbance reduction obtained for the whole river water. These totals correspond with the adsorption monitoring data for which 60 to 75% of the color or UV-absorbing species were observed to be isolated by the procedures used, indicating that similar magnitudes of absorbance reduction were achieved by the ozonation of each NOM fraction. For the total of the fractions of UV absorbance reduction obtained following ozonation to be similar to that present before ozonation (during adsorption monitoring) each of the NOM fractions, isolated and not isolated, must be similarly affected by ozonation indicating common functionalities among the fractions. This is not unexpected in that large molecules such as fulvic acids are known to contain many types of organic functional groups, some of which (e.g., sites of unsaturation) could be similar among different NOM fractions.

The slight deficit in the contributions from the lake water fractions (determined at 5 mg/l NPOC) relative to the lake water itself (16.7 mgC/l) indicate an influence of NOM concentration on UV absorbance removal by ozonation; however, this concentration effect as it affects UV absorbance reduction is relatively insignificant compared with its effect on DBP formation (as discussed further later). Still, the general agreement in the trends of the observed effects during ozonation of natural and synthetic waters implies that fulvic acid fractions can provide an accurate surrogate for the modeling of general color or UV absorbance reduction subsequent to ozonation, and could be used in experiments to evaluate the effectiveness of different processes and their operating parameters for color reduction.

Disinfection By-Product Formation

Ozonation Parameter Effects—The mean effects of pH, alkalinity, and ozone dosage on the formation of organic ozonation DBPs from NOM fractions and natural waters are shown in Figures 5 to 7. Data for aldehyde, oxoacid, and carboxylic acid formation were calculated on a molar basis and summed for each class of by-product to determine if class-specific effects were evident. The relative yields of the different classes of DBPs can be determined by comparing the scales of the y-axes for Figures 5 to 7. The significance of the each effect reported was calculated based on the standard deviation of the results of the corresponding replicated factorial midpoint samples. The effects were statistically significant at the 5% level when the difference in DBP production at the upper and lower parameter set points was greater than 507 μmol/g NPOC for aldehyde data, 32.5 μmol/g NPOC for the oxoacids, and 69.4 μmol/g NPOC for the carboxylic acids. For the following discussion, only statistically significant effects are discussed.

The parameter that was observed to affect ozonation by-product formation from NOM fractions to the greatest extent was pH. The effect of pH on aldehyde and carboxylic acid formation was different from that observed for oxoacid formation in that formation of aldehydes and carboxylic acids (Figures 5 and 7) was favored at low pH whereas formation of oxoacids proceeded with greater yield at pH 8 for all but one fraction (Figure 6). This effect can be explained by reference to the different mechanisms of ozone reaction with organic matter. At pH 6, molecular ozone reacts at sites of unsaturation to

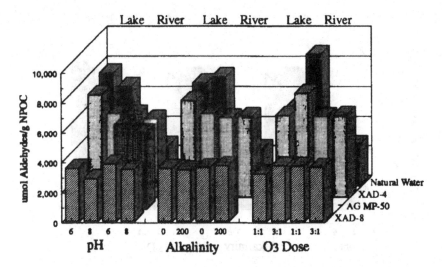

Figure 5 Mean effects of pH, alkalinity, and ozone dosage on the formation of aldehydes during ozonation of natural waters and respective NOM fractions (NOM fraction concentration 5 mg/l NPOC).

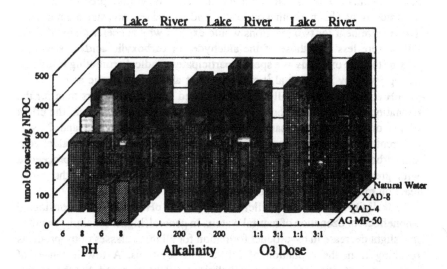

Figure 6 Mean effects of pH, alkalinity, and ozone dosage on the formation of oxoacids during ozonation of natural waters and respective NOM fractions (NOM fraction concentration 5 mg/l NPOC).

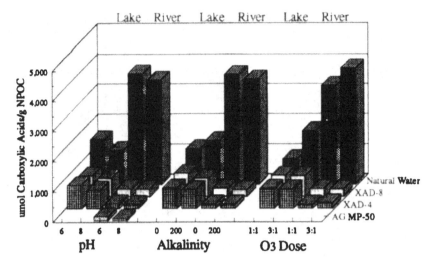

Figure 7 Mean effects of pH, alkalinity, and ozone dosage on the formation of carboxylic acids during ozonation of natural waters and respective NOM fractions (NOM fraction concentration 5 mg/l NPOC).

produce aldehydes directly which can be readily oxidized to form carboxylic acids.[17] At pH 8, radical reaction mechanisms become more predominant. For oxoacids, it may be that both a molecular and radical reaction step are required. The requirement for two reactions would explain why concentrations of these DBPs were less than those of the aldehydes or carboxylic acids. As well, it may be that the oxoacids as a species participate in radical promoting reactions, as suggested by Xiong et al.[21] and Caprio et al.[54] for glyoxalic acid.

pH effects were generally less pronounced for natural waters than for the ozonation of fulvic acids; however, the trends were similar. Again, the effect of pH on aldehyde and carboxylic acid formation (Figures 5 and 7) was different from that on oxoacid formation (Figure 6). Formation of aldehydes and carboxylic acids was favored at low pH while that of oxoacids proceeded with greater yield at pH 8, effects which can be explained by the ozone reaction mechanisms discussed above.

Alkalinity had a much smaller effect than did pH and could be considered a noneffect for ozonation of natural waters. In general, higher alkalinity resulted in a slight decrease in by-product formation for all three classes of by-products resulting from the ozonation of fulvic acid solutions. A few instances of increased production at increased alkalinity levels were noted, but the magnitudes of all alkalinity effects were such that they were usually within the experimental error associated with noneffects. The variability in alkalinity effects may be an indication of the noneffect of this parameter; however, it may also be that for some fulvic acids the experimental conditions represented those intermediate between causing fulvic acids to be scavengers or promoters of radical reactions. In the latter case, the alkalinity suppressed radical formation unless the promoting aspect of the fulvic acid was enough to overcome the inhibitory effects of the alkalinity with regard to DBP production.

Increasing the ozone dosage applied to the lake water fulvic acid fractions resulted in higher yields of all by-products except for carboxylic acids. In contrast, for the river water fulvic acids, changes in ozone dosage at the levels evaluated had little effect on DBP yields, except for those of the oxoacids which approximately doubled on increasing the ozone dose regardless of the water source or NOM fraction. The dosage effect was generally less pronounced for the formation of aldehydes and carboxylic acids than for oxoacids and was, in general, smaller for river water NOM than for lake water NOM. It may be that sites of unsaturation in the NOM react quickly with ozone to form aldehydes and carboxylic acids even at lower ozone doses. As has been stated, these sites react preferentially with molecular ozone. The greater ozone dosage effects observed for oxoacids formation may be due to a possible radical promoting ability of this class of compounds. This may also be related to the large ozone dose effects observed for UV absorbance reduction on ozonation of fulvic acids discussed previously. The lack of observed effect on changing ozone dose for river water may have been due to the high ozone concentrations used (1:1 and 3:1 ozone:NPOC ratios). Typical drinking water treatment employs ratios of 1:1 or less, and so greater effects would be expected to be observed in real situations under the lower dosages employed.

NOM Fraction Effects — The NOM fractions isolated were shown to produce similar normalized quantities of some DBPs regardless of the NOM source, and some fractions contributed more to the production of certain DBPs than did other fractions. For example, all XAD-8 fractions produced similar aldehyde yields, and the yields for the XAD-4 and AG MP-50 fractions were similar to each other (Figure 5). It is also evident that NOM isolated on XAD-4 and ion-exchange resins generally produced greater quantities of aldehydes per gram NPOC than did fractions isolated on XAD-8 resin. For either of the two water sources, XAD-4 fractions produced up to 1.5 times the aldehydes when compared to XAD-8 fractions, but only slightly higher oxoacids (Figure 6), and approximately the same amount of carboxylic acids (Figure 7). The ion-exchange fraction produced aldehydes in amounts similar to those derived from the XAD-4 fractions, but produced lower concentrations of oxoacids and carboxylic acids. While NOM isolated on XAD-8 resin generally produced greater quantities of carboxylic acid than did XAD-4, the differences were generally within 20%. The lowest observed yields for carboxylic acids resulted from ozonation of the hydrophilic basic NOM which had been isolated by ion exchange.

Oxoacids formation was similar for all NOM fractions and yields were dominated by ozone dose and pH (Figure 6). pH effects observed for oxoacids formation were the opposite to those observed for aldehyde and carboxylic acid production. Higher oxoacids yields generally resulted at higher pH. Ozone dosage had a marked effect on oxoacids production regardless of the NOM source or fraction. Increasing the ozone dose from 1:1 to 3:1 ozone:NPOC increased the oxoacids yield approximately twofold.

Some consideration should be given to the differences in the mean molecular weights of the NOM fractions when rationalizing the greater quantities of ozonation DBPs which were produced from the XAD-4 fractions than the XAD-8 fractions. Equal mass concentrations of the two types of NOM would be expected to result in higher molar concentrations of precursor material for the lower molecular weight fraction (XAD-4). For this molecular weight difference to be the cause of the yield discrepancies, each fulvic acid macromolecule, regardless of its size (molecular weight), would have to contain a similar number of reactive sites. For DBP formation, this may imply a surface-type phenomenon whereby only easily accessible portions of the molecule (those near an edge) react to form ozonation DBPs. This theory is possible because it is known that ozonation results in the disruption of large molecules to smaller ones (change in molecular weight) without DBP formation.

These results have implications with regard to water treatment methods and biological stability of water in distribution systems. Aldehydes and carboxylic acids, which are produced in appreciable yields upon ozonation of NOM, are important in microbial metabolism and likely contribute to microbial regrowth in distribution systems. These DBPs are not readily removed by traditional physical/chemical treatment methods; however, biological treatment has been shown to reduce the concentrations of biodegradable materials such as these. In addition, the lower molecular weight and higher carboxyl content of the DBP precursors isolated on XAD-4 resin may indicate that their removal during conventional treatment could be different from removals of NOM fractions isolated on XAD-8.

Water Source Influences — Some source-related effects were observed. In particular, carboxylic acid production was governed more by the NOM source than by NOM fraction or ozonation conditions. Lake water NOM fractions produced carboxylic acids in greater yields than river NOM fractions by up to three times (Figure 7); however, the opposite effect was observed in natural waters. Lake water fulvic acid fractions produced similar or slightly higher amounts of aldehydes compared with results using river water NOM fractions, and the effect was exaggerated for natural water supplies (Figure 5). This general trend was also observed for oxoacids production from NOM fractions; however, similar yields were obtained from both natural water sources (Figure 6).

It is interesting to note that the yields of aldehydes and carboxylic acids, normalized to NPOC, were different for the two water sources and that the source with the higher organic content appeared to produce lower normalized yield of these DBPs. This is another example of the concentration effect (discussed later) affecting DBP formation due to the very different NPOC concentrations associated with the two raw water samples.[43] The NPOC-normalized carboxylic acid yields from the fulvic acids experiments performed at equal NPOC concentrations were higher for the lake water fulvic acid fractions; however, with the natural water they were higher for the river water because of its much lower NPOC. Therefore, some consideration of concentration effects and the differences in NPOC concentrations employed or encountered in the factorial experiments should also be given in interpreting these results.

These source-related observations are likely a result of differences in DBP precursors in the two source water types and imply that, as has been observed in the past, while some NOM fractions produce consistent DBP yields regardless of NOM source, source-related differences in NOM character may negate the application of data obtained using NOM from one source to other sources.

DBP Quantitation with NOM Fractions vs. Natural Waters — In natural waters, aldehydes and carboxylic acids formed the majority of the by-product classes studied relative to the oxoacids, as is indicated by the scales on the y-axes of Figures 5 to 7 which illustrate DBP yields. This relative distribution is different from that obtained with most NOM fractions at 5 mg/l NPOC but is similar to that obtained when using an NPOC concentration of 20 mg/l (not shown; this concentration effect is discussed later). There are at least two possible explanations for these relative abundances. One is that, as portions of the fulvic acids are progressively oxidized from aldehydes to carboxylic acids, in the process some become multiply oxidized to oxoaldehydes, oxoacids, and others prior to being oxidized further to simple carboxylic acids. In this case, the relative abundances of the different classes may indicate the relative activation energies for formation of oxoacids, aldehydes, and carboxylic acids from precursor materials, e.g., fulvic acids. It is also possible that different types and amounts of precursor materials are available for oxidation, especially for ozonation of natural water. This second explanation seems most likely, as is discussed later.

As was shown for the UV absorbance reduction data, the contributions of each of the isolated NOM fractions to the overall NPOC were applied to their DBP yields, and their sum then compared to the DBP yields from the natural waters to obtain an indication of the contribution of these fractions to the overall DBP production.[53] DBP yield data collected from NOM fraction experiments conducted at NPOC concentrations near those of the natural waters (to minimize concentration effects) were averaged for each factorial experiment and multiplied by these proportions to estimate the respective contributions of each NOM fraction to the overall DBP production. These data are shown for river water, for which more NOM fractions could be obtained, in Figure 8.

Assuming that the similarity in NPOC concentrations employed adequately minimizes concentration effects, the data illustrated in Figure 8 show that the

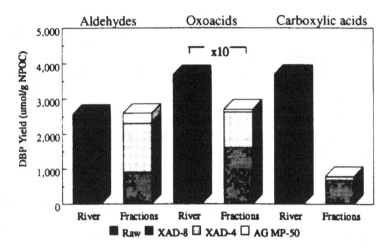

Figure 8 Contribution of NOM fractions to DBP production from ozonation of river water. (From Andrews, S. A. and Huck, P. M., *Ozone Sci. Eng.*, 16, 1, 1994a. With permission.[53])

major precursor materials for aldehydes are the fulvic acid and hydrophilic base fractions of NOM. Yields of oxoacids in natural water exceeded those predicted by summing the fractions (Figure 8). The river water NOM fractions accounted for approximately 74% of the natural oxoacid production. Therefore, it is probable that oxoacid precursors in addition to fulvic acids are significant. The yields of carboxylic acids in natural water far exceeded those predicted by summing the fractions. Clearly, in the case of carboxylic acid production, it is likely that the major carboxylic acid precursors are not the fulvic acids but some other chemical species.

UV-DBP Correlations

Initially, DBP production and UV absorbance reduction did not initially appear to correlate very well at the NPOC concentrations studied. The UV absorbances of XAD-8 fractions were those most affected by ozonation, whereas greater quantities of DBPs were produced by the XAD-4 fractions. Similarly, increased ozone dosages resulted in further reduced UV absorbance regardless of the fraction employed; however, changes in ozone dose had a variable or noneffect on the production of DBPs. Therefore, the initial conclusion was that UV data could not be used to quantitatively predict DBP formation.

However, on looking closely at the results obtained at 5 mg/l NPOC, similarities were observed for some of the effects of ozonation on UV absorbance reduction and DBP formation. For example, both UV absorbance reduction and oxoacid formation proceeded to greater extents at high pH whereas the opposite was observed for aldehyde and carboxylic acid formation.

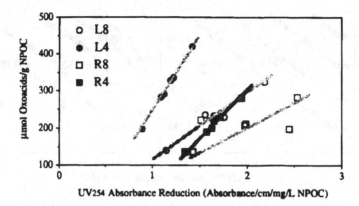

Figure 9 Correlations between UV absorbance reduction and oxoacids production on ozonation of various fractions of NOM at 5 mg/l NPOC (L = lake, R = river, 8 = XAD-8, 4 = XAD-4).[38]

In addition, while higher ozone doses resulted in only marginally increased aldehyde yields and did not affect carboxylic acid yields, the increase in oxoacids production was nearly double for a threefold increase in ozone:NPOC ratio.

In Figure 9, the correlation between UV absorbance reduction and oxoacid formation is shown under the conditions of ozonation employed in these experiments for all NOM fractions except one. The AG MP-50 resin could not be included in these comparisons because it had only been used in pH experiments and so there was insufficient data for this fraction. Good correlations were obtained, with r^2 values of greater than 0.95 being obtained for all fractions except one. Only the river water NOM isolated on XAD-8 resin showed a poor correlation (r^2 of 0.43). Similar correlations were made for aldehyde and carboxylic acid production, but the results were poor (r^2 of less than 0.4) and are not shown. These results indicate that the oxoacids may be formed by direct reaction with sites of unsaturation (which absorb UV light) on the NOM fractions.

In the experiments utilizing high NPOC concentrations (20 mgC/l), the correlation was much weaker ($r^2 < 0.42$), possibly because of the greater complexity of reactions occurring at these high concentrations. Statistical analysis of these data indicates that the correlation observed at the lower NPOC concentration (except for the one fraction) was significant at the 1% level whereas that at 20 mg/l NPOC was not significant at even the 5% level.

Ozonation Parameter Interactions

One of the advantages of using factorial design is that interactions between the effects of each of the parameters can be evaluated.[45] Since experimental parameters do not necessarily affect observed results independently, the facto-

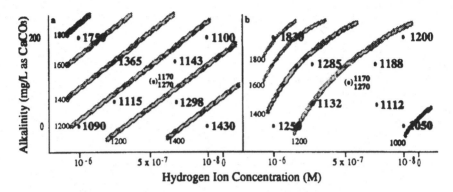

Figure 10   Response surface for interaction of alkalinity and pH on aldehydes formation during ozonation of river water NOM isolated on XAD-4 resin (values in bold are experimental aldehyde data as μmol/g NPOC; a = 1:1 ozone:NPOC mass ratio; b = 3:1 ozone:NPOC mass ratio). (From Andrews, S. A., Huck, P. M., and Coutts, R. T., *Vom Wasser*, 81, 151, 1993a. With permission.[50])

rial design can be used to determine the extent of the interaction or synergistic effects of each of the parameters investigated.

Illustrated in Figures 10 and 11 are the interactive effects of pH and alkalinity on aldehyde and oxoacid formation for NOM ozonated at two ozone dosages. Recall that some aldehydes and oxoacids have toxicological significance and that they may contribute to the biodegradable portion of NOM. Shown are data for the ozonation of 20 mg/l NPOC river water NOM which was isolated on XAD-4 resin.[50] These are response surfaces of experimentally determined DBP concentrations (μmol DBP/g NPOC, shown in bold) plotted at the conditions of pH (as hydrogen ion concentration H+) and alkalinity (mg/l CaCO₃) under which they were determined. Also shown are contours of equal DBP response, manually drawn to give a visual appreciation

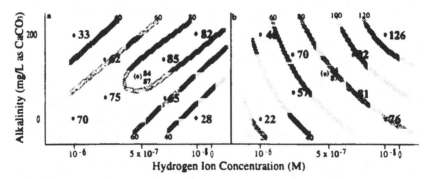

Figure 11   Response surface for interaction of alkalinity and pH on oxoacids formation during ozonation of river water NOM isolated on XAD-4 resin (values in bold are experimental aldehyde data as μmol/g NPOC; a = 1:1 ozone:NPOC mass ratio; b = 3:1 ozone:NPOC mass ratio).

of the effects. The midpoint values are shown in parentheses because they were obtained at an intermediate ozone dose (2:1 ozone:NPOC mass ratio). A more detailed examination of these data for model formulation purposes was not attempted due to the limited number of data points.

The effects of pH and alkalinity on aldehyde formation at the lower ozone dose of 1:1 ozone:NPOC mass ratio are shown in Figure 10a. Over the range of conditions studied, the quantity of aldehydes produced remained approximately constant for similar changes in alkalinity and hydrogen ion concentration. A minimum aldehyde formation of approximately 1100 µmol/g NPOC was apparent for this ozone dosage. Deviation from equivalent changes in hydrogen ion concentration and alkalinity resulted in increased aldehyde concentrations (note that the x-axis scale is pictured with the origin at the right). Therefore, conditions of high alkalinity and high hydrogen ion concentration (low pH) or low alkalinity and high pH produce greater quantities of aldehyde than at intermediate values of these parameters. Also illustrated in this figure is the tendency for greater quantities of aldehydes to be produced by molecular ozonation reactions (maximum concentration of 1750 µmol/g NPOC observed at low pH, high alkalinity) than by the free radical pathway (maximum concentration of 1430 µmol/g NPOC at high pH, low alkalinity) which is consistent with the literature.

The linearity of the contours in Figure 10a indicates that there were few significant interactions or synergisms between the effects of the two variables (pH and alkalinity) in the formation of aldehydes under the range of conditions employed.[45] However, the presence of a minimum suggests that, if the response surface were extended by expanding the range of the variables, it would probably be elliptical. If a mathematical model were to be formulated which describes the system it would be nonlinear, containing quadratic terms with respect to H^+ and alkalinity.

At the higher ozone dose of 3:1 ozone:NPOC (Figure 10b), the pH-alkalinity interaction appeared to increase, indicated by curvature in the contour lines of the response surface. These data show that higher concentrations of aldehyde were still produced at low pH and high alkalinity, favoring participation of a molecular ozone reaction as discussed previously. While the elliptical character of the response surface shown in Figure 10a, as evidenced by the minimum yield contour, was not evident in Figure 10b, concentration effects (discussed later) may have caused a shift in the contour of minimum yield, taking it out of range of the conditions employed in these tests.

pH- and alkalinity-related interactive effects for the formation of oxoacids are illustrated in Figure 11, again for the ozonation of river water NOM isolated on XAD-4. Maximum formation of oxoacids was observed at high pH and alkalinity. Contour lines shown in Figure 11a, which are generally parallel, indicate that, as described for aldehydes formation, (Figure 10a) there were few interactive effects at the lower ozone dose. Visualization of the effects of this proposed elliptical response surface are limited, however, by the number of data points.

Interactions were more important at the higher ozone dose (3:1 ozone:NPOC) as evidenced by the contours in Figure 11b. Oxoacid formation was greatly favored at high pH and alkalinity, conditions where neither discrete molecular nor discrete radical reaction mechanisms are present. This suggests that both molecular and radical reaction conditions are required for oxoacids formation. Therefore, during drinking water treatment, conditions of high pH and alkalinity would be expected to yield maximum concentrations of oxoacids at high ozone dosages relative to NPOC concentrations.

Effect of NPOC Concentration on DBP Yields

Typical experimental programs into DBP formation involve performing at least some initial experiments using reactant concentrations which are one or two orders of magnitude higher than would be encountered naturally in the systems under study to facilitate quantitation of products. This approach was also initially employed in the present research, in which the reactants (fractionated NOM and ozone) were employed at concentrations which represented approximately five to ten times those encountered in typical drinking water treatment practice. However, due to various competing reactions, the reactions being studied may occur differently in concentrated and dilute systems, making extrapolation of results obtained at high concentration to ambient concentrations difficult. For example, aldehyde formation was much more dominant relative to the other DBPs under lower NPOC conditions than at high NPOC, whereas at 20 mg/l NPOC aldehydes and carboxylic acids formed the majority of the by-product classes studied relative to the oxoacids. The ratio of aldehydes to oxoacids was similar at either NPOC concentration, but the ratio of carboxylic acids to the other DBP classes was substantially smaller for samples tested at the lower NPOC concentration. These results were not predicted from data collected initially. Therefore, to facilitate interpretation of comparison data obtained at different solute concentrations and perhaps allow inferences for even lower organic concentrations to be made, it was considered important to investigate the effect of NOM concentration on DBP formation.

Differences in NOM concentration did not affect the nature of the influence of ozonation parameters on UV absorbance reduction or DBP formation. At 20 mg/l NPOC, similar ozonation parameter effects were observed at 5 mg/l NPOC. pH was the major parameter affecting fulvic acids ozonation by-product formation, and formation of aldehydes and carboxylic acids was favored at low pH while UV absorbance reduction and the formation of oxoacids proceeded with greater yield at pH 8. Alkalinity had a much smaller effect than did pH, and in general, higher alkalinity resulted in a decrease in by-product formation for all three DBP classes. Increasing the ozone dosage

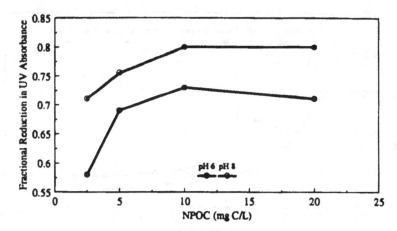

Figure 12 Reduction in lake water fulvic acid UV absorbance as a function of NPOC concentration and pH (3:1 applied ozone:NPOC ratio, no added alkalinity, 254 nm).

applied to the lake water fulvic acids resulted in higher yields of all by-products except for carboxylic acids. However, changes in ozone dosage at the levels employed had little effect on any of the DBP yields from river water NOM.

The influence of the type of NOM (XAD-4 vs. XAD-8) on by-product formation was also unaffected by NOM concentration. The yields of aldehydes and carboxylic acids were most strongly influenced by the source (lake vs. river) of the fulvic acids being ozonated.

The effects of varying NPOC concentrations on UV absorbance resulting from ozonation in the absence of bicarbonate at a constant ozone:NPOC ratio of 3:1 are illustrated in Figure 12 for ozonation of lake water fulvic acids. For the NPOC concentrations examined, greater changes in UV absorbance were observed at pH 8 than at pH 6, which was consistent with DBP results. However, the magnitude of the difference between the UV absorbance reduction determined at pH 6 and that determined at pH 8 was not the same for all NPOC concentrations tested. It is evident from these data that color reduction, measured by change in UV absorbance at 254 nm, was not a linear function of NPOC concentration, especially at concentrations less than 10 mgC/l. However, the removal efficiency appeared to plateau with increasing NPOC concentration at between 70 and 80%, depending on the pH employed. Part of the reason that higher reductions were not achieved may have been because many chemical species absorb UV radiation at or near 254 nm, including some that are by-products of ozonation. Therefore, the UV-absorbing species are being created during ozonation, limiting the extent of UV reduction possible.

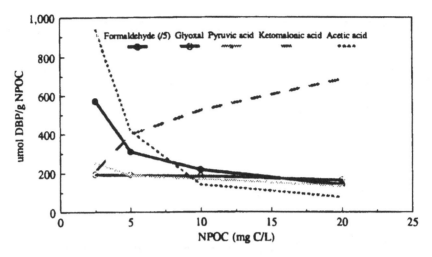

Figure 13 Effect of NPOC concentration on organic DBP formation at a constant
ozone:NPOC ratio of 3:1 (pH 6, no added alkalinity).

These observations were consistent with those described by Staehelin and
Hoigné[34] for the effects of humic acids and other solutes on ozone consumption.
They state that humic acids may be responsible for some of the initial "sponta-
neous ozone consumption" observed in water treatment plants by initially
scavenging it and/or its free radical decomposition products, and that interactions of various solutes with free radicals in solution will result in nonadditive
effects of various types of ozone reactions occurring. The observations made
in the present research indicate that the reaction of ozone at UV-absorbing
sites of the fulvic acids molecules becomes less prevalent at lower NPOC
concentrations supporting the idea of a competitive effect.

DBP production was not a linear function of NPOC concentration. Typical
DBP data normalized to their corresponding NPOC concentrations are shown
in Figure 13. For the various DBPs, the results obtained at pH 8 were qualitatively similar to those at pH 6;[43] therefore, only those obtained at pH 6
are discussed herein. While absolute DBP yields (μg/l, μM) increased with
increasing substrate concentration (not shown), the increases were not directly
proportional to the increases in fulvic acid concentration. If the absolute yields
had been directly proportional to NPOC concentration, then the curves shown
in Figure 13 would appear as horizontal lines. However, data presented in the
figure indicate that higher NPOC concentrations produced lower DBP yields
per gram NPOC, except for ketomalonic acid.

The reduced normalized yields observed at higher NPOC concentrations
may also be the result of the ozone-quenching ability of the fulvic acids
described by Staehelin and Hoigné,[34] as was discussed in relation to UV
absorbance reductions. Various interactions between ozone and solutes or
other radicals in solution may have prevented the ozone from reacting with

the fulvic acids at the same reactive sites as it did at low NPOC, and preventing DBP production.

An increase in absolute yields with increasing initial NPOC concentration was also observed by Xiong et al.[21] for the formation of glyoxylic acid from the ozonation of a similar fulvic acid at pH 3.4 and a 1:1 mass ratio of ozone:NPOC. However, it is not clear if the concentrations are expressed per mass of fulvic acid or are absolute values. While the present research and that of Xiong et al.[21] cannot be directly compared due to pH and ozone dose differences, the DBP yields appear to be similar to those for both NPOC concentrations reported, indicating that the results reported by Xiong et al.[21] are probably absolute yields. In their research, they identified glyoxylic acid as a precursor for radical reactions on the basis of the concentration effects noted (higher yield at higher NPOC, constant 1:1 ozone:NPOC mass ratio, pH 3.4). Since the same effect was observed for each of the ozonation DBPs studied in the present research, then according to the above rationale all of these DBPs could be precursors for radical reactions. So, too, could the NOM fractions in the present research since it has been recognized that humic and fulvic acids can behave as both quenchers and promoters of free radical reactions. Certainly, each of these species possesses active hydrogens that could participate in free radical reactions. The oxoacids and carboxylic acids in their ionized forms would also react with ozone to promote free radical reactions as described by Hoigné and Bader[55] and Staehelin and Hoigné[34] for formic acid, who also report glyoxylic acid as being an initiator and promoter of free radical reactions. They reported that the rate constant for ozone depletion in the presence of formic acid increased more than linearly with the concentration of formic acid, indicating that ozone reacts with formic acid to produce free radicals which can then react with ozone or with other species present to result in reaction termination. Similar results were observed for the reaction of ozone and glyoxylic acid.

This information emphasizes the problems involved in extrapolating bench or even pilot data to real situations. It has long been recognized that DBP formation cannot be predicted globally, i.e., for several source water types, from data not obtained at those sites. The data from this research show that even, if such data are available for a given site, it may be difficult to predict DBP formation for that site under varying NPOC concentrations. In particular, if such data are extrapolated to NPOC concentrations lower than those on which the database was formed, DBP formation can be seriously underestimated.

CONCLUSIONS

This work examined three major classes of ozonation by-products (aldehydes, oxoacids, and carboxylic acids) which are important in drinking water either from a microbiological or toxicological point of view. These were studied using both natural waters and solutions of NOM fractions obtained using resin isolation.

Modifications proposed by Malcolm[41] to the standard XAD-8 isolation procedure provided additional fractions of NOM to those previously isolable and allowed a more complete study of DBP precursors to be made. The use of Amberlite® XAD-4 resin in the NOM isolation procedure increased NOM recoveries by 100% and Bio Rad AG MP-50 increased NOM yields by another 25%. Fulvic and low molecular weight acids isolated by XAD-4 were significantly different from those isolated by XAD-8, both in directly measurable characteristics and by their DBP formation potentials resulting from ozonation. The hydrophilic bases isolated by ion exchange possessed higher aliphatic character and higher nitrogen content than did the other fractions.

All three types of NOM fractions reacted with ozone to form organic DBPs. Aldehydes were produced in greatest yield upon ozonation, particularly at lower NPOC. For both aldehydes and oxoacids, the lower molecular weight fulvic acid fractions, which were isolated on XAD-4, generally produced more DBPs per mass than did the fractions isolated on XAD-8. Hydrophilic bases produced quantities of aldehyde which were comparable to XAD-4 NOM, but fewer oxoacids and relatively small amounts carboxylic acid when compared to XAD-8 and XAD-4 fractions. DBP formation from these NOM fractions was somewhat site specific, especially in terms of quantities of DBPs formed. For example, lake water ozonation generally produced higher levels of carboxylic acids and slightly higher oxoacids concentrations.

The relative yields of some DBPs in river water differed from those obtained in fulvic acid experiments, particularly the carboxylic acids and oxoacids, indicating that there were likely precursors in the natural water for these compounds in addition to the fulvic acids. The yields of aldehydes were similar in both fulvic acid solutions and river waters, indicating that the fulvic acids may be their major precursor materials.

The effects of the ozonation parameters were similar for natural water to those observed for the isolated NOM fractions. Therefore, similar reaction mechanisms should apply. One parameter that was observed to affect organic ozonation by-product formation significantly was pH. The influence of pH on the formation of oxoacids and on the reduction of UV absorbance was different than for aldehydes and carboxylic acids, possibly because of the participation in different reaction mechanisms (molecular vs. radical). Greater UV absorbance reductions, loosely correlated with reduction in color, occurred at high pH, high alkalinity, and at the higher applied ozone dose. For the concentrations of NPOC used and the ozonation parameter values studied, it was apparent that the fulvic acids may have reacted both as radical scavengers and promoters, and that the transition between the two states might lie between the two ozone:NPOC dosage ratios investigated. Alkalinity had a smaller and more variable effect than did pH, and the magnitudes of all alkalinity effects were such that they were usually within the experimental error associated with noneffects. Increasing the ozone dosage applied to either natural water resulted in higher yields of all three classes of DBPs.

The effects of the ozonation parameters studied are not necessarily independent of each other. The interaction and synergistic effects of pH and alkalinity were most pronounced at the higher ozone dosage examined (3:1 ozone:NPOC mass ratio). Maximum aldehyde formation was observed at minimum pH and maximum alkalinity, whereas oxoacids were formed in greater quantities at high pH and alkalinity. At the lower ozone dose (1:1 ozone:NPOC), there was an apparent minimum aldehyde formation that could be maintained throughout the range of pH and alkalinity examined if certain pH/alkalinity ratios could be maintained. At this lower ozone dose, interactions between the effects caused by changing pH and alkalinity levels were almost a non-effect.

UV absorbance reduction correlated well with oxoacid formation, but it did not correlate with the production of the other ozonation DBPs studied. For natural waters, the reduction in UV absorbance on ozonation could be predicted using the NOM fractions employed, particularly the XAD-8 and XAD-4 fractions.

DBP production was not a linear function of NPOC concentration. These NOM concentration effects were particularly significant in DBP formation using NOM fractions, but were less evident with regard to UV absorbance reduction. The reactions being studied appeared to occur differently in concentrated and dilute systems, likely due to various competing reactions and possibly involving the ozone-quenching ability of some types of NOM. The data from this research show that it may be difficult to predict DBP formation for any given site under varying NPOC concentrations and that, if such data are extrapolated to NPOC concentrations lower than those on which the database was formed, DBP formation can be greatly underestimated.

All these data underscore the importance of the nature of the NOM in the raw water for the formation of ozonation by-products. Therefore, the global application of data obtained in a few locations may be difficult or impossible. To a lesser extent, the major seasonal changes occurring in some source waters may have significant nonlinear impacts on ozonation DBP production.

ACKNOWLEDGMENTS

This research was completed at the University of Alberta as part of the first author's doctoral dissertation and while the co-author was a professor in the Department of Civil Engineering at that university. Financial support for this research was provided by Health and Welfare Canada and by the Natural Sciences and Engineering Research Council of Canada. Especially acknowledged are contributions of Dr. R. T. Coutts and the efforts of Dr. D. T. Williams in supporting this research.

REFERENCES

1. Malcolm, R. L., Factors to be considered in the isolation and characterization of aquatic humic substances, in *Humic Substances in the Aquatic and Terrestrial Environment*, Boren, H. and Allard, B., Eds., John Wiley & Sons, London, 1991, 369.
2. Schnitzer, M. and Khan, S. V., *Humic Substance in the Environment*, Marcel Dekker, New York, 1972.
3. Christman, R. F. and Ghassemi, M., Chemical nature of organic color in water, *J. Am. Water Works Assoc.*, 58, 723, 1966.
4. Steelink, C., Implications of elemental characteristics of humic substances, in *Humic Substances in Soil, Sediment and Water: Geochemistry, Isolation and Characterization*, Aiken, G. R., McKnight, D. M., Wershaw, R. L., and MacCarthy, P., Eds., Wiley-Interscience, New York, 1985, 457.
5. Thurman, E. M. and Malcolm, R. L., Preparative isolation of aquatic humic substances, *Environ. Sci. Technol.*, 15(4), 463, 1981.
6. APHA-AWWA-WEF, *Standard Methods for the Examination of Water and Wastewater*, 18th ed., American Public Health Association, Washington, D.C., 1992.
7. Town, R. M. and Powell, H. K. J., Limitations of XAD resins for the isolation of the non-colloidal humic fraction in soil extracts and aquatic samples, *Anal. Chim. Acta*, 271, 195, 1993.
8. APHA-AWWA-WPCF, *Standard Methods for the Examination of Water and Wastewater*, 16th ed., American Public Health Association, Washington, D.C., 1989.
9. Smith, M. and Singer, P., Impact of ozonation on cyanogen chloride formation, Presented at the American Water Works Association Water Quality Technology Conference, San Francisco, 1994.
10. Cowman, G. et al., Presented at the American Water Works Association Water Quality Technology Conference, San Francisco, 1994.
11. Carlson, K. H., Elmund, G. K., and Gertig, K. R., Getting a jump on the information collection rule: Plant scale characterization of NOM and relationship to DBP formation. Presented at the American Water Works Association Water Quality Technology Conference, San Francisco, 1994.
12. Cook, P. L., Current perspectives on EPA's regulation of disinfection and disinfection by-products, in *Disinfection By-Products: Current Perspectives*, American Water Works Association, Denver, 1989, 171.
13. Rice, R. G., Implementation of the U.S. Safe Drinking Water Act amendments of 1986: their impacts on the use of ozone for drinking water treatment in the United States: 1991 update, in Proc. International Ozone Association, Toronto, September 1991, 1.
14. Anderson, L. J., Johnson, J. D., and Christman, R. F., Extent of ozone's reaction with aquatic fulvic acid, *Environ. Sci. Technol.*, 20, 739, 1986.
15. Bedessem, J., Price, D., Amy, G. L., and Sierka, R. A., Molecular weight fingerprints of dissolved organic matter (DOM) implications for treatment process selection and monitoring, in *Proc. American Water Works Association Water Quality Technology Conference*, American Water Works Association, Denver, 1990, 531.

16. Owen, D. M., Brennan, W. J., Chowdhury, Z. K., and Amy, G. A., A determination of the molecular weight fractionation of organic material and disinfection by-products, in *Proc. American Water Works Association Water Quality Technology Conference*, American Water Works Association, Denver, 1990, 957.

17. Glaze, W. H., Reaction products of ozone: a review, *Environ. Health Perspect.*, 69, 151, 1986.

18. Glaze, W. H., Ozone by-products workshop: summary and research needs, in Proc. of a Workshop Held in Los Angeles CA, January 12–15, American Water Works Association Research Foundation, Denver, 1988.

19. Matsuda, H., Ose, Y., Sato, T., Nagase, H., Kito, H., and Sumida, K., Mutagenicity from ozonation of humic substances, *Sci. Total Environ.*, 117/118, 521, 1992.

20. Coleman, W. E., Munch, J. W., Ringhand, H. P., Kaylor, W. H., and Mitchell, D. E., Ozonation/post-chlorination of humic acid: a model for predicting drinking water disinfection by-products, *Ozone Sci. Eng.*, 14, 51, 1992.

21. Xiong, F., Croué, J. -P., and Legube, B., Long-term ozone consumption by aquatic fulvic acids acting as precursors of radical chain reactions, *Environ. Sci. Technol.*, 2695, 1059, 1992.

22. Krasner, S. W., Sclimenti, M. J., and Hwang, C. J., Experiences with implementing a laboratory program to sample and analyze for disinfection by-products in a national study, in *Disinfection By-Products: Current Perspectives*, American Water Works Association, Denver, 1989, 129.

23. Huck, P. M., Anderson, W. B., Savage, E. A., von Borstel, R. C., Daignault, S. A., Rector, D. W., Irvine, G. A., and Williams, D. T., Pilot scale evaluation of ozone and other drinking water disinfectants using mutagenicity testing, *Ozone Sci. Eng.*, 11(3), 245, 1989.

24. Huck, P. M., Anderson, W. B., Rowley, S. M., and Daignault, S. A., Formation and removal of selected aldehydes in a biological drinking water treatment process, *J. Water Supply, Res. Technol. Aqua*, 39(5), 321, 1990.

25. Bull, R. J. and Kopfler, F. C., *Health Effects of Disinfectants and Disinfection By-Products*, American Water Works Association Research Foundation and American Water Works Association, Denver, 1991.

26. Cajelli, E., Cononero, R., Martelli, A., and Brambilla, G., Methylglyoxal-induced mutation to 6-thioguanine resistance in V79 cells, *Mutat. Res.*, 190, 47, 1987.

27. Sayato, Y., Nakamuro, K., and Ueno, H., Mutagenicity of products formed by ozonation of naphthoresorcinol in aqueous solutions, *Mutat. Res.*, 189, 217, 1987.

28. National Research Council (NRC), Drinking Water and Health, Vol. 2, National Academy Press, Washington, DC, 1980.

29. Xie, Y. and Reckhow, D. A., Research note: formation of ketoacids in ozonated drinking water, *Ozone Sci. Eng.*, 14, 269, 1992.

30. Glaze, W. H., Drinking water treatment with ozone, *Environ. Sci. Technol.*, 21(3), 224, 1987.

31. Reckhow, D. A., Legube, B., and Singer, P. C., The ozonation of organic halide precursors: effect of bicarbonate, *Water Res.*, 20, 987, 1986.

32. Legube, B., Croué, J. -P., Reckhow, D. A., and Doré, M., Ozonation of organic precursors: effects of bicarbonate and bromide, in *Proc. International Confer-*

ence on the Role of Ozone in Water and Wastewater Treatment, Perry, R. and McIntyre, A. E., Eds., Selper, London, 1985, 73.

33. Doré, M., Brunet, R., Legube, B., and Croué, J. -P., The role of alkalinity in the effectiveness of processes of oxidation by ozone, in *Proc. 2nd International Conference on the Role of Ozone in Water and Wastewater Treatment*, Smith, D. W. and Finch, G. R., Eds., Tektran International, Kitchener, Ontario, 1987, 1.

34. Staehelin, J. and Hoigné, J., Decomposition of ozone in water: rate of initiation by hydroxide ions and hydrogen peroxide, *Environ. Sci. Technol.*, 16, 676, 1982.

35. Bühler, R. E., Staehelin, J., and Hoigné, J., Ozone decomposition in water studied by pulse radiolysis. I. HO_2/O_2^- and HO_3/O_3^- as intermediates, *J. Phys. Chem.*, 88, 2560, 1984.

36. Staehelin, J., Bühler, R. E., and Hoigné, J., Ozone decomposition in water studied by pulse radiolysis: II. OH and OH_4 as chain intermediates, *J. Phys. Chem.*, 88, 5999, 1984.

37. Andrews, S. A., Huck, P. M., and Coutts, R. T., Ozonation and post-chlor(am)ination by-products of fractionated natural organic matter obtained using an improved isolation procedure, in Proc. American Water Works Association Water Quality Technology Conference, 1992.

38. Andrews, S. A., Huck, P. M., and Coutts, R. T., Identification of microbially important ozonation by-products of aquatic natural organic matter, in *Proc. of Wasser Berlin*, International Ozone Association, Berlin, submitted to *Ozone Sci. Eng.*

39. Andrews, S. A., Huck, P. M., Coutts, R. T., and Williams, D. T., Identification of ozonation by-products of fractionated natural organic matter, in *Disinfection Dilemma: Microbial Control Versus Byproducts*, Robertson, W., Tobin, R., and Kjartanson, K., Eds., AWWA, Denver, 1993c, 79.

40. Andrews, S. A. and Huck, P. M., Identification of ozonation by-product precursors using fractionated natural organic matter, in Proc. American Water Works Association Water Quality Technology Conference, 1993e.

41. Malcolm, R. L., personal communication, 1990b.

42. Leenheer, J. A., Comprehensive approach to preparative isolation and fractionation of dissolved organic carbon from natural waters and wastewaters, *Environ. Sci. Technol.*, 15(5), 578, 1981.

43. Andrews, S. A., Ph.D. dissertation, Department of Civil Engineering, University of Alberta, Edmonton, Alberta, 1993d.

44 Aiken, G. R., McKnight, D. M., Wershaw, R. L., and MacCarthy, P., *Humic Substances in Soil, Sediment and Water—Geochemistry, Isolation and Characterization*, John Wiley & Sons, New York, 1985.

45. Box, G. E., Hunter, W. G., and Hunter, J. S., *Statistics for Experimenters: An Introduction to Design, Data Analysis and Model Building*, John Wiley & Sons, New York, 1978.

46. Davies, O. L., *The Design and Analysis of Industrial Experiments*, 2nd ed., Longman Group Ltd., London, 1979.

47. Sclimenti, M. J., Krasner, S. W., Glaze, W. H., and Weinberg, H. S., Ozone disinfection by-products: Optimization of the PFBHA derivatization method for the analysis of aldehydes, in *Proc. AWWA Water Quality Technology Conference*, AWWA, Denver, 1990, 477.

48. Wong, J. T. F., Baker, G. B., and Coutts, R. T., Rapid and simple procedure for the determination of urinary phenylacetic acid using derivatization in aqueous medium followed by electron-capture gas chromatography, *J. Chromatog.*, 428, 140, 1988.
49. MacCarthy, P., personal communications, 1992.
50. Andrews, S. A., Huck, P. M., and Coutts, R. T., Quantitation of ozonation by-products of fractionated aquatic natural organic matter, *Vom Wasser,* 81, 151, 1993a.
51. Hayes, M. H., B., MacCarthy, P., Malcolm, R. L., and Swift, R. S., *Humic Substances II–In Search of Structure*, Wiley-Interscience, Chichester, Great Britain, 1989.
52. Malcolm, R. L., The uniqueness of humic substances in each of soil, stream and marine environments, *Anal. Chim. Acta*, 232, 19, 1990a.
53. Andrews, S. A. and Huck, P. M., Using fractionated natural organic matter to quantify organic by-products of ozonation, *Ozone Sci. Eng.*, 16, 1, 1994a.
54. Caprio, V., Insola, A., and Silvestre, A. M., The ozonation of glyoxylic acid in aqueous solution: chemical products and kinetics evolution, *Ozone Sci. Eng.*, 9, 13, 1987.
55. Hoigné, J. and Bader, H., Rate constants of reactions of ozone with organic and inorganic compounds in water. II. Dissociating organic compounds, *Water Res.*, 17, 185, 1983.

CHAPTER 21

Characterization of NOM Removal by Biofiltration: Impact of Coagulation, Ozonation, and Sand Media Coating

M. R. Collins and C. W. Vaughan

ABSTRACT

The overall objective of this paper is to characterize natural organic matter (NOM) removals by biologically active sand filters. Specific goals are to ascertain pretreatment effects, i.e., coagulation and ozonation, on NOM biodegradability and chlorine reactivity. Emphasis will be given to evaluating the influence of location within a conventional treatment process train on biofilter efficiency and to examine NOM removal mechanisms associated with biofiltration, especially with aged sand media.

As expected, preozonation enhances biofiltration removals of source water NOM. However, precoagulation reduces biofiltration efficiency and the positive treatment influence of preozonation on biofiltration performance. Alum coagulation removes significant quantities of higher molecular weight, hydrophobic fractions of NOM which are also susceptible to biofiltration removal. Finally, sand coating metal composition, which is generally associated with the age of the filter media, may have an impact on biofiltration removal of NOM.

INTRODUCTION

The most common approach used by water treatment utilities to minimize disinfection by-products (DBPs), such as trihalomethanes (THMs) and haloacetic acids (HAAs) in drinking water is to maximize NOM removals by conventional clarification processes preceding the chlorine disinfection stage. A variety of chemical coagulants are used in clarification; the most common of which is aluminum sulfate (alum). Advanced treatment techniques including ultrafiltration/nanofiltration, chemical oxidation, and activated carbon adsorption can be used to enhance removals of DBP organic precursors. Unfortunately, these physicochemical treatment processes are more costly and often less efficient when compared to biological treatment processes.[1,2]

Biological treatment can be used to remove organic carbon or nutrients from a source water; limit distribution system biofouling corrosion and taste and odor problems by reducing regrowth potential; and reduce the amount of chlorination required for disinfection purposes, thereby reducing DBP formation.[2] A number of practices employ biological treatment processes, most are fixed-film (biofilm) processes. They include fluidized bed filters, slow sand filters, biologically active GAC filters (biological activated carbon, BAC), and soil-aquifer treatment systems.[1] Many European treatment practices employ some form of biological treatment process within the overall treatment scheme to provide a more complete, stable finished water.[3]

The NOM of most source waters is comprised of humic substances (humic and fulvic acids), hydrophilic acids, carboxylic acids, amino acids, carbohydrates, and hydrocarbons in the approximate proportions of 50, 30, 6, 3, 10, and 1%, respectively.[4] However, in highly colored waters, the humic substance content may be as high as 50 to 90%.[4] The portion of NOM that can be biodegraded is sometimes defined as biodegradable dissolved organic carbon (BDOC) or assimilable organic carbon (AOC), which are measured by two distinct techniques.[5]

The amino acids and proteins, carbohydrates, and carboxylic acids found in natural waters vary in their susceptibility to microbial biodegradation. The nonhumic fraction usually requires little enzymatic hydrolysis or conversion before use in microbial catabolic and anabolic processes. At issue in biofiltration systems is the ability of the microbial consortia to degrade the bulk of the NOM present as humic and fulvic acids. The best estimates indicate that 5 to 50% of NOM in surface waters is biodegradable; 15 to 30% is average.[6]

Ozone treatment advancement has been focusing on decreasing the biodegradability of NOM by post-treatment with biofiltration. Biological growth rates on ozonated dissolved organic carbon were up to four times higher than controls.[7] Ozonation of NOM, particularly humic substances, will partially oxidize and break apart larger molecules[8,9] and aromatics.[10,11] The resulting compounds, which are reduced in size and polarity, are believed to be more readily biodegradable.[12,13] Other than slow sand filtration, removals of biodegradable NOM fractions have not been well documented in typical water treatment plant operations. A fundamental understanding of biological filtration needs to be advanced, especially given the growing importance of biological drinking water treatment.

The overall objective of this research program was to characterize NOM removals by biologically active sand filters. Specific goals of the research

Table 1   Typical Parameter Ranges of Portsmouth (NH) Raw Water Quality

Parameter	Range
Dissolved oxygen (mg/l)	8.0–12.0
Temperature (°C)	8–16
pH	5.5–6.1
Iron (mg/l)	0.1–1.0
Turbidity (ntu)	1.0–2.3
DOC (mg/l)	4.4–8.2
UV absorbance (cm^{-1})	0.21–0.34
THMFP (μg/l)	290–460

presented herein were to evaluate the influence of coagulation and ozonation on NOM biodegradability with respect to biofilter location within a conventional treatment scheme, and examine NOM removal mechanisms associated with ripened or aged biofilters.

METHODS AND MATERIALS

A global approach taken in this study was to evaluate only one source water vs. many sources of NOM and to utilize only one BDOC method, after an initial comparison phase. The reasoning behind this approach was to maintain consistency between biodegradability comparisons. Any observed differences between BDOC measurements would not be confounded by varying inorganic nutrient levels or biofilm ripening conditions.

NOM SOURCE WATER

The natural water for all biofilter runs was the raw water source for the Portsmouth (NH) Water Treatment Plant, the Bellamy Reservoir. Selected water quality parameters analyzed on the raw water during the course of the investigation are summarized in Table 1. The Bellamy Reservoir water is a moderate-to-heavy colored, low turbidity, low alkalinity source water. The raw water has a high aromaticity content with significant iron concentrations ranging from 0.1 to 1.0 mg/l. Typical pH and alkalinity values averaged near 6 and 4 mg/l as $CaCO_3$, respectively. Concentrations of Br^- were typically below 30 μg/l and that chloroform accounted for over 85 to 95% of the total THMs during this study.

BDOC METHODOLOGY

Three different methods of determining BDOC were compared during the initial stages of this investigation. The principal BDOC method used was the recirculating acclimated-sand column technique. The other BDOC determination methods explored, at least in the early phases, included a modified version of the Servais method using a raw water bacterial inoculant, and a shaker acclimated-sand method conducted at the University of Cincinnati. Brief descriptions of the three different BDOC techniques follow.

Recirculating Sand Column

The BDOC analysis was determined by a variation of a column technique used by Mogren et al.[14] A 2.54-cm diameter glass column, with a 245-ml volume partially filled with biologically active sand (170 ml), served as the main body of the reactor. A 1-l recirculation flask with two inlets and one outlet was used as a reservoir where most of the sample resides during recirculation. One of the inlets allows air to enter the system after passing through a bacterial removal filter. All aqueous samples were filtered through a 0.45-μm filter membrane prior to recirculation. Prior to sample application to the bioreactor, approximately 10 l of type-II water was circulated through the column and sent to waste until the effluent dissolved organic carbon (DOC) equals the influent DOC. Approximately 200 ml of sample passed through the bioreactor and sent to waste. The recirculation flasks were refilled to 1 l and recirculated through the media at a flow rate of 49 ml/min (12 bed volumes per hour or 6 m/h) for 5 to 7 days. The biofilter empty-bed-contact time (EBCT) was 4 ± 0.5 min throughout the run. Samples for analysis were periodically withdrawn from the recirculation flasks. The difference between the initial and final DOC was operationally defined as the BDOC. The reactor was kept in darkness at all times to prevent algal growth. Temperature was held constant between 19 and 21°C. All sand media used in the biofilters were acclimated with the Portsmouth (NH) raw water for *at least* 3 to 4 weeks before the commencement of a biofiltration run.

Modified Servais Method

A modified method of Servais et al.[15] was also used to determine the BDOC during the initial phase of investigation. The method involves taking a filtered (0.22 μm membrane) water sample and reinoculating the water with an inoculum of raw bacteria that does not contain protozoan grazers. The DOC of the batch solution was monitored over a 30-d incubation period. Changes in DOC were attributed to bacterial uptake and mineralization of the biodegradable natural organic matter. The difference in the final steady-state DOC value and the initial DOC concentration was termed the BDOC. Modifications to the Servais method included the following: utilization of 200-ml filter-sterilized samples, use of Waltman GFC (washed) prefilters, use of 0.22-μm Micron sep membrane filters rather than 0.2-μm cellulose acetate membranes, and reinoculation of the sample with 2 ml of raw water.

Selected samples not inoculated with raw water bacteria were used to check for glassware adsorption and volatilization of the dissolved organic matter. The sample blanks used for the 28-d BDOC determinations revealed unaccountable DOC losses of less than 0.1 mg/l. Moreover, glucose samples were inoculated to quantify the mineralization potential of the raw water bacterial inoculum.

Shaker Sand Method

Bench-scale batch reactors modified after the Joret-Levi reactor[16] and others was used by Dr. Scott Summers at the University of Cincinnati to provide another independent assessment of the BDOC for the Portsmouth (NH) raw water. Duplicate water samples with a volume of about 500 ml each were incubated at 20°C for 5 d in reactors that contain roughly 100 ml of bioacclimated sand. The BDOC values of samples were determined by the difference between initial DOC concentration and that after 5 d of incubation. All of the samples collected for DOC analysis were filtered through 0.45-μm pore membrane filters. The BDOC detection limit is considered to be 0.1 mg/l.[17]

Ozonation and Coagulation Pretreatment

At the times of sampling, the Portsmouth (NH) conventional water treatment plant used alum and caustic soda additions of 32 to 38 and 7 to 10 mg/l, respectively, at pHs ranging from 5.8 to 6.2. Settling basin detention times averaged 4 to 5 h. Typical DOC, UV absorbance, and THMFP reductions achieved by the WTP varied from 50 to 70, 70 to 90, and 70 to 90%, respectively.

The influence of ozonation and conventional alum coagulation on biofiltration removals were assessed using the Portsmouth (NH) source water and the settling tank effluent as outlined in Figure 1. Four raw water samples and three coagulated/settled samples were applied to the recirculating biofilters. The raw water samples included a nonozonated sample, a low ozone dose sample, a high ozone dose sample, and a high ozone dose duplicate sample. The settling tank effluent samples included a nonozonated sample, a low ozone dose sample, and a high ozone dose sample. A glucose/glutamic acid mixture was also applied to the biofilters to confirm biological activity. The mixture was made from high purity reagent stock, and included a nutrient seed addition.

The eight recirculating biofilter columns, as described in the BDOC method, were filled with Portsmouth pilot slow sand filter media and acclimated *in situ* an additional 30 d with the Portsmouth raw water. The settling tank effluent followed conventional sweep-floc coagulation treatment using alum (36 mg/l). Both the raw and settling tank effluent target waters were filtered through 0.45-μm filters prior to ozonation or biofiltration.

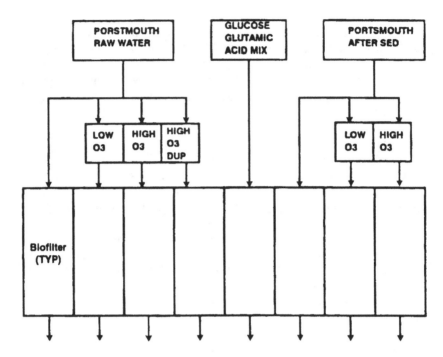

Figure 1   Experimental design used to access influence of coagulation/sedimentation and ozonation pretreatment on biofiltration removals of natural organic matter.

The ozonation procedure and experimental setup consisted of parallel trains of a 2-l reactor followed by two 500-ml potassium iodide traps. The potassium iodide traps were used to measure ozone that was not consumed in the reactor vessels. The target milligrams of ozone per milligrams of DOC dosages consumed for the low and high ozone doses were 0.5 and 1.0, respectively. The actual ozone/DOC doses consumed for the raw water low dose, high dose, and high dose duplicate were 0.44, 0.7, and 0.78 mg ozone per mg of DOC, respectively. The actual doses consumed for the coagulated/settled water low dose and high dose were 0.54 and 1.18 mg ozone per mg DOC, respectively. Exact ozone dosages applied to samples could not be controlled precisely because of limited ozonator sensitivity control, but measurement of the actual dosage consumed was determined using proven techniques.[10]

NOM CHARACTERIZATION

All samples before and after ozonation and before and after biofiltration were fractionated for hydrophilic/hydrophobic content and apparent molecular weight (AMW) distribution using DOC and UV absorbance (254 nm at pH 7) as analytical parameters. THM formation potential (THMFP) was analyzed for all samples before and after ozonation, and before and after biofiltration, as well as all hydrophilic and <3000 AMW fractions.

AMW distributions were determined by a ultrafiltration (UF) membrane technique as described by Fenstermacher.[18] Molecular weight separations were performed using a pressurized stirred cell apparatus (Amicon Corp., Danvers, MA) with membranes characterized by nominal molecular weight cutoffs (MW) of 500, 3000, and 10,000 AMW units. The ultrafiltrations were performed in parallel; therefore, the procedure did not provide a discrete AMW fraction, but rather the accumulation of organics below the given AMW. Prior to sample filtration, 100 ml of type II water was filtered and tested for DOC to assure that there was no contamination. The clean water flux rates at 55 psi were frequently analyzed and compared with manufacturer's specifications to check the membrane integrity.

Hydrophilic/hydrophobic separations were performed by nonionic resin adsorption (XAD-8, Amberlite-Supelco, Bellefonte, PA). NOM that is adsorbed to the XAD-8 resin is operationally defined as hydrophobic (humic substances) material.[19] A 31.3-ml column (12-mm diameter) was filled with XAD-8 resin and connected to a recirculation pump and flask. One liter of filtered (<0.45 μm) sample was acidified to a pH of 2 and applied to the resin. The application rate was 12 resin bed volumes per hour (6.3 ml/min) for 15 h. The fraction that did not adsorb to the resin was defined as hydrophilic. The hydrophobic compounds was eluted from the resin by applying 0.1 N NaOH at a rate of five resin bed volumes per hour (2.6 ml/min). The first 28 ml of the eluate (void volume) was sent to waste. The next 50 ml of yellowish brown eluate was collected with a resulting concentration factor of 20:1. The resin was then cleaned with 320 ml of 0.1 N NaOH and 500 ml of type II water. At the end of the type II water rinse a sample for DOC was analyzed to assure no contamination prior to the next sample application.

ANALYTICAL MEASUREMENTS

Dissolved organic carbon was determined by the UV-persulfate oxidation method using a DC-80 Dohrmann TOC analyzer (Santa Clara, CA). Two drops of concentrated phosphoric acid was added to 10 ml of sample and purged with N_2 for 5 min to remove CO_2. Three 1000-μl injections for every sample were analyzed and the average reported as milligrams per liter DOC. Standard deviations of replicated injections were below 0.1 mg/l and usually below 0.05 mg/l.

UV absorbance was measured utilizing the Spectronic 2000 Bausch & Lomb Spectrometer (Rochester, NY) using a path length of 1 cm and a pH of 7. Triplicate absorbance readings were analyzed at a wavelength of 254 nm. Standard deviations of replicated readings were below 0.002 cm^{-1}. The sample and reference cells were matched quartz curvettes. The sample and reference blank were adjusted to pH 7 using two drops of a weak phosphate

Table 2   Assessment of the Reproducibility of the NOM Fractionation and BDOC Techniques Using Replicated Portsmouth (NH) Source Water after Biofiltration

Parameter	No. of replicates	DOC (mg/l)		UV absorbance @ 254 nm (cm^{-1})		THMFP (µg/l)	
		$\bar{X}$	± S.D.	$\bar{X}$	± S.D.	$\bar{X}$	± S.D.
After Biofiltration[a]	6	3.52	0.15	0.179	0.007	173	9
<0.5 k AMW	3	0.53	0.05	0.002	0.002	10	1
<3.0 k AMW	3	2.16	0.50	0.035	0.014	55	12
<10.0 k AMW	3	2.38	0.33	0.083	0.011	116	15
Hydrophilic	3	1.73	0.10	0.046	0.002	79	12
Hydrophobic[b]	3	1.79	0.18	0.133	0.007	94	15

[a] Initial DOC, UV absorbance, and THMFP for Portsmouth, NH source water: 6.62 ± 0.22 mg/l, 0.265 ± 0.002 cm^{-1}, and 468 ± 26 µg/l, respectively.
[b] Hydrophobic content was determined as the difference between the initial concentration and the hydrophilic fraction. Base extractions from the XAD resin typically recovered within 10% of the calculated value.

buffer. A blank sample was analyzed every ten samples to check for instrument drift. All samples were filtered through a 0.45-µm filter prior to analysis.

Trihalomethane formation potential (THMFP) was determined by quantifying the total concentration of all THM species present after a 7-d incubation with sodium hypochlorite at 20°C and pH 7 with a chlorine dose-to-carbon mass ratio of 4:1. All samples after incubation were tested for residual chlorine and pH. THM species were determined by a liquid/liquid extraction gas chromatograph electron capture detection method, EPA and Standard Methods.

Metal analyses were conducted by commercial laboratory (NH Materials Laboratory, Dover, NH) using inductively coupled argon plasma emission spectrometry (ICP) (USEPA method 6010, 1986). Concentrations of calcium, aluminum, iron, and manganese were determined from composite samples filtered through a prerinsed 0.45-µm membrane filter and acidified using concentrated reagent grade nitric acid. The average precision and accuracy measured for metals analyses were acceptable according to EPA method 6010 (1986).

CHARACTERIZATION REPEATABILITY

The reproducibility of the analytical and NOM characterization procedures was assessed at the beginning of the research program. The standard deviations for various characterization methods when quantified by either DOC, UV absorbance, or THMFP from replicated samples are summarized in Table 2. In many cases, the reproducibility of the NOM fraction or BDOC determination for a given sample approached the minimum detection level of the analytical instrument. Because of the confidence in the analytical protocol, a decision was made to replicate only one sample during an experimental run rather than replicate all individual samples to maximize the number of comparisons. A

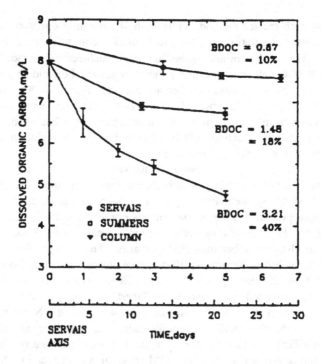

Figure 2   Comparison between methods used to determine BDOC on Portsmouth (NH) raw water.

random replicated sample was used as a check to insure that the reproducibilities initially established for the NOM characterization and BDOC procedure are being maintained.

RESULTS AND DISCUSSION

BDOC METHOD COMPARISONS

An opportunity was taken to compare three different methods of determining BDOC for the same source water. The three methods that were compared include the Servais method using an unattached bacteria inoculant, the Summers/batch-reactor sand method conducted at the University of Cincinnati, and the recirculating sand column method as previously described. All analyses were performed in duplicate (at least) to give a level of statistical confidence to all comparisons.

There was a significant difference in the BDOC content of the Portsmouth (NH) raw water as determined by each method as noted in Figure 2. In general, the Servais method underpredicted the BDOC content as determined by the

Summers/batch-reactor method, while both significantly underpredicted the BDOC content as determined by the recirculation column method. A BDOC content of 18% as determined by the Summers/batch-reactor method is in the range of BDOC values determined previously for the Portsmouth raw water source using the same modified Servais method. Typical BDOC values as determined by the Servais method for the Portsmouth raw water have been known to vary from 9 to 17% over a four-month period.[20] What was unexpected was the significant reduction in DOC observed in the recirculating sand columns after 5 d of continuous operation.

Usually, significant reductions of dissolved organic carbon (>35%) by biological treatment can rarely be achieved without increasing the biodegradability of NOM by pretreatment processes such as ozonation.[21] Since no preoxidation processes were utilized in this comparison, the DOC reduction achieved by the recirculating sand column appears to be associated with removal mechanisms other than biodegradation. The sand medium placed in the biofilter columns before the 30-d source water acclimation period was collected from 20-year ripened slow sand filters located in West Hartford. The possibility existed that the extremely "aged" or "ripened" sand media used may have accounted for this significant increase in biofilter NOM removals. According to the slow sand filtration literature,[22] aged or ripened sand media is a more efficient collector than new or fresh sand; consequently, the impact of sand media age or condition on NOM removal was included in this study. Because of the reproducibility and the ease by which the biodegraded sample could be collected for additional analysis and characterization, a decision was made to continue the study using the recirculating sand column method for determining BDOC. Emphasis was given to using a sand medium that has been acclimated to the Portsmouth (NH) source water for months via pilot slow sand filters.

INFLUENCE OF OZONATION AND COAGULATION PRETREATMENT

Although drinking water treatment is often viewed as a series of individual processes placed together in a rational sequence, the performance of a particular process may be directly related to the type and performance of upstream processes.[23] A major goal of this research was to assess the influence of preozonation on biofiltration efficiently as a function of treatment plant location. Preozonation is usually practiced either at the headworks of a conventional treatment plant or after chemical coagulation and sedimentation, but prior to granular media filtration.

Both biofilter location within a conventional water treatment plant and ozone dose exerted a significant impact on biofiltration efficiency as noted in Figure 3. A review of graphical trends suggests that alum coagulation in the sweep floc zone was more influential than ozonation (utilized in this study)

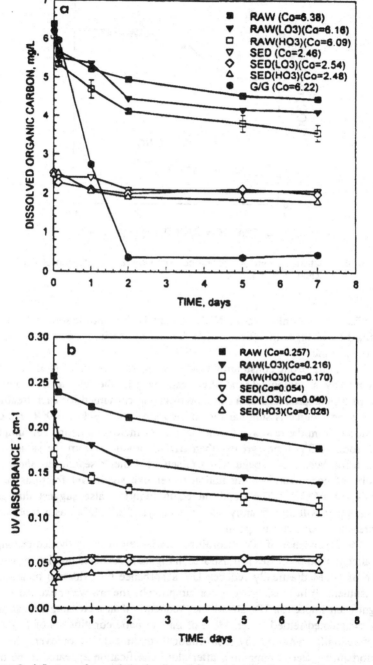

Figure 3   Influence of coagulation/sedimentation and ozonation on biofiltration performance as quantified by (a) dissolved organic carbon and (b) UV absorbance.

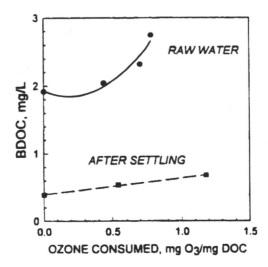

Figure 4 Relationship between ozone consumption and production of BDOC as a function of WTP location.

in effecting not only overall NOM removals, but biofiltration kinetics as defined by changes in concentration divided by the initial concentration (C/Co) over time.

Increasing ozone consumption did increase the amount of NOM that was removed by the biological sand filters, especially for the raw water. A comparison of BDOC produced as a function of ozone consumed for each treatment plant water source is summarized in Figure 4. The yield of BDOC material produced from the raw water as a function of increasing ozone consumption was roughly $3^1/_2$ times greater than BDOC produced from the same water which has been alum coagulated and settled. Other research has shown the ability of coagulation/sedimentation to remove significant fractions (up to 100%) of BDOC.[24] The trends noted in Figure 4 also suggest that alum coagulation will significantly decrease the yield of BDOC as a function of increasing ozone consumption.

The dependence of UV absorbing material removals on the sequencing of treatment plant processes is illustrated in Figure 5. As noted previously, ozonation of NOM drastically reduces UV absorbance by oxidation of aromatic constituents.[10] Increasing ozone consumption by the raw water exerted a more significant impact on UV absorbance reduction than increasing ozonation of the coagulated/settled water. Most of the aromatic constituents of NOM are preferentially removed by conventional treatment;[25] moreover, the UV absorbance material remaining after alum clarification appeared to be more resistant to ozonation influences as noted by the significant differences in slopes between the two waters.

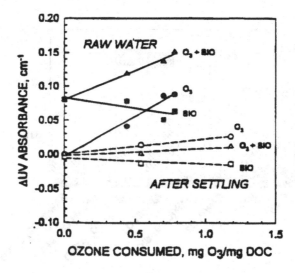

Figure 5   Relationship between ozone consumption and change in UV absorbance as a function of treatment process.

The ability of the biofilters to reduce UV absorbance, as shown by the negative slopes in Figure 5, was diminished by preozonation of either raw water or coagulated/settled water. It is not clear whether this negative influence of ozonation was due to the preferential attraction of ozone for the same UV absorbing material that would be also amenable to biofiltration or to a diminutive release of UV-absorbing material from an active biofilm that was acclimated to a nonozonated raw water.

The combined influence of ozonation and biofiltration resulted in a diverging UV absorbance removal trend with increasing ozone consumption between the raw water and the coagulated/settled water. The combination of ozonation and biofiltration resulted in little change in UV absorbance for the coagulated/settled water even with increasing ozone dosages. The same treatment combination resulted in increased UV absorbing material removals for the raw water although the contribution of the biological filter to these removals diminished with increasing ozone consumption.

Because ozone consumption at the levels used in this study (0.44 to 1.18 mg O_3/mg DOC) should not have appreciably mineralized the reduced organic carbon to carbon dioxide,[10] most of the changes noted in UV absorbance reflected changes in the composition and characterization of NOM. An assessment of these NOM transformations can be reflected in the AMW and hydrophobic content distributions for each target water. The raw water changes are depicted in Figures 6 and 7 while the coagulated/settled changes are shown in Figures 8 and 9. Although DOC, UV absorbance, and THMFP were performed as analytical parameters, the focus here will be on DOC.

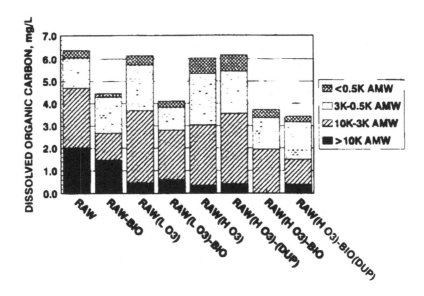

Figure 6   Change in AMW distribution of raw water as a function of ozonation and biofiltration.

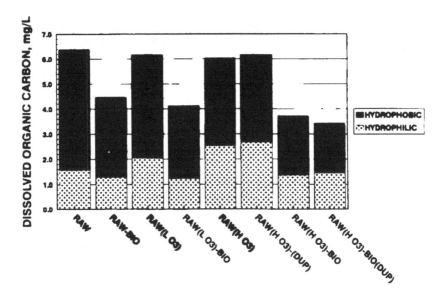

Figure 7   Change in hydrophobic/hydrophilic content of raw water as a function of ozonation and biofiltration.

AFTER SEDIMENTATION

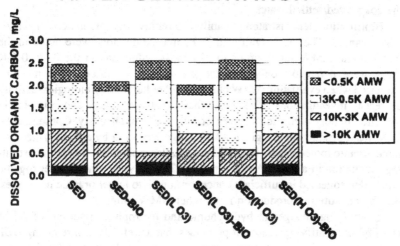

Figure 8   Change in AMW distribution of coagulated/settled water as a function of ozonation and biofiltration.

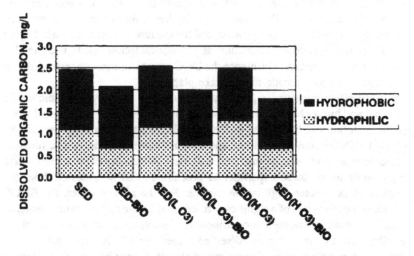

Figure 9   Change in hydrophobic/hydrophilic content of coagulated/settled water as a function of ozonation and biofiltration.

As expected, ozonation did not significantly affect the total organic carbon concentration but did change the AMW distribution for both the raw and coagulated/settled waters as noted in Figures 6 and 8, respectively. In all cases, ozonation increased organic material in the lower MW ranges at the expense of the higher MW ranges, a trend that has been observed by many others.[9,26]

However, the shift into the lowest AMW range (<0.5 K) was minimal for the coagulated/settled water.

Biofiltration demonstrated the ability to remove organic material from all AMW ranges. The most significant DOC mass reductions were noted in the intermediate AMW range of 10 to 3K. Mass reductions in the 3- to 0.5-K range were noticeable only with the preozonated biofilters. Care must be exercised when evaluating AMW changes and reductions by biological activity as recent investigations[27,28] suggest that repolymerization and microbially mediated depolymerization may be occurring simultaneously. Consequently, DOC mass production in a given AMW range can occur by either (1) cleaving large organic molecules by bacteria into smaller molecules, thereby constituting production in a lower AMW range or (2) repolymerizing smaller organic molecules collected in sufficient concentrations into larger organic molecules, thereby constituting production in a higher AMW range.[29]

Quantifying changes in hydrophobic and hydrophilic fractions of NOM is useful in identifying treatment processes that are effective in reducing DBP precursors as well as producing or removing the most biodegradable fractions of organic carbon. Both ozonation and biofiltration exerted a significant impact on hydrophobic/hydrophilic fractionations of the raw and coagulated/settled waters as shown in Figures 7 and 9, respectively. Ozonation *increased* the hydrophilic material at the expense of the hydrophobic or humic material although this trend was more pronounced for the raw water as shown in Figure 10. As noted previously, ozonation at the consumption levels used in this study did not significantly influence the DOC concentration, suggesting mineralization to carbon dioxide did not take place.[10]

Biofiltration removed significant fractions of both hydrophobic/humic material and hydrophilic/nonhumic material in the raw water while only hydrophilic material was biofiltered in the coagulated/settled water. A breakdown of the BDOC content of both raw and coagulated/settled waters as a function of ozone consumption is summarized in Table 3. A measure of the accumulative reproducibility of these experimental procedures can be assessed from the duplicated raw water, high ozone sample. For the natural water, the BDOC composition consisted of a significant fraction of humic and nonhumic material with ozonation increasing the contribution of nonhumic material amenable to biofiltration. For the coagulated/settled water, the BDOC composition was comprised mostly from nonhumic material with ozonation having a minimal impact since the humic material amenable to biofiltration and ozonation was already removed by alum coagulation. According to the literature, large hydrophobic/humic organic molecules are predominantly removed by adsorption[30,31] and alum coagulation,[25,32,33] whereas smaller organic molecules are removed by both adsorption and biodegradation[34,35] but are more difficult to remove by coagulation.[25,36,37]

Another way of viewing the impact of biofilter location within a conventional water treatment facility on organic precursor removal can be made by

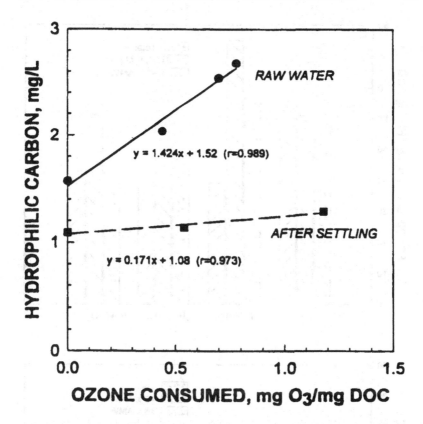

Figure 10 Relationship between ozone consumption and production of hydrophilic carbon as a function of WTP location.

Table 3 Hydrophobic/Humic and Hydrophilic/Nonhumic Composition of BDOC as a Function of Coagulation and Ozonation Pretreatment

Biode- gradable DOC fraction	Raw water ozone consumption (mg O_3/mg DOC)				Coagulated/settled water ozone consumption (mg O_3/mg DOC)		
	0	0.4	0.70	0.78 (dup)	0	0.54	1.18
Hydrophobic	1.64	1.25	1.15	1.51	−0.04	0.14	0.05
Hydrophilic	0.28	0.80	1.17	1.23	0.43	0.40	0.63
Total BDOC	1.92	2.05	2.32	2.74	0.39	0.54	0.71

examining biofilter enduced changes in specific yield (μg THM/mg DOC) of various NOM fractions for the raw water and after coagulation/sedimentation water (AS). The calculated specific yields for the various treated waters are summarized in Figure 11. Both the raw water, high ozone [RW (HO3)] and the raw water, high ozone, biofiltered [RW (HO3)-Bio] samples were duplicated to give an indication of specific yield reproducibility. The low ozone samples are not shown because of space limitations.

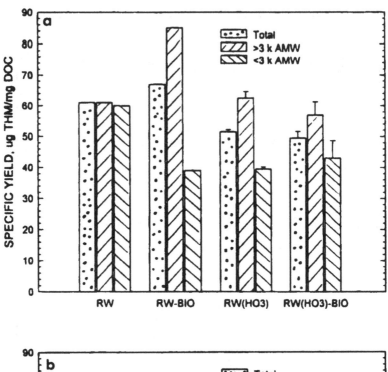

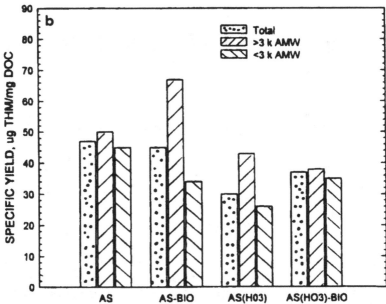

Figure 11   Change in specific yield (μg THMFP/mg DOC) of various AMW fractions as a function of coagulation and biofiltration. (a) Raw water and (b) after coagulation/sedimentation.

Several observations are evident from a review of the trends depicted in Figure 11. As expected, alum coagulation/sedimentation removed the most chlorine-reactive species of NOM.[25] Ozonation exerted a tendency in both raw and coagulated waters to reduce NOM reactivity with chlorine, especially in the lower MW range. Biofiltration preferentially removed in both raw and coagulated waters (1) the less-reactive material from the larger MW organic molecules and (2) the more-reactive material from the smaller MW organic molecules. Biofiltration did not exert a noticeable preference for chlorine-reactive species of NOM following ozonation.

It is important to note that over 90% of the total trihalomethanes in all MW-fractionated samples were chloroforms ($CHCl_3$). The highest recorded bromodichloromethane concentrations was less than 8 μg/l. Unfortunately, hydrophobic/hydrophilic separations using the XAD-8 resin resulted in the hydrophilic fraction containing a significant amount of bromodichloromethane (up to 45% of the total THMs on a weight basis) after chlorine addition. Either the hydrophobic fractionation technique significantly changed the Br/DOC ratio in the hydrophilic fraction thereby promoting bromated THM species[38] or the hydrochloric acid used to lower the pH to facilitate hydrophobic adsorption to the nonionic resin may have been contaminated with bromide. The latter possibility was more likely to have occurred since the source water had low bromide concentrations (typically <30 μg/l) and the bromated THM species were low for all other nonfractionated and MW-fractionated samples.

SAND MEDIA COATING

An experimental design used to assess different sand media coatings on NOM removals by biologically active filters is depicted in Figure 12. Each sand represents a different degree of aging or ripening under various treatment conditions. The West Hartford slow sand filter (SSF) medium was aged *in situ* for roughly 20 years prior to collection. The Portsmouth rapid sand filter (RSF) medium was ripened *in situ* at the conventional alum clarification plant for roughly 30 years prior to collection. The Portsmouth SSF medium was aged in pilot SSFs for roughly 6 months prior to collection. The sand media were all biologically ripened simultaneously for at least 3 weeks on the Portsmouth (NH) raw water prior to the BDOC run. The glucose/glutamic acid feed was to monitor for any differences in biokinetics.

As depicted in Figure 13, biofiltration removals of NOM appear to be a function of sand media source. This observation was confirmed in a previous comparison between the Portsmouth pilot SSF medium and the West Hartford SSF medium as shown in Figure 14. The standard deviations from the replicated samples suggest these differences are significant. Both Figures 13 and 14 suggest that the West Hartford SSF was a more efficient biofilter using DOC as the analytical parameter. Using UV absorbance as the measuring parameter, the Portsmouth RSF medium was a more efficient collector than

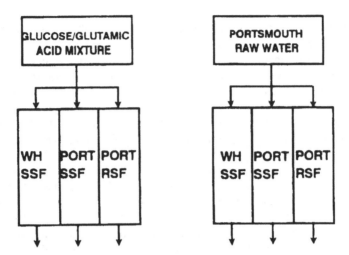

Figure 12 Experimental design used to access influence of sand media coating on biofiltration performance.

the West Hartford SSF medium which was more efficient than the Portsmouth SSF medium. The Portsmouth pilot SSF medium was the least efficient sand collector as determined by either parameter.

An obvious explanation of why one filter media would be more efficient than the other cannot be easily given. All sand media were acclimated for at least 3 weeks with the target water before the commencement of a biofilter run. As noted in Figure 13a, almost identical removals of glucose/glutamic acid by the three different biofilters suggested that the biological activity was the same on each sand media source. Epifluorescent direct bacterial counts of the various sand coatings using a phosphate detergent to extract the bacteria are presented in Table 4. The bacterial counts on each sand media source were considered to be in the same order of magnitude. Interestingly, the bacteria counts was inversely related to the UV absorbance removal trend depicted in Figure 13b.

A revealing correlation may be found by relating the metal content of the sand coating to the removal trends. Following a hot nitric acid/hydrochloric acid extraction technique, sand coating metal concentrations were determined on a gram dry weight basis and are summarized in Table 5. The West Hartford slow sand filter medium contained the highest concentrations of iron and manganese, which is not surprising since biological filters are noted for their exceptional removals of iron and manganese.[39] The Portsmouth RSF contained low concentrations of iron, manganese, and calcium, but significantly high concentrations of aluminum, possibly as aluminum oxide from the aging of $Al(OH)_3$ floc carryover. Apparently significant aluminum concentrations can build up on rapid sand media over long periods of time. Other research has

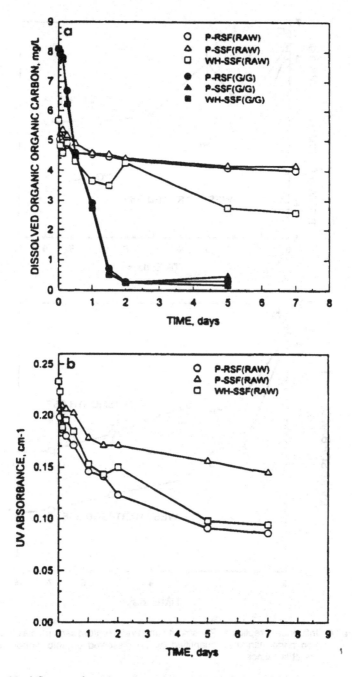

Figure 13   Influence of sand media source on biofiltration performance as quantified by (a) dissolved organic carbon and (b) UV absorbance.

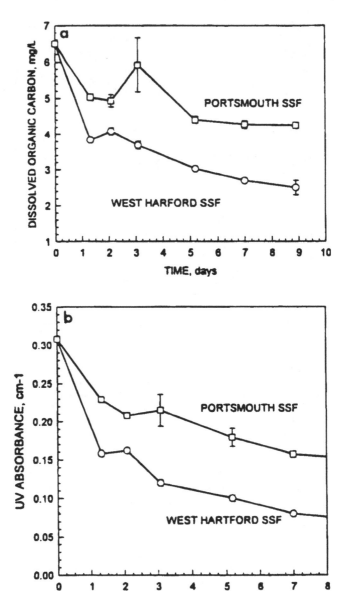

Figure 14 Influence of replicated Portsmouth and West Hartford sand media on biofiltration performance as quantified by (a) dissolved organic carbon and (b) UV absorbance.

Table 4   Epifluorescence Bacterial Direct
          Counts of the Various Ripened
          Sand Coatings

Sand source	Counts/g dry wt
Portsmouth SSF	9.42 E7 ± 1.04 E6
Portsmouth SSF (G/G)	9.03 E7 ± 2.04 E7
West Hartford SSF	8.73 E7 ± 1.96 E7
Portsmouth RSF	5.18 E7 ± 1.86 E7

shown significant increases of Fe_2O_3, MnO_2, and Al_2O_3 content in the coatings of the rapid sand filter medium over time, 2 to 34 years.[40]

Overall, the equivalent metal concentrations ranked from highest to lowest were Portsmouth RSF, West Hartford SSF, and the Portsmouth SSF. This trend also corresponds to UV absorbing matter removals observed in the biological filters as noted in Figure 13b. Reasonable explanations to support the biofilter correlation between UV absorbance reductions and equivalent metal concentrations include (1) the UV absorbance components of DOC were preferentially removed (as noted by comparing percent reductions in Figures 13 and 14), (2) the dominant UV-absorbing components of NOM are generally associated with the hydrophobic/humic material,[23] and (3) the hydrophobic/humic materials are preferentially adsorbed.[31] An investigation using a raw water inoculum to determine BDOC (modified method of Servais et al.[15]) found the hydrophilic material fractionated from the Portsmouth (NH) source water to be more biodegradable than the hydrophobic material.[41]

A recent study attempted to distinguish between adsorption and biodegradation by using two metabolic toxins, sodium azide and mercuric chloride, and temperature (5°C).[42] The results from using 100 mg/l of mercuric chloride in the feedwater following an 8-h static application of the inhibitor to the biofilm and from comparative runs at two different temperatures (5 and 20°C) are shown in Figure 15. The impact of each inhibitor depended upon the biodegradability of the target organic compound. For glucose/glutamic acid, the rate of removal by biofiltration was greatly reduced at 5°C while mercuric chloride addition resulted in a significant increase or release of organic material from the biofilm beginning within the first few hours of application. Removals of DOC from the Portsmouth (NH) raw water were also affected by temperature although nowhere near the extent demonstrated for a more readily biodegraded compound. The colder temperatures resulted in roughly a 10% reduction in

Table 5   Metal Concentrations from the Various Sand Coatings mg/kg dry wt

	WH-SSF	Port-SSF	Port-RSF
Fe	3732 ± 257	2986 ± 452	641 ± 67
Mn	200 ± 13	45 ± 1	10 ± 1
Ca	270 ± 19	283 ± 39	5 ± 1
Al	1695 ± 117	1148 ± 91	7149 ± 60
Average metal milliequivalents/kg dry wt	343	250	818

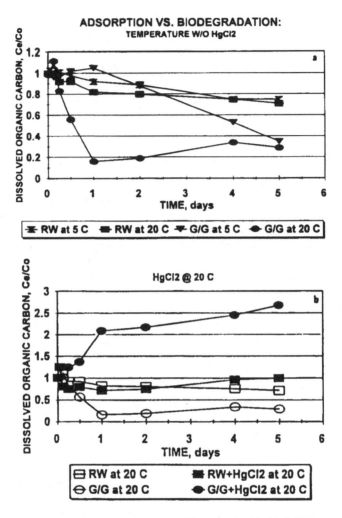

Figure 15 Influence of (a) temperature and (b) mercuric chloride inhibitor on biofilter performance.

removal efficiency through the first 2 d of run time. Removals were similar at both temperatures after 4 d. The mercuric chloride inhibitor did not appear to exert a significant influence on NOM until the later stages of the BDOC run. These reductions in DOC with the inhibited biological filters support the role that adsorption may play during biofiltration. Additional studies that can distinguish between adsorption and biodegradation removals of NOM would be useful.

Research studies have looked more closely at the retention of NOM by artificially coated iron oxide coated media[43] and naturally occurring aluminum and iron oxides.[44-47] Each study has demonstrated the efficacy of NOM adsorption on the metal oxide surface. The major mechanisms by which NOM adsorb onto mineral surfaces have been proposed to involve[44] (1) anion exchange or electrostatic interactions, (2) ligand exchange-surface complexation, (3) hydrophobic interaction, (4) entropic effect, (5) hydrogen bonding, and (6) cation bridging. In general, hydrophobic and macromolecular NOM[48] and organic molecules with greater contents of aromatic moieties, carboxylic acid groups, and amino acid residues[43] are preferentially sorbed onto mineral oxide surfaces. Consequently, total metal content and composition in the surface coatings of sand media may contribute to the natural aging or ripening of filter media. An aged or ripened sand media has always been known to be a better collector in granular media filtration.[22,49]

Biodegradation is considered to be the dominant removal mechanism of NOM by biofiltration. However, there is enough evidence from this study and others to suggest that adsorption, especially of hydrophobic/humic material, may also play a role, a role that has not yet been fully defined.

CONCLUSION

The overall objective of this research was to characterize NOM removals by biologically active sand filters with emphasis given to ascertaining the influence of coagulation and ozonation pretreatment on biofiltration efficiency. Specific conclusions deduced from the investigation are summarized below.

Increasing ozone consumption enhanced (1) the production of BDOC with the raw water being roughly $3^1/_2$ times more productive than the alum coagulated/settled water, (2) the production of lower MW, hydrophilic organic molecules at the expense of higher MW, hydrophobic organic molecules, and (3) biofiltration removals of hydrophilic material, especially in the raw water compared to coagulated/settled water.

Alum coagulation/sedimentation reduced biofiltration removals of NOM by (1) removing higher MW, hydrophobic organic carbon molecules that are amenable to removal by ripening biofilters and (2) reducing beneficial influence of preozonation on biofiltration performance.

Biofiltration removal efficiency of NOM was influenced by alum coagulation, ozonation, the order of coagulation and ozonation pretreatment, and the extent of "aging" or "ripening" of the granular medium.

ACKNOWLEDGMENTS

The authors would like to thank Peter Dwyer, Graduate Research assistant, and Nan Collins for their assistance in the manuscript preparation.

REFERENCES

1. Bouwer, E. J. and Crowe, P. B., Biological processes in drinking water treatment, *J. AWWA*, 80(9), 82, 1988.
2. Rittmann, B. E. and Huck, P. M., Biological treatment of public water supplies, *Crit. Rev. Environ. Control*, 19(2), 119, 1989.
3. Huck, P. M., *Use of Biological Processes in Drinking Water Treatment: Review of European Technology*, Vol. 2, Biotechnology Research Institute, Montreal, 1988.
4. Thurman, E. M., *Organic Geochemistry of Natural Waters*, Martinus Nijhoff, Dordrecht, The Netherlands, 1985.
5. Huck, P. M., Measurement of biodegradable organic matter and bacterial growth potential in drinking water, *J. AWWA*, 82(7), 78, 1990.
6. van der Kooij, D., Orange, J. P., and Hijnen, W. A., Growth of *Pseudomonas aeruginosa* in tap water in relation to utilization of substrates at concentrations of a few micrograms per liter, *Appl. Environ. Microbiol.*, 44, 1086, 1982.
7. Sontheimer, H. and Hubele, C., The use of ozone and granular activated carbon in drinking water treatment, in *Proc. 2nd National Conference on Drinking Water—Treatment of Drinking Water for Organic Contaminants*, Huck, P. M. and Toft, P., Eds., Pergamon Press, Elmsford, NY, 1986.
8. Gilbert, E., Investigations on the changes of biological degradability of single substances induced by ozonation, *Ozone Sci. Eng.*, 5, 137, 1983.
9. Langlais, D., Reckhow, D. A., and Brink, D. R., *Ozone in Water Treatment— Application and Engineering*, Lewis Publishers, Chelsea, MI, 1991.
10. Reckhow, D. A. and Singer, P. C., The removal of organic halide precursors by preozonation and alum coagulation, *J. AWWA*, 76(4), 151, 1984.
11. Gilbert, E., Biodegradability of ozonation products as a function of COD and DOC elimination by example of substituted aromatic substances, *Water Res.*, 21, 1273, 1987.
12. Kuo, P. P. K., Chain, E. S. K., and Chang, B. J., Identification of end products resulting from ozonation and chlorination of organic compounds commonly found in water, *Environ. Sci. Technol.*, 11, 1177, 1977.
13. Sontheimer, H., Process engineering aspects in the combination of chemical and biological oxidation, in *Oxidation Techniques in Drinking Water Treatment*, Kuhn, W. and Sontheimer, H., Eds., EPA-57019-79-020, 1979.
14. Mogren, E. M., Scarpino, P., and Summers, R. S., Measurement of biodegradable dissolved organic carbon in drinking water, *AWWA Annual Conference Proceedings*, Cincinnati, OH, 1990, 573.
15. Servais, P., Bellen, G., and Hascoet, M.-C., Determination of the biodegradable fraction of dissolved organic matter in waters, *Water Res.*, 21, 445, 1987.
16. Joret, J. C., Levi, Y., Dupin, T., and Gilbert, M., Rapid method for estimating bioeliminable organic carbon in water, AWWA Annual Conference Proceedings, Orlando, FL, June 19–23, 1988, 1715.
17. Wang, J. Z. and Summers, R. S., The evaluation of organic matter and disinfection by-product control in biofilters with biomass and bioactivity analyses, AWWA Proceedings Water Quality Technology Conference, November 7–11, Miami, 1993, 1341.
18. Fenstermacher, J. M., Modifications to Slow-Rate Filtration to Improve

Removal of Trihalomethane Precursors, M.S. thesis, University of New Hampshire, Durham, 1988.

19. Thurman, E. M. and Malcolm, R. L., Preparation isolation of aquatic humic substances, *Environ. Sci. Technol.*, 15(4), 463, 1981.

20. Eighmy, T. T. et al., Biologically enhanced slow sand filtration for removal of natural organic matter, AWWARF Final Report, Denver, 1993.

21. Summers, R. S., University of Cincinnati, personal communication, 1993.

22. Hendricks, D. et al., *Manual of Design for Slow Sand Filtration*, Denver, AWWARF, 1991.

23. Randtke, S. J., Organic contaminant removal by coagulation and related process combinations, *J. AWWA*, 80(5), 40, 1988.

24. Amy, G. L. et al., Biodegradability of natural organic matter: a comparison of methods (BDOC and AOC) and correlations with chemical surrogates, AWWA Annual Conference Proceeding, Vancouver, BC, June 18–22, 1992, 523.

25. Collins, M. R., Amy, G. L., and Steelink, C., Molecular weight distribution, carboxylic acidity, and humic substances content of aquatic organic matter: implications for removal during water treatment, *Environ. Sci. Technol.*, 20, 1028, 1986.

26. Leinhard, H., Einfluss der Verfahrens-bedingungen des Ozoneintrags auf die Ozonwirksamkeit bei der Wasserraufbereitung, Dissertation, Universität Karlsruhe, 1980.

27. Kaplan, L. A. and Bott, T. L., Microbial heterotrophic utilization of dissolved organic matter in a Piedmont stream, *Fresh Water Biol.*, 13, 363, 1983.

28. Geller, A., Degradation and formation of refractory DOM by bacteria during simultaneous growth on labile substrates and persistent lake water constituents, *Schweis. Z. Hydrol.*, 47, 27, 1985.

29. Hedges, J. I., Polymerization of humic substances in natural environments, in *Humic Substances and their Role in the Environment*, Frimmel, F. H. and Christman, R. F., Eds., John Wiley & Sons, Chichester, U.K., 1988, 45.

30. Davis, J. A. and Gloor, R., Adsorption of dissolved organics in lake water by aluminum oxide: effect of molecular weight, *Environ. Sci. Technol.*, 15, 1223, 1981.

31. Schneider, J. K. et al., Analytical fractionation of dissolved organic matter in water using on-line carbon detection, *Water Res.*, 18(12), 1515, 1984.

32. Rest, C. et al., The removal of specific molecular weight fractions of trihalomethane precursors by alum coagulation AWWA Annual Conference Proceedings, Las Vegas, NV, 1983.

33. Semmens, M. J. and Staples, A. B., The nature of organics removed during treatment of Mississippi River water, *J. AWWA*, 78(2), 76, 1986.

34. DeHaan, H., Effect of benzoate on microbial decomposition of fulvic acids in Tjeukemeer (the Netherlands), *Limnol. Oceanogr.*, 22(1), 38, 1977.

35. Meyer, J. L. et al., Bacterial growth on dissolved organic carbon from a blackwater river, *Microb. Ecol.*, 13, 13, 1987.

36. Sinsabaugh, R. L. et al., Removal of dissolved organic carbon by coagulation with iron sulfate, *J. AWWA*, 78(5), 74, 1986.

37. Vik, E. et al., Removing aquatic humus from Norwegian lakes, *J. AWWA*, 77(3), 58, 1985.

38. Shukairy, H. M., Koeching, M., and Summers, R. S., Fractionated organic

matter DBP formation: kinetics, reactivity and speciation, *AWWA Annual Conference Proceedings*, New York, June 19–23, 1994.

39. Degremont, *Water Treatment Handbook*, Lavoisier, Paris, 1991.

40. Galvin, R. M., Ripening of silica sand used for filtration, *Water, Res.*, 26(5), 683, 1992.

41. Shaw, J., Masters thesis, University of New Hampshire, Durham, 1994.

42. Collins, M. R. and Vaughan, C. W., Assessing biofilter treatability of natural organic matter, AWWA Proceedings Water Quality Technology Conference, November 7–11, Miami, 1993, 1249.

43. Benjamin, M. M. et al., NOM adsorption onto iron-oxide-coated sand, AWWARF Final Report, Denver, 1993.

44. Gu, B. et al., Adsorption and desorption of natural organic matter on iron oxide: mechanisms and models, *Environ. Sci. Technol.*, 28(1), 38, 1994.

45. McKnight, D. M. et al., Sorption of dissolved organic carbon by hydrous aluminum and iron oxides occurring at the confluence of Deer Creek with the Snake River, Summit County, Colorado, *Environ. Sci. Technol.*, 26(7), 1388, 1992.

46. Tipping, E., The adsorption of aquatic humic substances by iron oxides, *Geochim. Cosmochim. Acta*, 45(2), 191, 1981.

47. Davis, J. A., Adsorption of natural dissolved organic matter at the oxide/water interface, *Geochim. Cosmochim. Acta*, 46(11), 2381, 1982.

48. McCarthy, J. F. et al., Mobility of natural organic matter in a sandy aquifer, *Environ. Sci. Technol.*, 27(4), 667, 1993.

49. O'Melia, C. R., Practice, theory and solid-liquid separation, *Aqua*, 40(6), 371, 1991.

CHAPTER 22

Inhibition of GAC-Induced Phenol Coupling: Reactions by Humic Materials and Model Compounds

Richard A. Larson, Lina S. Chin, and Vernon L. Snoeyink

INTRODUCTION

Granular activated carbon (GAC) has been widely used to remove hydrophobic organic compounds from water and wastewater due to its high adsorptive capacity. Phenolic compounds, however, are quite sensitive to one-electron oxidation to form phenoxy radicals, which can couple to form dimers and polymers. GAC is well known to promote oxidative reactions of both inorganic and organic species, particularly phenols.[1] These reaction products may be health hazards; chlorinated phenols, for example, can be oxidized by various pathways to chlorohydroxybiphenyls, chlorinated biphenyl ethers, chlorinated dibenzofurans, chlorinated dibenzodioxins, chloroquinones, and many other compounds. Phenolic dimers have been observed in GAC extracts after phenols were adsorbed on virgin or disinfectant-treated GAC surfaces.[2-4] We report that coupling reactions of phenol are partially suppressed when GAC is treated with humic materials or the phenolic acid, syringic acid.

MATERIALS AND METHODS

GRANULAR ACTIVATED CARBON

The carbons used included F400 GAC and BPL (Calgon Carbon Corp., Pittsburgh) and Nuchar WV-G (Westvaco, Covington, VA). Before use all the carbons were washed in deionized water several times to remove water soluble ash and fines, and baked at 175°C for 1 week to vaporize volatile impurities. The washed and baked carbons were kept continuously at 103°C. Carbons

1-56670-136-8/96/$0.00+$.50

were cooled to room temperature in a desiccator shortly before preparing a GAC column.

ORGANIC COMPOUNDS

The structures of the model organic compounds used in this study are shown in Figure 1. Phenol (1) (Mallinckrodt, Inc., St. Louis) was dissolved in CH_2Cl_2 and its purity was checked by GC. Phenol solutions were prepared within the concentration range 0.3 to 50 mg/l. The quantity of phenol adsorbed on GAC was calculated by analyzing the phenol breakthrough curve of the GAC column; samples were taken periodically from the column effluent and phenol concentrations were measured by ultraviolet absorption spectrophotometry at 270 nm.

Syringic acid (4-hydroxy-3,5-dimethoxybenzoic acid, 2: Aldrich Chemical Co., Milwaukee) and 3,4,5-trimethoxybenzoic acid (3: Aldrich) are both slightly soluble in water at pH 8. To prepare their solutions, syringic acid or 3,4,5-trimethoxybenzoic acid was first dissolved in about 30 ml of ethanol, and then added to 10 l of phenol solution. After agitation and pH adjustment to 8, the combined solution was pumped to a column containing 2 g of F400 GAC at 10 ml/min. Molar ratios of syringic acid to phenol varied from 1 to 0.08; phenol concentration was kept at 30 mg/l. 3,4,5-Trimethoxybenzoic acid was applied with phenol at equimolar concentrations. The purity of syringic acid and 3,4,5-trimethoxybenzoic acid were checked by GC/MS. In the sample of syringic acid (m/z 198), an impurity of m/z 212 was identified as methyl-4-hydroxy-3,5-dimethoxybenzoate. In the sample of 3,4,5-trimethoxybenzoic acid (m/z 212), a minor impurity with m/z 226 was tentatively identified as methyl-3,4,5-trimethoxybenzoate.

Commercially available dimers were used as authentic standards for the analysis of the products from phenol coupling: 2,2'-dihydroxybiphenyl (Aldrich Chemical Co., Milwaukee), 4-hydroxydiphenyl ether (American Tokyo Kasei, Portland, OR), 4,4'-dihydroxybiphenyl (Aldrich), and 4,4'-dimethoxybiphenyl (Aldrich). The compounds in the sample were identified by comparing their mass spectra with literature, or authentic standard spectra. The quantitative analyses were performed by addition of an internal standard (anthracene, 20 or 40 ppm) to the sample. The peak area ratios were then compared with the calibration data provided by mixtures of authentic and internal standards.

In general, all chemicals employed were reagent grade or better and were used without further purification. If not stated otherwise solutions to be studied were freshly prepared in deionized-distilled water and buffered with 0.005 M phosphate mixture to the desired pH value. Dilute HCl and NaOH solutions were used to adjust pH when necessary. Solutions were placed in glass bottles covered with black plastic and aluminum caps to avoid photochemical reaction.

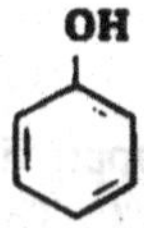

Figure 1 Structures of compounds mentioned in the text.

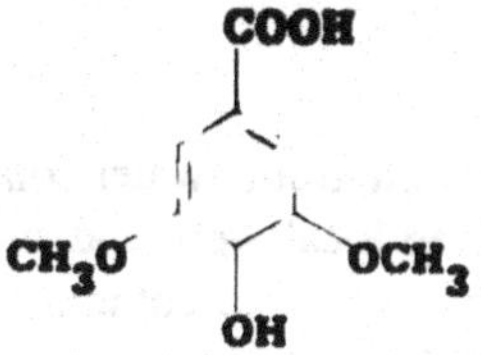

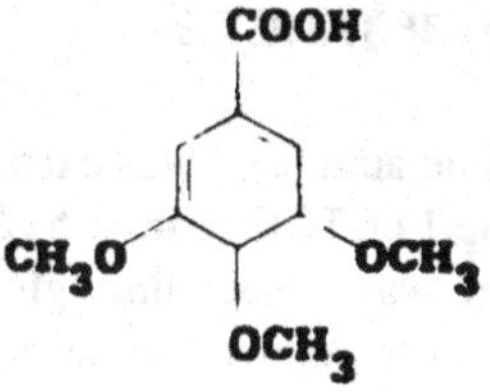

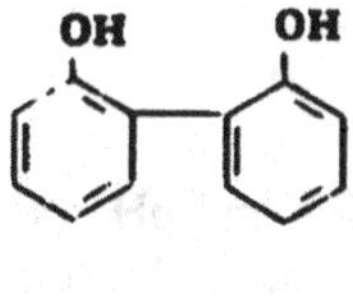

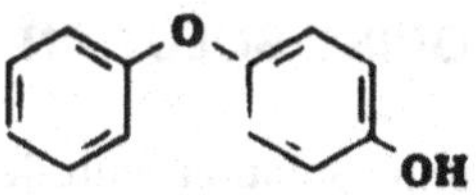

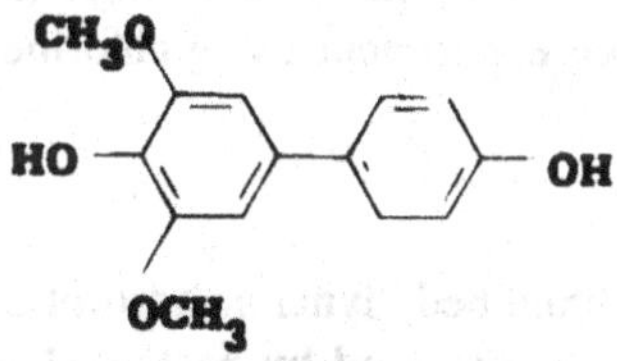

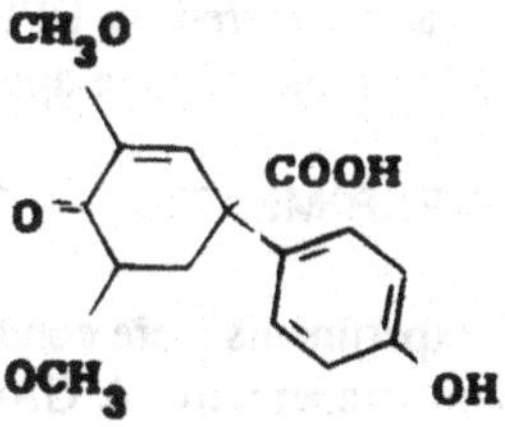

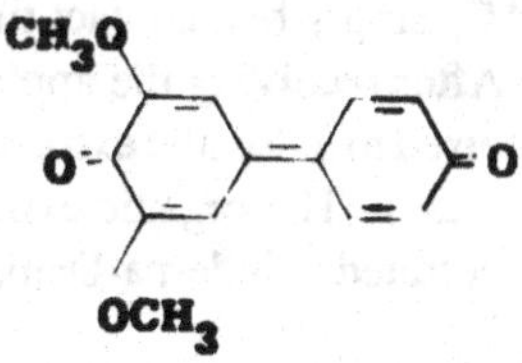

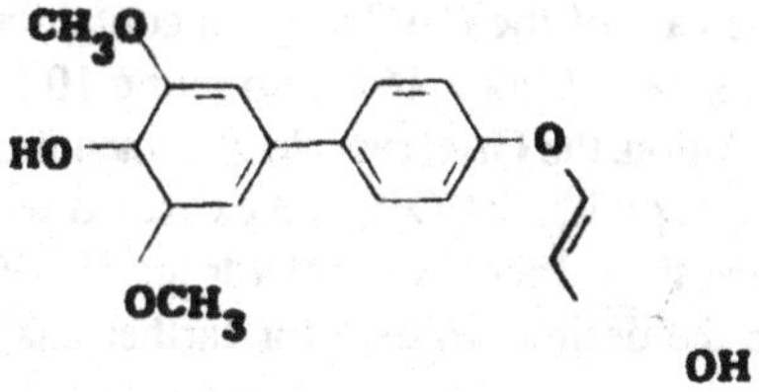

HUMIC SUBSTANCES

One fulvic acid used was extracted from groundwater containing approximately 3 mg/l of TOC from a 512-ft-deep local well, and stored at 4°C. The groundwater was passed through an ion-exchange column containing green sand (zeolite) to remove some dissolved metal ions before use. A different fulvic acid was isolated from peat soil.

The fulvic acid solution used to prepare a phenol solution was made up by adding an appropriate amount of stock solution to 10 l of deionized-distilled water containing a 0.005 M phosphate buffer. TOC measurements of fulvic acid stock solution and groundwater were performed with a Dohrmann DC-80 carbon analyzer (Xertex Corporation, Santa Clara, CA).

FREE CHLORINE SOLUTION

Free chlorine solution with desired concentration and pH was prepared by diluting its stock solution with 0.005 M phosphate buffer. The stock solution was obtained by bubbling free chlorine gas into deionized-distilled water and stored at 4°C. Sodium hydroxide pellets were added to neutralize the acid produced during the dissolution of free chlorine gas. The concentration of free chlorine was measured by DPD titration.[5] Twenty liters of 14 mg/l (as Cl_2) free chlorine solution were applied for each experiment using chlorine.

COLUMN EXPERIMENTS

All GAC experiments were conducted in a fixed bed, dynamic adsorption system at room temperature. A GAC column was prepared by packing 1 to 2 g GAC into a 1 cm 8 cm glass tube. The GAC was secured in the column by the use of pyrex wool. Solutions were continuously pumped at a flow rate of 10 ml/min (7600 l/m^2 · h) through the GAC column from the bottom. In the case of the GAC column containing 2 g of GAC, empty bed contact time was 38 s. It took 16.7 h to pump 10 l of solution. After receiving the applied solution, the GAC sample was vacuum-dried, transferred to a Soxhlet extractor, wetted with CH_3OH, and extracted with CH_2Cl_2 for 24 h. The organic extract was then dried over anhydrous Na_2SO_4 and concentrated (Kuderna-Danish) to the desired volume for further analysis.

AQUEOUS REACTIONS

Ten liters of solution containing desired reactants were mixed for 10 min. Excess sodium bisulfite (Na_2SO_3) was then added to destroy residual free chlorine if necessary. The solution was acidified (dilute sulfuric acid) and pumped through an XAD-2 resin column at 10 ml/min. The resin column was dried under vacuum to remove water and wetted with methanol; the adsorbed compounds were eluted by about 100 ml of methylene chloride into a flask containing anhydrous sodium sulfate. The methylene chloride extract was then treated similarly to a carbon extract (Kuderna-Danish concentration, gas chromatography, and gas-chromatography-mass spectrometry analyses).

GAS CHROMATOGRAPHY (GC) AND GAS CHROMATOGRAPHY/MASS SPECTROMETRY (GC-MS)

The quantitative and qualitative analysis of compounds of interest in methylene chloride extracts were accomplished by GC and GC-MS, respectively. A gas chromatograph (5830 A, Hewlett-Packard, Avondale, PA) with a splitless injector system was equipped with a fused silica capillary column (liquid phase: DB-1; film thickness: 0.25 mm; column dimensions: 30 M × 0.322 mm; J and W Scientific, Inc., Folsom, CA) connected to a flame ionization detector. The analysis conditions were: helium carrier, 1.2 ml/min; column pressure, 10 psi; auxiliary gas (He), 30 ml/min; injector temperature, 200°C; detector temperature, 300°C; oven temperature, 40°C for 5 min, 4°min to 240°C, then hold for 30 min. Some column extracts were methylated before analysis using DIAZALD® (Aldrich), according to the manufacturer's instructions.

A quadrupole mass spectrometer (5985A, Hewlett Packard, Avondale, PA) was used in the electron-impact, positive-ion mode. Samples were introduced to the mass spectrometer via a gas chromatograph also equipped with a DB-1 fused silica capillary column. The exit of the column was led directly into the ion source (which was maintained at 202°C). GC-MS samples were temperature programmed from 60 to 240°C at 4°C/min. The oven temperature was held at the final temperature until all the components were eluted. Conventional EI spectra were recorded with an electron energy of 70 eV. Total ion chromatograms were recorded for the mass-to-charge (m/z) range of 35 to 500. The mass spectrometer was tuned before the first injection of each day.

RESULTS AND DISCUSSION

DIMER FORMATION ON THE GAC SURFACE

When 10 l of phenol solution, 30 mg/l, was pumped into a F400 GAC column at pH 8, a light yellow Soxhlet extract was obtained from the carbon. Its gas chromatogram is shown in Figure 2. Three phenolic products were detected from the phenol reaction. They are 2,2'-dihydroxybiphenyl ("C-C dimer", 4), 4-hydroxydiphenyl ether ("C-O dimer", 5), and a compound with MW 276. These three products were absent in blank experiments. The first two compounds were identified by comparing their mass spectra with those of authentic standards. To demonstrate that the activity toward adsorbed phenol

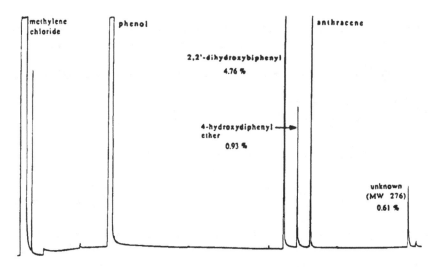

Figure 2   Gas chromatogram of the Soxhlet extract of GAC after phenol adsorption.

was not a unique property of the selected F400 GAC, three other types of commercial GAC (F400 with different dates of manufacture, BPL, and Nuchar WV-G) from two sources were tested. The same products were observed for all the carbons tested.

Different results were observed by Voudrias[6] under similar experimental conditions in previous studies. The virgin carbon he used showed less activity toward adsorbed phenol, and no phenolic dimer was detected without surface activation by free or combined chlorine. Since GAC of the same batch was no longer available, we could not repeat these experiments. Vidic and Suidan[7] have shown that unidentified oligomers were formed when the more readily oxidized phenol, *o*-cresol, was adsorbed to GAC for several weeks.

In the usual phenol coupling mechanism, phenoxy free radicals are essential intermediates to form dimers.[8] Hence, it is most probable that phenol reacted with active one-electron oxidizing sites on the GAC surface and generated phenoxy radicals that dimerized to give the observed products.

EFFECTS OF HUMIC MATERIALS

Soil Fulvic Acid

A fulvic acid, extracted from soil, was dissolved in the phenol solution at two different concentrations, 7.8 and 33.3 mg C/l and then was applied to a GAC column. The results, shown in Figure 3, indicate that fulvic acid had two significant effects on the phenol-GAC reaction. At high concentrations, fulvic acid inhibited the formation of all products. In keeping with the free radical mechanism of phenol coupling, the inhibition behavior of fulvic acid

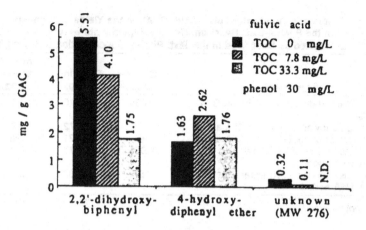

Figure 3 Effects of soil fulvic acid concentration on yields of products from the GAC-phenol reaction.

can be explained by its properties as a free radical scavenger. Furthermore, the presence of fulvic acid markedly favored C-O coupling relative to C-C coupling. A plot of the molar ratio of C-O to C-C dimer vs. fulvic acid concentration is shown in Figure 4. Similar phenomena were noticed in 2,6-dimethylphenol oxidative coupling catalyzed by copper(I) chloride and pyridine;[6] C-O coupling was promoted more than C-C coupling by an increase

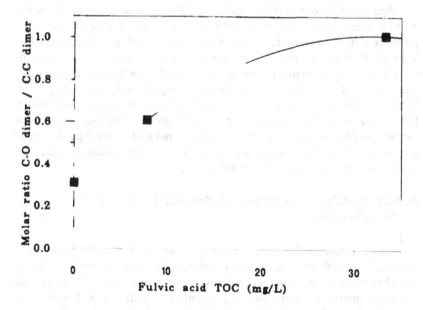

Figure 4 Effects of increasing concentrations of soil fulvic acid on the C-O/C-C dimer ratio.

**Table 1   Effects of Groundwater Fulvic Acid (GFA) on the Yields of Products
from the Phenol-GAC Reaction. Column Experiments were
Conducted as Described in the Text. Phenol Concentration was 1 mg/L**

Experiment	C-C dimer (mg)	C-O dimer (mg)	mw 276 product (mg)
Phenol, distilled-deionized water, no GFA added	0.034	0.070	0.013
Phenol, groundwater (10 l), no GFA added	0.030	0.072	0.014
Phenol, groundwater (48 l), no GFA added	ND	0.021	ND
Phenol, distilled-deionized water with 4.3 mg/L GFA added	0.029	0.077	0.028
Phenol, distilled-deionized water with 36.3 mg/L GFA added	ND	0.032	0.002

ND = not detected.

of the concentration of pyridine or the ligand ratio (N/Cu ratio). Analogously, humic substances (which contain 1 to 6% nitrogen) are able to form complexes with metal ions.[9] However, more studies with better-defined materials will be necessary to understand the roles fulvic acid plays in the phenol-GAC reaction.

Humic Substances from Groundwater

The effects of groundwater humic substances on the yields of phenolic products are summarized in Table 1. An experiment with virgin carbon receiving phenol in deionized-distilled water was taken to be a zero-humic material control. When deionized-distilled water was replaced with groundwater in the phenol solution, no significant change of product yield was observed unless a greatly increased volume of groundwater was passed over the column. At the TOC level detected in groundwater, fulvic acid extracted from groundwater and added to deionized-distilled water did not significantly affect the phenol-GAC reaction either. However, increasing the concentration of groundwater fulvic acid to 36 mg/l, comparable to the high concentration used in the previous soil fulvic acid experiment, markedly reduced the total product yield and increased the relative fraction of the C-O dimer. The results are consistent with those obtained using the soil fulvic acid.

SYRINGIC ACID AS A MODEL COMPOUND OF HUMIC SUBSTANCES

Because humic substances are complex and polymeric mixtures, it is difficult to identify the substructural elements responsible for activity. Phenols and phenolic acids, however, which are well known to be present in decomposing plant matter, are important constituents of freshwater and soil humic materials. By analogy with the structural characteristics of synthetic antioxidants, it is likely that these compounds could exhibit antioxidant activity.

Table 2 Compounds Tentatively Identified in the GAC-Promoted Reaction of
Phenol in the Presence of Syringic Acid. Product Structures are
Shown in the Text as Unmethylated Isomers, but were Methylated
during Workup

Compound	Retention time (methylated derivative)	Molecular weight (methylated deriv.)	Relative peak area (phenol = 1.0)
1	4.5 min	108	1.0
5	31.8	200	0.01
2	34.1	226	0.59
"hybrid dimer" (6 or isomer)	41.2	260	0.36
"hybrid dimer"	41.7	274	0.02
"hybrid dimer"	42.0	260	0.01
"hybrid dimer"	42.3	260	0.10
quinone 8 (?)	42.7	244	0.04
"hybrid dimer"	47.0	274	trace
dimer precursor (7 or isomer)	50.2	318	trace
dimer precursor	51.5	304	trace
"hybrid trimer" (9 or isomer)	54.5	352	0.03
"hybrid trimer"	56.1	352	0.06
"hybrid trimer"	57.4	338	0.03

Syringic acid, a lignin-derived, potential constituent of humic substances, was
chosen as a model compound to begin to probe the inhibition mechanism of
the phenol coupling reaction. Syringic acid (3.38 mmol) was dissolved in
ethanol and added to 10 l of 0.005 M phosphate buffer solution (pH 8)
containing phenol equal in concentration to syringic acid. The extract of the
carbon receiving syringic acid and phenol was dark red after 24-h Soxhlet
extraction indicating possible formation of syringic acid reaction products.
Analysis by GC/MS indicated a complex mixture of products, including unre-
acted phenol and syringic acid, the C-O dimer (but not the C-C dimer), and
several cross-linked "hybrid" dimers and trimers (Table 2).

As observed in the previous experiments employing humic material, both
C-C and C-O dimers decreased markedly and the C-O dimer became the
major dimer product from phenol coupling when syringic acid was applied
together with phenol. The decreased formation of dimers may have been due
to decreased formation of phenoxy radicals or the destruction of such radicals
by syringic acid acting as a scavenger. Competition for active sites between
phenol and syringic acid is possible, even though at pH 8 syringic acid is not
highly hydrophilic. Its competition with phenol for adsorption on GAC was
not negligible, but an intense GC peak for phenol indicated that strong phenol
adsorption still occurred in the presence of syringic acid. Because phenol
adsorption on the GAC surface could result in the formation of phenoxy
radicals for further coupling reaction to form dimers, the significantly lower
yield of C-C and C-O dimers in the presence of syringic acid suggested that
syringic acid scavenged phenoxy radicals to inhibit phenol coupling. The
variety of "hybrid" (cross-coupled) dimers and trimers observed suggests that

this pathway is important. A high yield of a compound with m/z 246 was obtained. It appears to be a "hybrid dimer", consisting of a phenol and syringic acid coupling product with loss of a carboxyl group. Decarboxylation of syringic acid during oxidative coupling has also been observed in the linkage of syringic acid with itself,[10] substituted anilines,[11] and 2,4-dichlorophenol.[12] The mass spectrum of the methylated derivative of the hybrid dimer has the following characteristics: largest ion, m/z 245 (M-CH$_3$, 100%); m/z 260 (M, 72%); m/z77 (23%); and m/z 217 (M-CH$_3$-CO, 16%). A possible structure for the compound is 6. Several additional isomers of the hybrid dimer were also detected at lower concentrations. Trace amounts of possible precursors to the hybrid dimers were also detected; their mass spectra indicated that they retained the carboxyl group of syringic acid. A structure such as 7 would be consistent with their mass spectra.

A possible quinone form of the phenol-syringic acid structure was detected in low yield; it may have a structure like 8. Three isomers of a hybrid trimer (possible structure, 9) composed of two molecules of phenol and one of syringic acid (also with loss of the carboxyl group) were also found in the carbon extract.

Because the phenoxy radical is resonance stabilized, there are three possible decarboxylated dimers resulting from reaction with syringic acid (coupling could occur at the phenol oxygen, *ortho*, or *para* position). The addition of only one methyl group during derivatization of the major coupling product could indicate either that carbon-oxygen cross coupling between phenol and syringic acid resulted in only the hydroxyl group on syringic acid being available for methylation or that carbon-carbon cross-linkage occurred, but methylation of the two available phenolic hydroxyl groups was incomplete. However, the occurrence of two isomers with m/z 274 (m/z 246-2H+2CH$_3$) supports carbon-carbon cross-linkage at both ortho and para positions of phenol.

Cross-coupling between phenol and syringic acid was strongly preferred over homolytic coupling of phenol or syringic acid on the GAC surface. The preference for the cross-linkage between syringic acid and another phenolic compound was also observed in the reaction between syringic acid and 2,4-dichlorophenol in the presence of fungal laccase.[13] The C-C dimer yield was reduced significantly even in the presence of a small amount of syringic acid. Relatively, the decrease of the C-O dimer was much slower. When the molar ratio of syringic acid to phenol was greater than 0.26, the C-O dimer was the only detectable phenol dimer in the carbon extract (Table 3).

The compounds extracted from an aqueous reaction (no GAC present) between equimolar syringic acid and phenol at pH 8 included phenol, syringic acid, the impurity in the commercial syringic acid, and relatively small amounts of demethylated syringic acid and hybrid dimer. It is rather unusual to observe the occurrence of cross-linkage without the presence of GAC. No phenol dimers were detected.

Table 3   Yields of Products in Syringic Acid-Phenol-GAC Experiments. See text for details

Starting Materials			Products (mg/g GAC)		
Concentration, 10^{-4} M		Molar ratio (syringic acid to phenol)			
syringic acid	phenol		C-C	C-O	MW 276
N.A.	3.19	—	5.51	1.63	0.31
0.25	3.19	0.08	3.02	1.25	0.37
0.83	3.19	0.26	N.D.	0.39	N.D.
1.67	3.19	0.52	N.D.	0.10	N.D.
3.38	3.19	1.06	N.D.	0.02	N.D.

N.A. = not applicable. N.D. = not detectable.

It is proposed that syringic acid reacts with an active site on the GAC surface, leading to the removal of one electron and proton from the hydroxyl group and the formation of a carbon-associated quinonoid radical. By reacting with a molecule of phenol or a phenoxy radical, and concomitant loss of CO_2, the intermediate could be converted to one of the observed hybrid dimers. A proposed mechanism accounting for the formation of the observed products is shown in Figure 5.

The structures of compounds tentatively identified in these experiments were determined mostly on mechanistic grounds. Authentic standards for the suspected products were not available and thus the identifications are equivocal. Synthesis of the derivatives of syringic acid and phenol will be necessary for final identification.

If the mechanism proposed above for such a reaction is correct, then the formation of hybrid dimers would be inhibited if the hydroxyl group on syringic acid is replaced by a methoxy group. In keeping with this hypothesis, when 3,4,5-trimethoxybenzoic acid (3) was applied together with phenol (1:1 molar ratio), phenol coupling dimers were the only reaction products. No hybrid dimers of phenol and 3,4,5-trimethoxybenzoic acid were found. A high yield of 3,4,5-trimethoxybenzoic acid was recovered in carbon extract, which also indicated negligible chemical reaction between phenol and 3,4,5-trimethoxybenzoic acid. However, the yield of phenol dimers in the presence of 3,4,5-trimethoxybenzoic acid (0.16 mg/g GAC of C-C dimer and 1.17 mg/g GAC of C-O dimer) decreased compared to the case when only phenol was applied (5.51 mg/g GAC of C-C dimer and 1.63 mg/g GAC of C-O dimer). The observed high decrease in the yield of the C-C dimer is especially notable. The diminished yield of phenol dimers may be due to decreased phenol adsorption due to the competition for active sites on GAC surface between phenol and 3,4,5-trimethoxybenzoic acid, or to some other antioxidant mechanism.

The activity of syringic acid in scavenging a phenoxy radical may parallel the ability of humic material to act as a free radical inhibitor in the phenol-GAC reaction; the dimer decrease in the presence of humic substances was

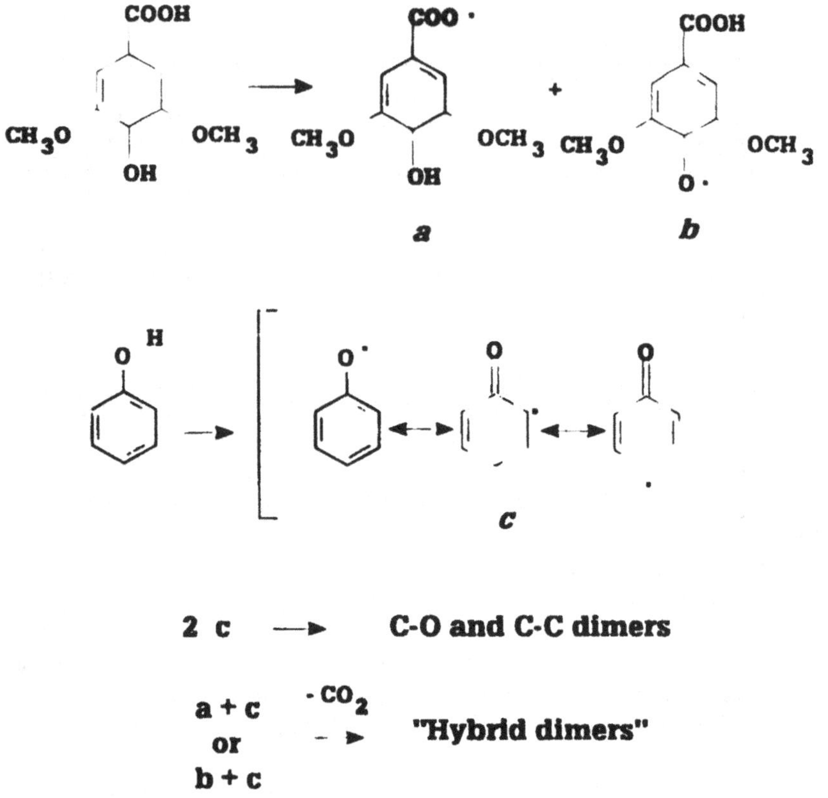

Figure 5 Proposed mechanism of syringic acid inhibition of GAC-induced phenol coupling reactions.

probably not only due to the competitive adsorption but also to free radical inhibition. Cross-linking of phenoxy radicals with humic substances may have formed inextractable polymeric adducts.

CONCLUSIONS

GACs of various types are capable of promoting phenol coupling reactions both in the presence and absence of disinfectants. The evidence suggests that free-radical oxidizing sites are present on the GAC surface. However, some constituents of natural waters, such as humic or fulvic acids, act to inhibit these reactions. The possibility that this was due to an antioxidant effect of the humic materials was supported by the results of a model compound, syringic acid. This compound, or a free radical derived from it, appears to react with phenol to form cross-coupling products. Similar reactions in water treatment practice could lead to the formation of adducts between humic materials and added phenols or other readily oxidizable compounds. The potential health significance of these adducts remain to be assessed.

ACKNOWLEDGMENT

This research was supported by the National Science Foundation (Grant #CEE-84-18647).

REFERENCES

1. Safe Drinking Water Committee, National Research Council, *Drinking Water and Health*, National Academy Press, Washington, DC, 1980, 316.
2. Voudrias, E. A., Larson, R. A., and Snoeyink, V. L., Effects of activated carbon on the reactions of free chlorine with phenols, *Environ. Sci. Technol.*, 19, 441, 1985.
3. Voudrias, E. A., Larson, R. A., Snoeyink, V. L., Chen, A. S.-C., and Stapleton, P. L., Hypochlorous acid-activated carbon: an oxidizing agent capable of producing hydroxylated polychlorinated biphenyls, *Environ. Health Perspect.*, 69, 97, 1986.
4. Chin, L. S., Larson, R. A., and Snoeyink, V. L., Oxidation of phenol on granular activated carbon, in *Biohazards of Drinking Water Treatment*, Larson, R. A., Ed., Lewis Publishers, Chelsea, MI, 1988, chap. 18.
5. American Public Health Association, *Standard Methods for the Examination of Water and Wastewater*, 15th ed. APHA, AWWA, and WPCF, Washington, DC, 1981.
6. Voudrias, E. A., Effects of Activated Carbon on Reactions of Free and Combined Chlorine with Phenols, Doctoral thesis, University of Illinois, Urbana, 1985.
7. Vidic, R. D. and Suidan, M. T., Role of dissolved oxygen on the adsorptive capacity of activated carbon for synthetic and natural organic matter, *Environ. Sci. Technol.*, 25, 1612, 1991.
8. Taylor, W. I. and Battersby, A. R., *Oxidative Coupling of Phenols*, Marcel Dekker, New York, 1967.
9. Endres, G. F., Hay, A. S., and Eustance, J. W., Polymerization by oxidative coupling. V. Catalytic specificity in the copper-amine-catalyzed oxidation of 2,6-dimethylphenol, *J. Org. Chem.*, 28, 1300, 1963.
10. Schnitzer, M. and Khan, S. U., *Humic Substances in the Environment*, Marcel Dekker, New York, 1972.
11. Liu, S.-Y., Minard, R. D., and Bollag, J.-M., Oligomerization of syringic acid, a lignin derivative, by a phenoloxidase, *Soil Sci. Soc. Am. J.*, 45, 1100, 1981.
12. Bollag, J.-M., Minard, R. D., and Liu, S.-Y., Cross linkage between anilines and phenolic humus constituents, *Environ. Sci. Technol.*, 17, 72, 1983.
13. Bollag, J.-M., Liu, S.-Y., and Minard, R. D., Cross-coupling of phenolic humus constituents and 2,4-dichlorophenol., *Soil Sci. Soc. Am. J.*, 44, 52, 1980.

Index

HAN, *see* Haloacetonitrile
HCO₃ radicals, 194
Health effects, 60, 86, 88
Health risks, *see* Health effects
HOBr, *see* Hypobromous acid
HOCl, *see* Hypochlorous acid
HRT, *see* Hydraulic residence time
Humic acids, 168, 228, 411, 421
Humic materials, 477
Hydraulic residence time (HRT), 214
Hydrogen
  atom, 134–135, 154
  peroxide, 203, 277
    effects of, 250
    production of OH radicals and, 204
    role of, 201
Hydrolysis, 74, 78
Hydrophilic acids, 411
Hydroxylation reactions, 323
Hydroxyl (HO) radicals, 134–135, 154
  concentration, estimating, 269
  mechanism, 193
  ozone decomposition induced by, 323
  produced by sunlight, 257
Hypobromite, 190, 208
Hypobromous acid (HOBr), 66, 69, 74, 76, 78, 81, 88, 173
Hypochlorous acid (HOCl), 66, 69, 74, 78, *see also* Chlorine

IC, *see* Ion chromatography
ICP, *see* Inductively coupled argon plasma emission spectrometry
ICR, *see* Information collection rule
ICT, *see* Insulated-core transformer
Inductively coupled argon plasma emission spectrometry (ICP), 456
Influent gas phase, 375
Information collection rule (ICR), 4, 60
Initial [Br⁻]/average [Cl⁺] molar ratio, 93
Initial [Br⁻]/[DOC] ratio, 93
Inorganic by-products, 17, 53, *see also* Drinking water treatment, use of ClO₂ in
Inorganic precursor concentration, 321
Insulated-core transformer (ICT), 133
Intermolecular changes, production of, 228
Internal standard, 1,1,2-trichlorethane, 134
Ion chromatography (IC), 170, 211, 236, 294
Ion-exchange resin, 423
IR spectra, 423
Iron, 216
Irradiation of water, 134
Isoelectric points, 217

Kaolin clay, 134–135
Kerosene, 10
Keto acids, 166
Keto-malonic acid, 175, 179
Ketones, 413
Kilorads, dose, 139
Kinetic calculations, 203
Kinetic simulation, 189, 195

Lactonic group, 211
Lake Austin water, 325
LARKIN, 196
Ligands, 216
Liquid-liquid extraction, 153

Maximum contaminant level (MCL), 59, 66, 72
  for bromate, 236
  current, 5
  establishment of, 3
  exceeding, 151
  in finished water, 173
  goal (MCLG), 60
  lowering of, 132
  regulated, 256
  set by EPA, 363, 371
  stage 1, 86
  use of for THMs, 74
Maximum residual disinfectant level, 4
MBAA, *see* Monobromoacetic acid
MCAA, *see* Monochloroacetic acid
MCL, *see* Maximum contaminant level
MCLG, *see* Maximum contaminant level goal
Membranes, 60
Mercury lamp, low-pressure, 210
Microflocculation, 152
Mineralization by ozonation, 318
Minimum reporting levels (MRLs), 63
Mississippi River water, 62, 82, 85
Mob, halogenating agent as, 95
Molecular weights, 371, 422
Monobromamine
  amino group in, 193
  oxidation, 197
Monobromoacetic acid (MBAA), 93, 96
Monochloramine, 349, 351
  concentration, linear function of, 367
  formation, 345
  inorganic, 356
  residual, 355
  solutions, nitrate formed by photolysis of, 367

Printed in the United States

Printed in the United States
by Baker & Taylor Publisher Services